Introduction to Dislocations

Introduction to Dislocations

Fourth Edition

by D. Hull and D. J. Bacon
Department of Engineering,
Materials Science and Engineering,
University of Liverpool, UK

OXFORD AUCKLAND BOSTON JOHANNESBURG MELBOURNE NEW DELHI

Butterworth-Heinemann
Linacre House, Jordan Hill, Oxford OX2 8DP
225 Wildwood Avenue, Woburn, MA 01801-2041
A division of Reed Educational and Professional Publishing Ltd

A member of the Reed Elsevier plc group

First published 1965
Second edition 1975
Third edition 1984
Fourth edition 2001

British Library Cataloguing in Publication Data
A catalogue record for this book is available from the British Library

Library of Congress Cataloguing in Publication Data
A catalogue record for this book is available from the Library of Congress

ISBN 0 7506 4681 0

For information on all Butterworth-Heinemann publications visit our website at www.bh.com

Typeset in India at Integra Software Services Pvt Ltd, Pondicherry, India 605005; www.integra-india.com

Contents

Preface

The subject of dislocations is essential for an understanding of many of the physical and mechanical properties of crystalline solids. It is therefore reassuring to learn that this book is still widely used in undergraduate courses and as an introduction to postgraduate study. When the first edition was prepared nearly 40 years ago, it was stated in the Preface that 'it seems unlikely that the ground work of the subject will be changed or modified appreciably, particularly at an introductory level'. This has indeed proved to be so, and thus we have found no reason to modify the main structure of the book or method of approach used. However, the present edition includes some substantial new sections. To avoid making the book significantly larger we have removed some material that has become a little dated as techniques and understanding have developed. We have also taken the opportunity to rewrite some sections to clarify the content and make it conform to a more modern standpoint.

The emphasis on a physical understanding of the geometry and properties of dislocations remains throughout, but the slightly more rigorous approach used to specify the mathematical content of the subject in the third edition has been retained. This will be of value to those who need to apply the principles of dislocations but it should not deter those who only require the briefest of introductions. No attempt has been made to change the book into a research text.

Although not normally recommended in a student textbook, we have included a fairly extensive bibliography at the end of each chapter. The books and research papers that are listed were chosen for those who wish to take the subject one step further. They introduce many of the more speculative aspects of the subject and demonstrate the way dislocation theory is applied in practice.

We acknowledge the helpful comments of Professor R.C. Pond on our first draft of sections 9.6 and 9.7, and of other colleagues who pointed out errors in the third edition and made suggestions for the fourth. We would also thank our publishers for encouraging us to prepare this new edition and agreeing to produce it in a completely new format.

Liverpool, March 2001

Derek Hull
David Bacon

1 Defects in Crystals

1.1 Crystalline Materials

Dislocations are an important class of defect in crystalline solids and so an elementary understanding of crystallinity is required before dislocations can be introduced. Metals and many important classes of non-metallic solids are crystalline, i.e. the constituent atoms are arranged in a pattern that repeats itself periodically in three dimensions. The actual arrangement of the atoms is described by the *crystal structure*. The crystal structures of most pure metals are simple, the three most common being the body-centred cubic, face-centred cubic and close-packed hexagonal structures and are described in section 1.2. In contrast, the structures of alloys and non-metallic compounds are often complex.

The arrangement of atoms in a crystal can be described with respect to a three-dimensional net formed by three sets of straight, parallel lines as in Fig. 1.1(a). The lines divide space into equal sized parallelepipeds and the points at the intersection of the lines define a *space lattice*. Every point of a space lattice has identical surroundings. Each parallelepiped is called a *unit cell* and the crystal is constructed by stacking identical unit cells face to face in perfect alignment in three dimensions. By placing a *motif unit* of one or more atoms at every lattice site the regular structure of a perfect crystal is obtained.

The positions of the *planes, directions* and *point sites* in a lattice are described by reference to the unit cell and the three principal axes, x, y and z (Fig. 1.1(b)). The cell dimensions $OA = a$, $OB = b$ and $OC = c$ are the lattice parameters, and these along with the angles $\angle BOC = \alpha$, $\angle COA = \beta$ and $\angle AOB = \gamma$ completely define the size and shape of

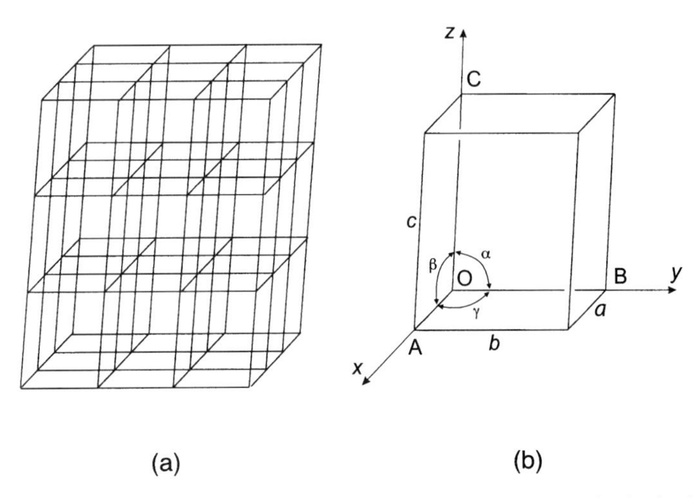

Figure 1.1 (a) A space lattice, (b) unit cell showing positions of principal axes.

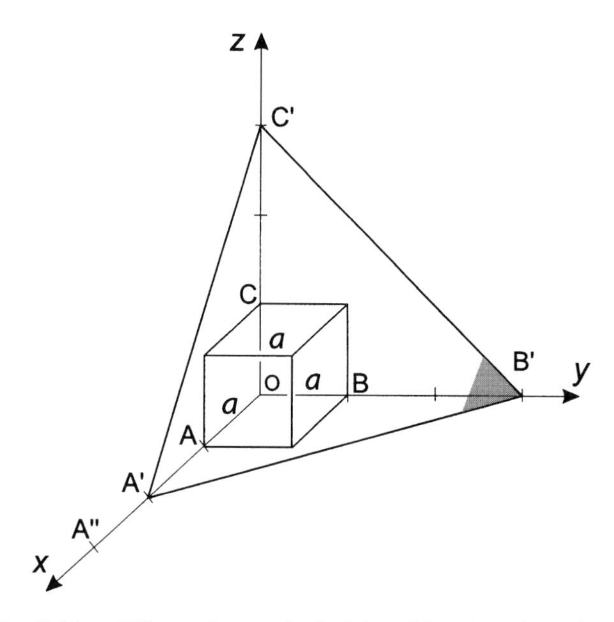

Figure 1.2 Cubic cell illustrating method of describing the orientation of planes.

the cell. For simplicity the discussion here will be restricted to cubic and hexagonal crystal structures. In cubic crystals $a = b = c$ and $\alpha = \beta = \gamma = 90°$, and the definition of planes and directions is straight-forward. In hexagonal crystals it is convenient to use a different approach, and this is described in section 1.2.

Any *plane* $A'B'C'$ in Fig. 1.2 can be defined by the intercepts OA', OB' and OC' with the three principal axes. The usual notation (*Miller indices*) is to take the reciprocals of the ratios of the intercepts to the corresponding unit cell dimensions. Thus $A'B'C'$ is given by

$$\left(\frac{OA}{OA'}, \frac{OB}{OB'}, \frac{OC}{OC'}\right)$$

and the numbers are then reduced to the three smallest integers in these ratios.

Thus from Fig. 1.2 $OA' = 2a$, $OB' = 3a$, and $OC' = 3a$; the reciprocal intercepts are

$$\left(\frac{a}{2a}, \frac{a}{3a}, \frac{a}{3a}\right)$$

and so the Miller indices of the $A'B'C'$ plane are (322). A plane with intercepts OA, OB, and OC has Miller indices

$$\left(\frac{a}{a}, \frac{a}{a}, \frac{a}{a}\right)$$

or, more simply, (111). Similarly, a plane $DFBA$ in Fig. 1.3 is

$$\left(\frac{a}{a}, \frac{a}{a}, \frac{a}{\infty}\right)$$

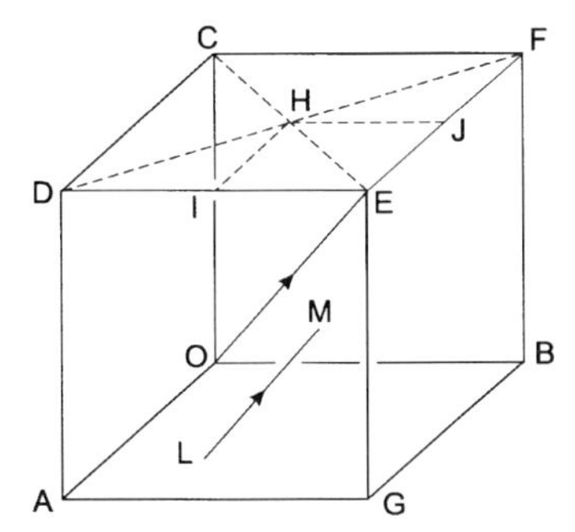

Figure 1.3 Cubic cell illustrating the method of describing directions and point sites. *LM* is parallel to *OE*.

or (110); a plane *DEGA* is

$$\left(\frac{a}{a}, \frac{a}{\infty}, \frac{a}{\infty}\right)$$

or (100); and a plane $AB'C'$ in Fig. 1.2 is

$$\left(\frac{a}{a}, \frac{a}{3a}, \frac{a}{3a}\right)$$

or (311). In determining the indices of any plane it is most convenient to identify the plane of lattice points parallel to the plane which is closest to the origin *O* and intersects the principal axis close to the origin. Thus plane $A''B'C'$ in Fig. 1.2 is parallel to *ABC* and it is clear that the indices are (111). Using this approach it will be seen that the planes *ABC*, *ABE*, *CEA* and *CEB* in Fig. 1.3 are (111), $(11\bar{1})$, $(1\bar{1}1)$, and $(\bar{1}11)$ respectively. The minus sign above an index indicates that the plane cuts the axis on the negative side of the origin. In a cubic crystal structure, these planes constitute a group of the same crystallographic type and are described collectively by $\{111\}$.

Any *direction LM* in Fig. 1.3 is described by the line parallel to *LM* through the origin *O*, in this case *OE*. The direction is given by the three smallest integers in the ratios of the lengths of the projections of *OE* resolved along the three principal axes, namely *OA*, *OB* and *OC*, to the corresponding lattice parameters of the unit cell. Thus, if the cubic unit cell is given by *OA*, *OB* and *OC* the direction *LM* is

$$\left[\frac{OA}{OA}, \frac{OB}{OB}, \frac{OC}{OC}\right]$$

or

$$\left[\frac{a}{a}, \frac{a}{a}, \frac{a}{a}\right]$$

or [111]. Square brackets are used for directions. The directions *CG*, *AF*, *DB* and *EO* are $[11\bar{1}]$, $[\bar{1}11]$, $[\bar{1}1\bar{1}]$ and $[\bar{1}\bar{1}\bar{1}]$ respectively and are a group of directions of the same crystallographic type described collectively by $\langle 111 \rangle$. Similarly, direction *CE* is

$$\left[\frac{a}{a}, \frac{a}{a}, \frac{O}{a}\right]$$

or [110]; direction *AG* is

$$\left[\frac{O}{a}, \frac{a}{a}, \frac{O}{a}\right]$$

or [010]; and direction *GH* is

$$\left[\frac{-a/2}{a}, \frac{-a/2}{a}, \frac{a}{a}\right]$$

or $[\bar{1}\bar{1}2]$. The rule that brackets [ ] and ( ) imply specific directions and planes respectively, and that $\langle\ \rangle$ and { } refer respectively to directions and planes of the same type, will be used throughout this text.

In cubic crystals the Miller indices of a plane are the same as the indices of the direction normal to that plane. Thus in Fig. 1.3 the indices of the plane *EFBG* are (010) and the indices of the direction *AG* which is normal to *EFBG* are [010]. Similarly, direction *OE* [111] is normal to plane *CBA* (111).

The coordinates of any *point* in a crystal relative to a chosen origin site are described by the fractional displacements of the point along the three principal axes divided by the corresponding lattice parameters of the unit cell. The centre of the cell in Fig. 1.3 is $\frac{1}{2}, \frac{1}{2}, \frac{1}{2}$ relative to the origin *O*; and the points *F*, *E*, *H* and *I* are $0, 1, 1$; $1, 1, 1$; $\frac{1}{2}, \frac{1}{2}, 1$; and $1, \frac{1}{2}, 1$ respectively.

1.2 Simple Crystal Structures

In this section the atoms are considered as hard spheres which vary in size from element to element. From the hard sphere model the parameters of the unit cell can be described directly in terms of the radius of the atomic sphere, *r*. In the diagrams illustrating the crystal structures the atoms are shown as small circles in the three-dimensional drawings and as large circles representing the full hard sphere sizes in the two-dimensional diagrams. It will be shown that crystal structures can be described as a stack of lattice planes in which the arrangement of lattice sites within each layer is identical. To see this clearly in two-dimensional figures, the atoms in one layer represented by the plane of the paper are shown as full circles, whereas those in layers above and below the first are shown as small shaded circles. The order or sequence of the atom layers in the stack, i.e. the *stacking sequence*, is referred to by fixing one layer as an *A* layer and all other layers with atoms in identical positions as *A* layers also. Layers of atoms in other positions in the stack are referred to as *B*, *C*, *D* layers, etc.

In the *simple cubic structure* with one atom at each lattice site, illustrated in Fig. 1.4, the atoms are situated at the corners of the unit cell. (Note that no real crystals have such a simple atomic arrangement.) Figures 1.4(b) and (c) show the arrangements of atoms in the (100) and (110) planes respectively. The atoms touch along $\langle 001 \rangle$ directions and therefore the lattice parameter *a* is twice the atomic radius *r* ($a = 2r$). The atoms in adjacent (100) planes are in identical atomic sites when projected along the direction normal to this plane, so that the stacking sequence of (100) planes is *AAA*... The atoms in adjacent (110) planes are displaced $\frac{1}{2}a\sqrt{2}$ along $[\bar{1}10]$ relative to each other and the spacing of atoms along $[\bar{1}10]$ is $a\sqrt{2}$. It follows that alternate planes have atoms in the same atomic sites relative to the direction normal to (110) and the stacking sequence of (110) planes is *ABABAB*... The spacing between successive (110) planes is $\frac{1}{2}a\sqrt{2}$.

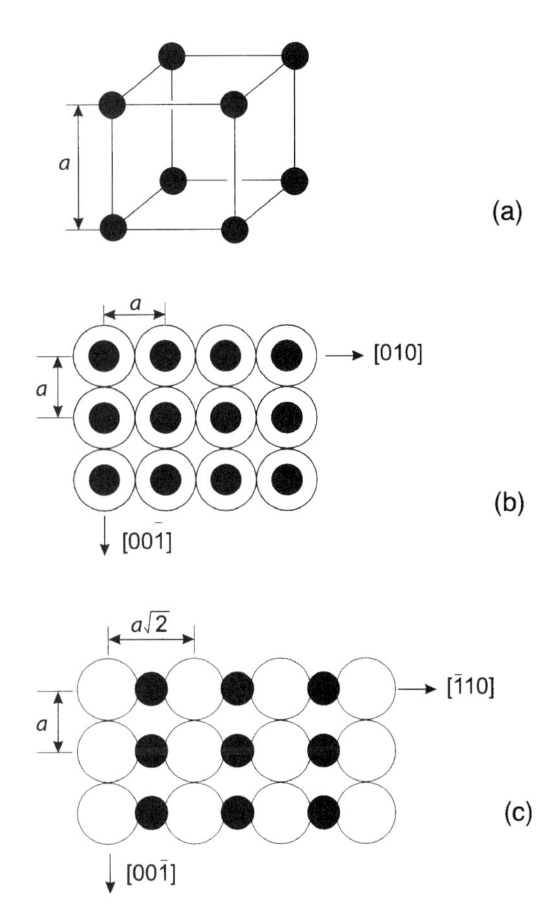

Figure 1.4 Simple cubic structure: (a) unit cell, (b) arrangement of atoms in (100) layers, (c) arrangement of atoms in (110) layers.

In the *body-centred cubic structure* (bcc), which is exhibited by many metals and is shown in Fig. 1.5, the atoms are situated at the corners of the unit cell and at the centre site $\frac{1}{2}, \frac{1}{2}, \frac{1}{2}$. The atoms touch along a $\langle 111 \rangle$ direction and this is referred to as the *close-packed direction*. The lattice parameter $a = 4r/\sqrt{3}$ and the spacing of atoms along $\langle 110 \rangle$ directions is $a\sqrt{2}$. The stacking sequence of $\{100\}$ and $\{110\}$ planes is $ABABAB\ldots$ (Fig. 1.5(b)). There is particular interest in the stacking of $\{112\}$ type planes. Figure 1.6 shows two body-centred cubic cells and the positions of a set of $(1\bar{1}2)$ planes. From the diagrams it is seen that the stacking sequence of these planes is $ABCDEFAB\ldots$, and the spacing between the planes is $a/\sqrt{6}$.

In the *face-centred cubic structure* (fcc), which is also common among the metals and is shown in Fig. 1.7, the atoms are situated at the corners of the unit cell and at the centres of all the cube faces in sites of the type $0, \frac{1}{2}, \frac{1}{2}$. The atoms touch along the $\langle 011 \rangle$ close-packed directions. The lattice parameter $a = 2r\sqrt{2}$. The stacking sequence of $\{100\}$ and $\{110\}$

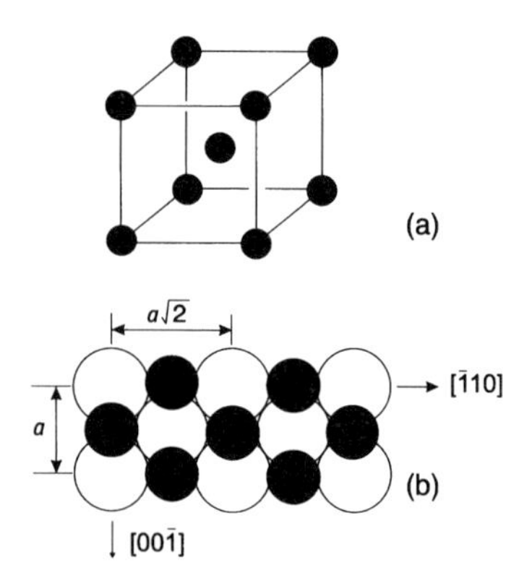

Figure 1.5 Body-centred cubic structures: (a) unit cell, (b) arrangement of atoms in (110) layers.

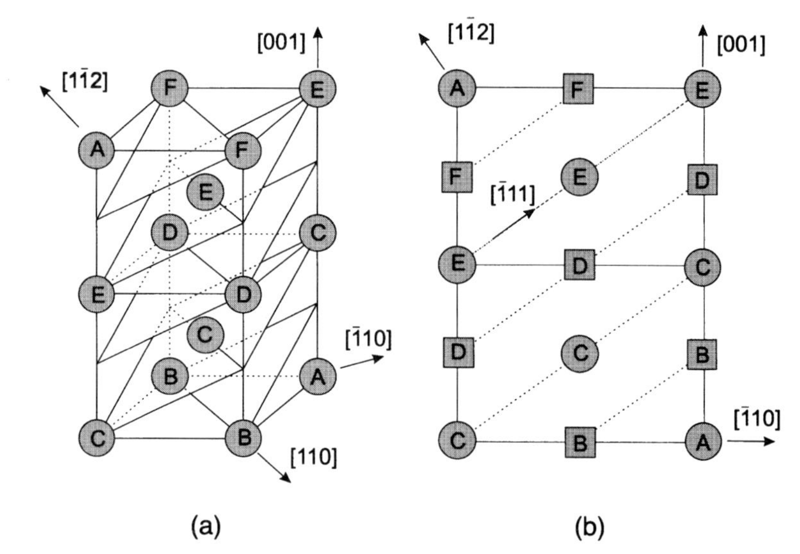

Figure 1.6 Stacking sequence of {112} planes in a body-centred cubic crystal. (a) Two unit cells showing positions of atoms in (1$\bar{1}$2) planes. (b) Traces of the (1$\bar{1}$2) planes on a (110) projection: atom sites marked by circles lie in the plane of the diagram; those marked by squares lie $a/\sqrt{2}$ above and below.

planes is *ABABAB*..., and the stacking sequence of {111} planes is *ABCABC*... The latter is of considerable importance and is illustrated in Figs 1.7(b) and (c). The atoms in the {111} planes are in the most close-packed arrangement possible for spheres and contain three close-packed directions 60° apart.

The *close-packed hexagonal structure* (cph or hcp) is also common in metals. It is more complex than the cubic structures but can be described very simply with reference to the stacking sequence. The unit cell with lattice parameters a, a, c is shown in Fig. 1.8(a), together with the hexagonal cell constructed from three unit cells. There are two atoms per lattice site, i.e. at 0, 0, 0 and $\frac{2}{3}, \frac{1}{3}, \frac{1}{2}$ with respect to the axes a_1, a_2, c. The atomic planes perpendicular to the c axis are close-packed, as in the fcc case, but the stacking sequence is now *ABABAB*..., as shown in Fig. 1.8(b).

For a hard sphere model the ratio of the length of the c and a axes (axial ratio) of the hexagonal structure is 1.633. In practice, the axial ratio varies between 1.57 and 1.89 in close-packed hexagonal metals. The variations arise because the hard sphere model gives only an approximate value of the interatomic distances and requires modification depending on the electronic structure of the atoms.

If Miller indices of three numbers based on axes a_1, a_2, c are used to define planes and directions in the hexagonal structure, it is found that crystallographically equivalent sets can have combinations of different numbers. For example, the three close-packed directions in the basal plane (001) are [100], [010] and [110]. Indexing in hexagonal crystals is therefore usually based on *Miller–Bravais* indices, which are referred to

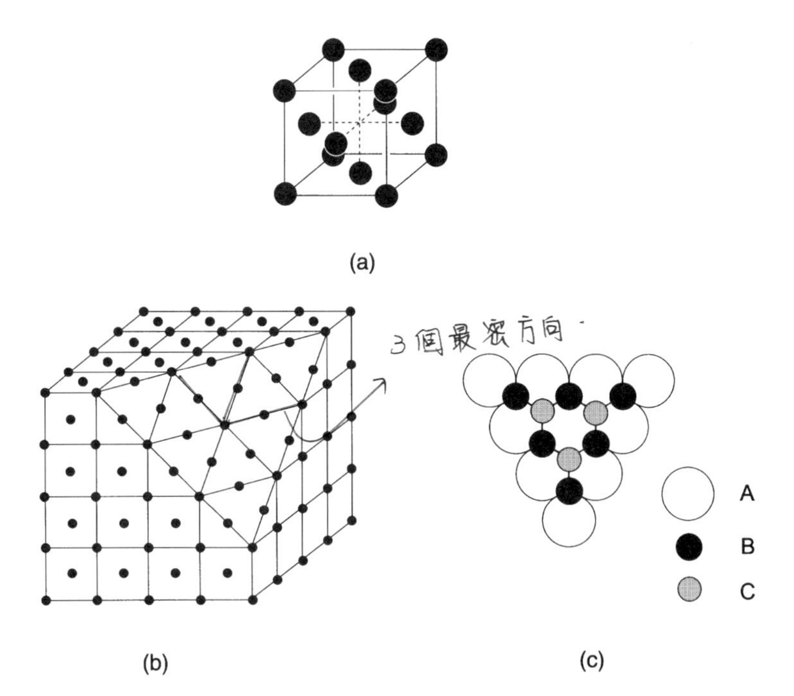

Figure 1.7 Face-centred cubic structure: (a) unit cell, (b) arrangement of atoms in a (111) close-packed plane, (c) stacking sequence of {111} planes.

the four axes a_1, a_2, a_3 and c indicated in Fig. 1.8(a). When the reciprocal intercepts of a plane on all four axes are found and reduced to the smallest integers, the indices are of the type (h, k, i, l), and the first three indices are related by

$$i = -(h + k) \tag{1.1}$$

Equivalent planes are obtained by interchanging the position and sign of the first three indices. A number of planes in the hexagonal lattice have been given specific names. For example:

Basal plane	(0001)
Prism plane: first order	$(1\bar{1}00)$ $(\bar{1}100)$, etc.
Prism plane: second order	$(11\bar{2}0)$ $(\bar{2}110)$, etc.
Pyramidal plane: first order	$(10\bar{1}1)$ $(\bar{1}011)$, etc.
Pyramidal plane: second order	$(11\bar{2}2)$ $(\bar{1}\bar{1}22)$, etc.

Some of these planes are indicated in Fig. 6.3. Direction indices in hexagonal structures are defined by the components of the direction parallel to the four axes. The numbers must be reduced to the smallest integers and the third index is the negative of the sum of the first two. To satisfy this condition the directions along axes a_1, a_2 and a_3 are of the type $\langle\bar{1}2\bar{1}0\rangle$ as illustrated in Fig. 1.9.

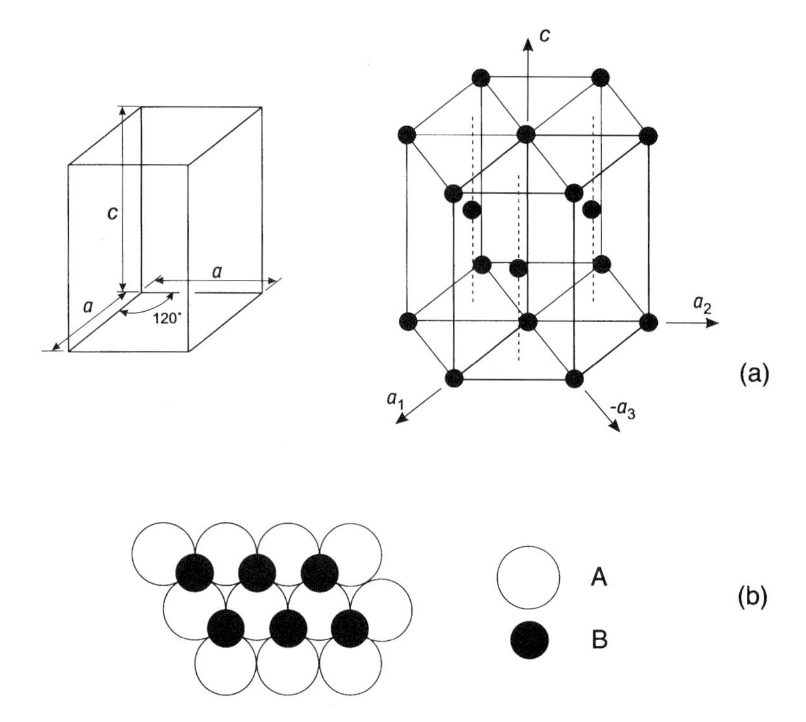

Figure 1.8 Close-packed hexagonal structure: (a) the unit cell of the lattice and the hexagonal cell showing the arrangement of atoms, (b) *ABAB*... stacking sequence of the atomic planes perpendicular to the *c* axis.

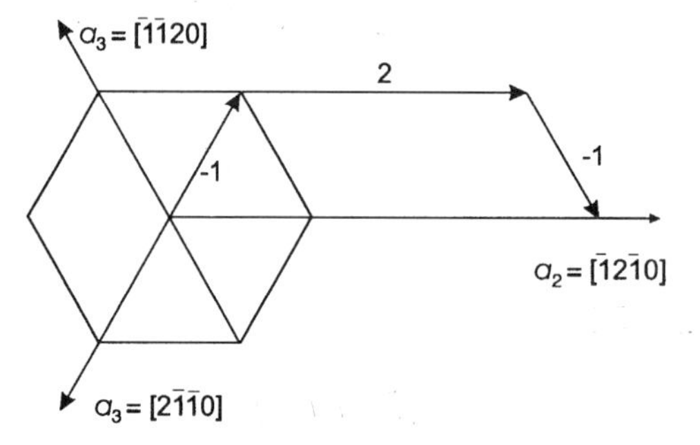

Figure 1.9 Determination of direction indices in the basal plane of an hexagonal crystal. The translations giving rise to $[\bar{1}2\bar{1}0]$ are shown explicitly.

1.3 Defects in Crystalline Materials

All real crystals contain *imperfections* which may be *point*, *line*, *surface* or *volume* defects, and which disturb locally the regular arrangement of the atoms. Their presence can significantly modify the properties of crystalline solids, and although this text is primarily concerned with the line defects called *dislocations*, it will be seen that the behaviour and effects of all these imperfections are intimately related.

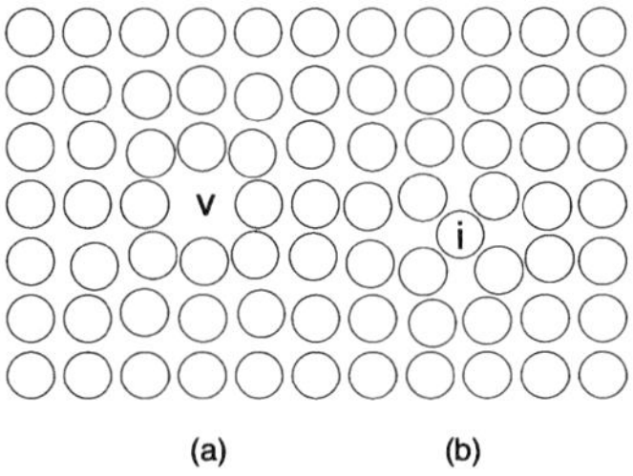

Figure 1.10 (a) Vacancy, (b) self-interstitial atom in an (001) plane of a simple cubic lattice.

Point Defects

All the atoms in a perfect lattice are at specific atomic sites (ignoring thermal vibrations). In a pure metal two types of point defect are possible, namely a *vacant atomic site* or *vacancy*, and a *self-interstitial atom*. These *intrinsic* defects are illustrated for a simple cubic structure in Fig. 1.10. The vacancy has been formed by the removal of an atom from an atomic site (labelled v) and the interstitial by the introduction of an atom into a non-lattice site at a $\frac{1}{2}, \frac{1}{2}, 0$ position (labelled i). It is known that vacancies and interstitials can be produced in materials by plastic deformation and high-energy particle irradiation. The latter process is particularly important in materials in nuclear reactor installations. Furthermore, intrinsic point defects are introduced into crystals simply by virtue of temperature, for at all temperatures above 0 K there is a thermodynamically stable concentration. The change in free energy ΔF associated with the introduction of n vacancies or self-interstitials in the lattice is

$$\Delta F = nE_f - T\Delta S \tag{1.2}$$

where E_f is the energy of formation of one defect and ΔS is the change in the entropy of the crystal. nE_f represents a considerable positive energy, but this is offset by an increase in the configurational entropy due to the presence of the defects. The equilibrium concentration of defects, given by the ratio of the number of defects to the number of atomic sites, corresponding to the condition of minimum free energy is

$$c_0 = \exp\left(-\frac{E_f}{kT}\right) \tag{1.3}$$

where k is Boltzmann's constant and T is the temperature (in deg K).

For the vacancy, the formation energy, E_f^v, is that required to remove one atom from its lattice site to a place on the surface of the crystal. Experimental values fall in the range $\sim$1–3 eV, i.e. 0.16–0.48 aJ. They scale with the melting temperature, T_m, and an approximate rule is $E_f^v \simeq 8kT_m$. Thus for copper, for which $T_m = 1356$ K and $E_f^v = 1.3$ eV, the fraction of atom sites vacant at 1300 K is $\sim 10^{-5}$ and at 300 K is $\sim 10^{-22}$. The self-interstitial is created by removing one atom from the surface and inserting it into an interstitial site, and the formation energy E_f^i is typically two to four times E_f^v. Consequently, the concentration given by equation (1.3) is many orders of magnitude smaller for interstitials, and so in metals in thermal equilibrium the concentration of interstitials may be neglected in comparison with that of vacancies. In non-metals, ionic charge and valence effects may modify this conclusion.

The rate at which a point defect moves from site to site in the lattice is proportional to $\exp(-E_m/kT)$, where E_m is the defect *migration energy* and is typically $\sim$0.1–1.0 eV. The rate decreases exponentially with

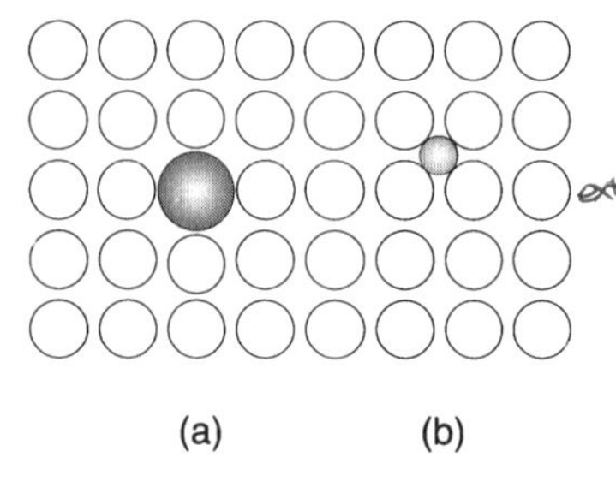

Figure 1.11 (a) Substitutional impurity atom, (b) interstitial impurity atom.

decreasing temperature and consequently in many metals it is possible to retain a high vacancy concentration at room temperature by rapidly quenching from a high equilibrating temperature.

Impurity atoms in a crystal can be considered as *extrinsic* point defects and they play a very important role in the physical and mechanical properties of all materials. Impurity atoms can take up two different types of site, as illustrated in Fig. 1.11: (a) *substitutional*, in which an atom of the parent lattice lying in a lattice site is replaced by the impurity atom, and (b) *interstitial*, in which the impurity atom is at a non-lattice site similar to the self-interstitial atoms referred to above.

All the point defects mentioned produce a local distortion in the otherwise perfect lattice. The amount of distortion and hence the amount of additional energy in the lattice due to the defects depends on the amount of 'space' between the atoms in the lattice and the 'size' of the atoms introduced.

The interstice sites between atoms generally have volumes of less than one atomic volume, and the interstitial atoms therefore tend to produce large distortions among the surrounding atoms. This accounts for the relatively large values of E_f^i referred to above, and can result in crystal volume increases as large as several atomic volumes per interstitial atom.

Additional effects are important when the removal or addition of atoms changes the local electric charge in the lattice. This is relatively unimportant in crystals with metallic binding, but can be demonstrated particularly well in crystals in which the binding is ionic. The structure of sodium chloride is shown in Fig. 1.12. Each negatively charged chlorine ion is surrounded by six nearest neighbours of positively charged sodium ions and vice versa. The removal of a sodium or a chlorine ion produces a local negative or positive charge as well as a vacant lattice site. These are called *cation* and *anion* vacancies respectively. To conserve an overall neutral charge the vacancies must occur either (a) in pairs of opposite sign, forming divacancies known as *Schottky defects*, or (b) in association with interstitials of the same ion, *Frenkel defects*.

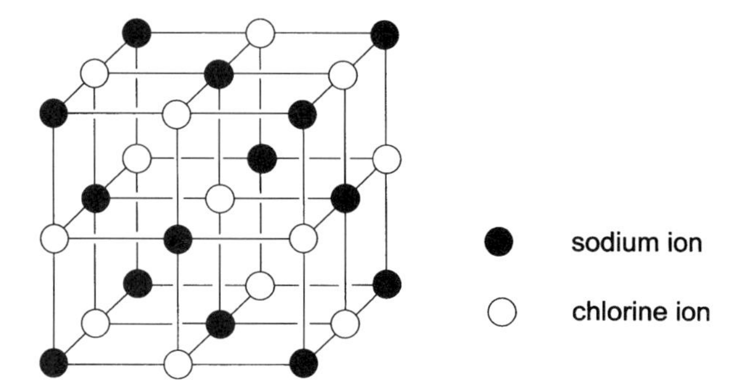

Figure 1.12 Sodium chloride structure which consists of two interpenetrating face-centred cubic lattices of the two types of atom, with the corner of one located at the point $\frac{1}{2}$, 0, 0 of the other.

Stacking Faults

In section 1.2 it was emphasised that perfect crystals can be described as a stack of atom layers arranged in a regular sequence. For the simple metallic structures discussed in section 1.2, the atomic layers are identical. A *stacking fault* is a *planar defect* and, as its name implies, it is a local region in the crystal where the regular sequence has been interrupted. Stacking faults are not expected in planes with *ABABAB*... sequences in body-centred or face-centred cubic metals because there is no alternative site for an *A* layer resting on a *B* layer. However, for *ABC ABC*... or *ABABAB*... stacking of the close-packed planes in close-packed structures there are two possible positions of one layer resting on another (Fig. 1.7). According to the hard sphere model, a close-packed layer of atoms resting on an *A* layer can rest equally well in either a *B* or a *C* position and geometrically there is no reason for the selection of a particular position. In a face-centred cubic lattice two types of stacking fault are possible, referred to as *intrinsic* and *extrinsic*. These are best described by considering the change in sequence resulting from the removal or introduction of an extra layer. In Fig. 1.13(a) part of a *C* layer has been removed which results in a break in the stacking sequence. This is an *intrinsic fault* and it can be seen that the stacking sequences above and below the fault plane are continuous right up to the fault itself. In Fig. 1.13(b) an extra *A* layer has been introduced between a *B* and a *C* layer. There are two breaks in the stacking sequence and it is referred to as an *extrinsic fault*. The extra layer does not belong to the continuing patterns of the lattice either above or below the fault.

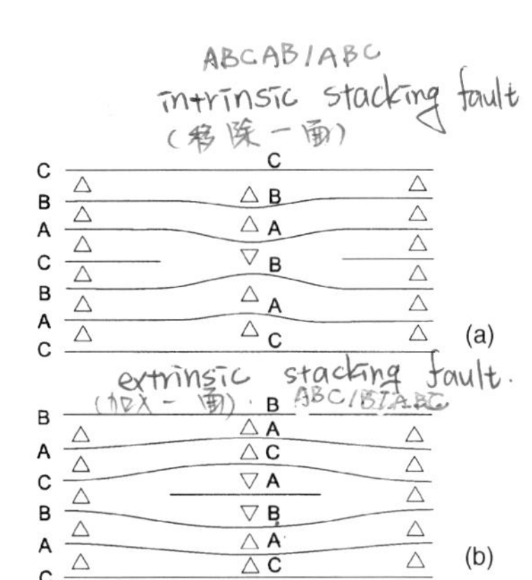

Figure 1.13 Stacking faults in the face-centred cubic structure. The normal stacking sequence of (111) planes is denoted by *ABCA*... Planes in normal relation to one another are separated by Δ, those with a stacking error by ∇: (a) intrinsic stacking fault, (b) extrinsic stacking fault.

The presence of stacking faults can play an important role in the plasticity of crystals. It should be noted, for example, that the intrinsic fault in the face-centred cubic structure can be produced by a sliding process. If, say, an *A* layer is slid into a *B* position and all the layers above are moved in the same way, i.e. *B* to *C*, *C* to *A*, *A* to *B*, etc., then the new sequence *ABCBCAB*... is identical to that discussed above. This aspect will be discussed in more detail in Chapter 5.

Stacking faults have been reported in many crystal structures. (Those which occur in the close-packed hexagonal metals are described in Chapter 6.) They destroy the perfection of the host crystal, and the associated energy per unit area of fault is known as the *stacking-fault energy*. Typical values lie in the range 1–1000 mJ/m^2. In the faults described above, the first- and second-nearest-neighbour atomic bonds in the close-packed structure are preserved, so that only bonds to more-distant neighbours and electronic effects contribute to the energy. In other structures, covalent bonding and ionic effects can be important.

Grain Boundaries

Crystalline solids usually consist of a large number of randomly oriented *grains* separated by grain boundaries. Each grain is a single crystal and

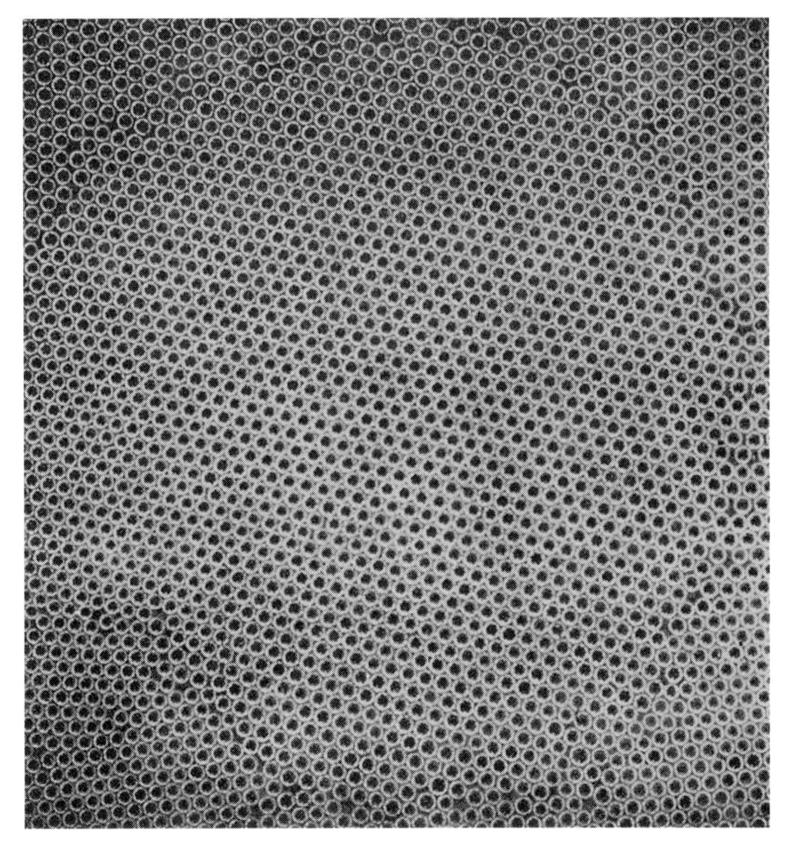

Figure 1.14 Crystal grains simulated by a bubble raft. (From *Scientific American*, Sept. 1967.)

contains the defects already described. When the misorientation between the grains is large, the atomic arrangement at the boundary is complicated and varies significantly with the angle of misorientation. An easy way to visualise the atomic arrangement is to use bubble models (Fig. 1.14) in which a two-dimensional raft of equal sized bubbles floats on the surface of a liquid. Figure 1.14 shows a grain or 'crystal' surrounded by grains of different orientation. A notable feature of the boundary structure is that the region of disorder is very narrow, being limited to one or two 'atoms' on each side of the boundary. For certain misorientation relationships between grains, the structure of the boundary can be described as an array of dislocations, as discussed in Chapter 9.

Twin Boundaries

Deformation twinning is a process in which a region of a crystal undergoes a homogeneous shear that produces the original crystal structure in a new orientation. In the simplest cases, this results in the atoms of the original crystal ('parent') and those of the product crystal ('twin') being mirror images of each other by reflection in a *composition plane*, as illustrated in Fig. 1.15. The open circles represent the positions of the atoms before twinning and the black circles the positions after twinning. The atoms above $x-y$ are mirror images of the atoms below it and therefore $x-y$ represents the trace of the twin composition plane in the plane of the paper. The homogeneous shear of the lattice parallel to the composition plane is denoted by arrows. Deformation twinning can be induced by plastic deformation and is particularly important in body-centred cubic and close-packed hexagonal metals and many non-metallic crystals (Chapter 6). When a growing twin meets a flat surface it produces a well-defined tilt, and this can readily be detected in an optical microscope. Figure 1.16 shows the tilts produced by deformation twins in a 3.25 per cent silicon iron crystal deformed at 20 K. Although the twinning process differs from slip (Chapter 3), in which there is no rotation of the lattice, the sequential shear of atoms in planes parallel

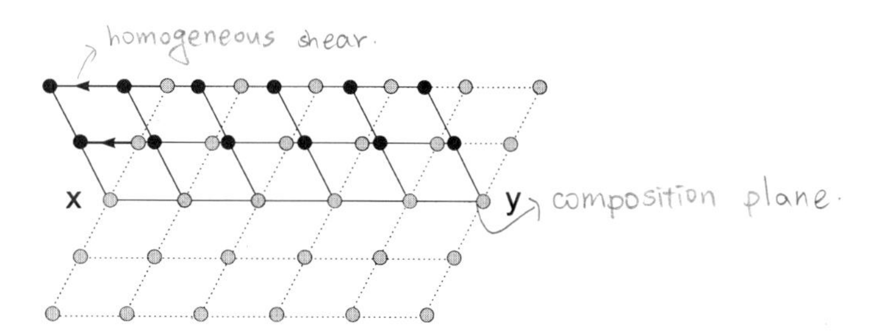

Figure 1.15 Arrangement of atoms in a twin related structure; $x-y$ is the trace of the twin composition plane.

Figure 1.16 Deformation twins in 3.25 per cent silicon iron. The surface at the twins is tilted so light is reflected in a different direction.

to the composition plane (Fig. 1.15) occurs by the movement of twinning dislocations (Chapter 9).

Volume Defects

Crystal defects such as precipitates, voids and bubbles can occur under certain circumstances and have important effects on the properties of crystalline solids. As an example, it will be seen in Chapter 10 how the interaction of dislocations with precipitates has played a vital role in the development of high-strength alloys.

1.4 Dislocations

Although there are many techniques now available for the direct observation of dislocations (Chapter 2), the existence of these line defects was deduced by inference in the early stages of dislocation study (1934 to the early 1950s). Strong evidence arose from attempts to reconcile theoretical and experimental values of the applied shear stress required to *plastically deform* a single crystal. As explained in section 3.1, this deformation occurs by atomic planes sliding over each other. In a perfect crystal, i.e. in the absence of dislocations, the sliding of one plane past an adjacent plane would have to be a rigid co-operative movement of all the atoms from one position of perfect registry to another. The shear stress required for this process was first calculated by Frenkel in 1926. The situation is illustrated in Fig. 1.17. It is assumed that there is a periodic shearing force required to move the top row of atoms across the bottom row which is given by the sinusoidal relation:

$$\tau = \frac{Gb}{2\pi a} \sin \frac{2\pi x}{b} \tag{1.4}$$

where τ is the applied shear stress, G is the shear modulus, b the spacing between atoms in the direction of the shear stress, a the spacing of the rows of atoms and x is the shear translation of the two rows away from the low-energy equilibrium position.

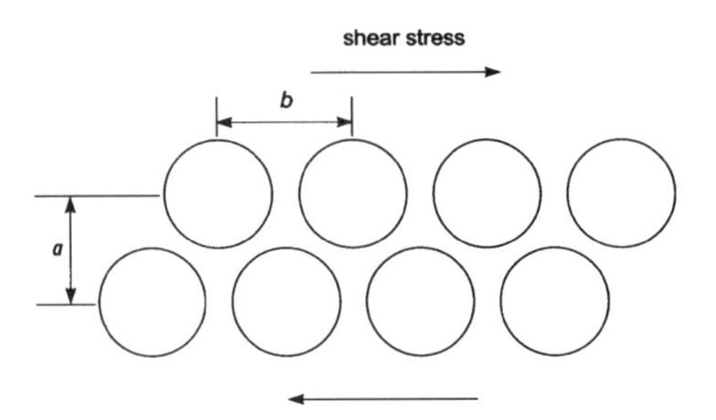

Figure 1.17 Representation of atom positions used to estimate the theoretical critical shear stress for slip.

The right-hand side of equation (1.4) is periodic in b and reduces to Hooke's law for small strains x/a, i.e. in the small-strain limit, $\sin(2\pi x/b) \simeq (2\pi x/b)$. The maximum value of τ is then the *theoretical critical shear stress* and is

$$\tau_{th} = \frac{b}{a}\frac{G}{2\pi} \tag{1.5}$$

Since $b \approx a$, the theoretical shear strength is a sizeable fraction of the shear modulus. Using more realistic expressions for the force as a function of shear displacement, values of $\tau_{th} \approx G/30$ have been obtained. Although these are approximate calculations, they show that τ_{th} is many orders of magnitude greater than the observed values (10^{-4} to $10^{-8}\,G$) of the resolved shear stress for slip measured in real, well-annealed crystals. This striking difference between prediction and experiment was accounted for by the presence of dislocations independently by Orowan, Polanyi and Taylor in 1934. Since then, it has been possible to produce crystals in the form of fibres of a small diameter, called whiskers, which have a very high degree of perfection. These whiskers are sometimes entirely free of dislocations and their strength is close to the theoretical strength.

Other evidence which contributed appreciably to the universal acceptance of the existence of dislocations in crystals, was the reconciliation of the classical *theory of crystal growth* with the experimental observations of growth rates. Consider a perfect crystal having irregular facets growing in a supersaturated vapour. At a low degree of supersaturation, growth occurs by the deposition of atoms on the irregular or imperfect regions of the crystal. The preferential deposition in imperfect regions results in the formation of more perfect faces consisting of close-packed arrays of atoms. Further growth then requires the nucleation of a new layer of atoms on a smooth face. This is a much more difficult process, and nucleation theory predicts that for growth to occur at the observed rates a degree of supersaturation of approximately 50 per cent would be required. This is contrary to many experimental observations which show that growth occurs readily at a supersaturation of only 1 per cent. The difficulty was resolved when it was demonstrated that the presence of dislocations in the crystal during growth could result in the formation of steps on the crystal faces which are not removed by preferential deposition, as in a perfect crystal. These steps provide sites for deposition and thus eliminate the difficult nucleation process.

Geometry of Dislocations

The role of dislocations in plastic deformation is explained in Chapter 3. At this stage it will be sufficient to describe the basic geometry of an *edge* and a *screw* dislocation and introduce the appropriate definitions and terminology.

Figure 1.18(a) represents an elementary descriptive model of the atomic arrangement and bonding in a simple cubic structure. For con-

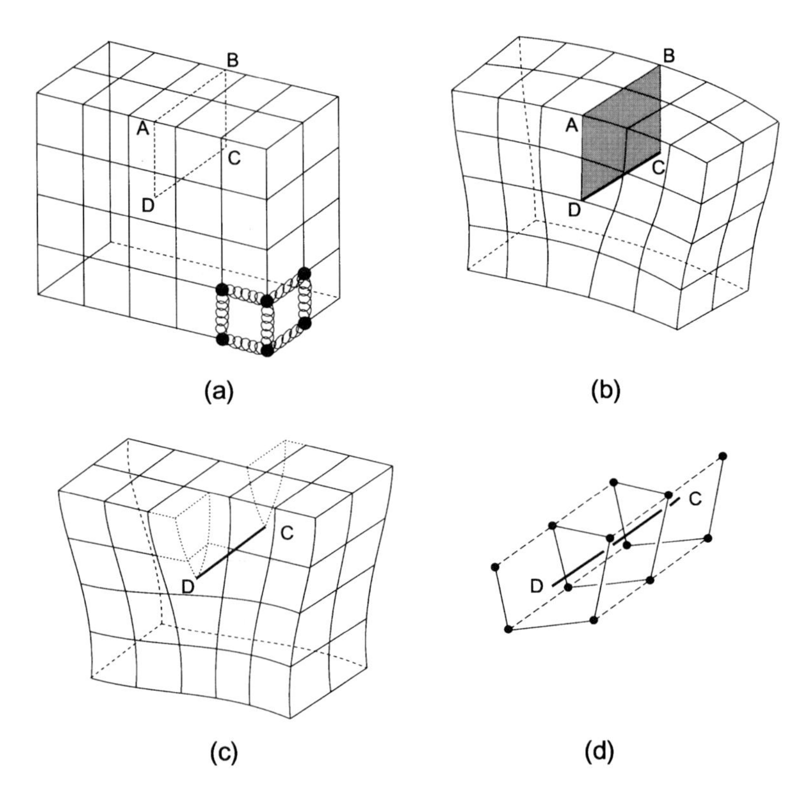

Figure 1.18 (a) Model of a simple cubic lattice; the atoms are represented by filled circles, and the bonds between atoms by springs, only a few of which are shown; (b) positive edge dislocation *DC* formed by inserting an extra half-plane of atoms in *ABCD*; (c) left-handed screw dislocation *DC* formed by displacing the faces *ABCD* relative to each other in direction *AB*; (d) spiral of atoms adjacent to the line *DC* in (c).

venience it is assumed that the bonds can be represented by flexible springs between adjacent atoms. It must be emphasised that bonding in real solids is complex and, in fact, the nature of the bonding determines the fine detail of the arrangement of the atoms around the dislocation. The arrangement of atoms around an edge dislocation can be simulated by the following sequence of operations. Suppose that all the bonds across the surface *ABCD* are broken and the faces of the crystal are separated so that an *extra half-plane of atoms* can be inserted in the slot, as illustrated in Fig. 1.18(b). The faces of the slot will have been displaced by one atom spacing, but the only large disturbance of the atoms from their normal positions relative to their neighbours is close to the line *DC*. The deflection and distortion of the interatomic bonds decrease with increasing distance from the line. This line *DC* is called a *positive edge dislocation* and is represented symbolically by ⊥. A *negative edge dislocation* would be obtained by inserting the extra plane of atoms below plane *ABCD* and is represented by ⊤.

The arrangement of atoms round a screw dislocation can be simulated by displacing the crystal on one side of *ABCD* relative to the other side in the direction *AB* as in Fig. 1.18(c). Examination of this model shows that it can be described as a *single surface helicoid*, rather like a spiral staircase. The set of parallel planes initially perpendicular to *DC* have been transformed into a single surface, and the spiral nature is clearly demonstrated by the atom positions shown in Fig. 1.18(d). *DC* is a screw dislocation. Looking down the dislocation line, if the helix advances one plane when a *clockwise circuit* is made round it, it is referred to as a *right-handed screw* dislocation, and if the reverse is true it is *left-handed.*

It is important to realise that in both the edge and the screw dislocations described, the registry of atoms across the interface *ABCD* is identical to that before the bonds were broken.

Burgers Vector and Burgers Circuit

The most useful definition of a dislocation is given in terms of the *Burgers circuit*. A Burgers circuit is any atom-to-atom path taken in a crystal containing dislocations which forms a closed loop. Such a path is illustrated in Fig. 1.19(a), i.e. *MNOPQ*. If the same atom-to-atom sequence is made in a dislocation-free crystal, Fig. 1.19(b), and the circuit does not close, then the first circuit, Fig. 1.19(a), must enclose one or more dislocations. The vector required to complete the circuit is called the *Burgers vector*. It is essential that the circuit in the real crystal passes entirely through 'good' parts of the crystal. For simplicity consider the Burgers circuit to enclose one dislocation as in Fig. 1.19(a). The sequence of atom-to-atom movements in the perfect crystal is the same as for the circuit *MNOPQ* in Fig. 1.19(a). The closure failure *QM* is the Burgers vector and is at *right angles* to the dislocation line (cf. Fig. 1.18(b)). When the Burgers circuit is drawn round a screw dislocation (Fig. 1.20), again with a closed circuit in the crystal containing the dislocation, the Burgers vector *QM* is parallel to the dislocation line. This leads to two important rules:

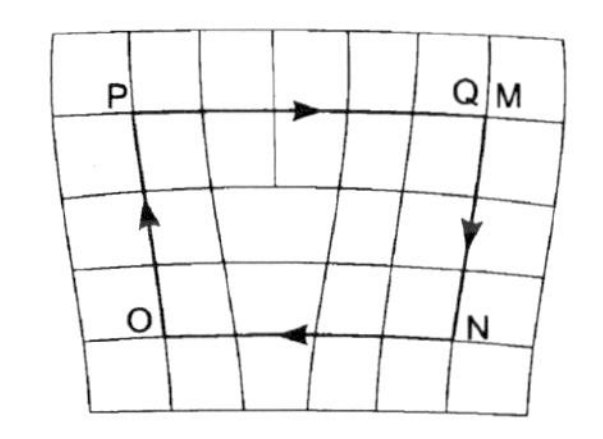

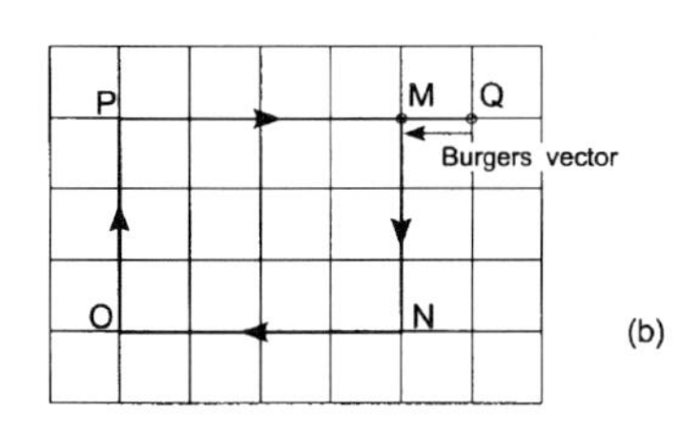

Figure 1.19 (a) Burgers circuit round an edge dislocation with positive line sense into the paper (see text), (b) the same circuit in a perfect crystal; the closure failure is the Burgers vector.

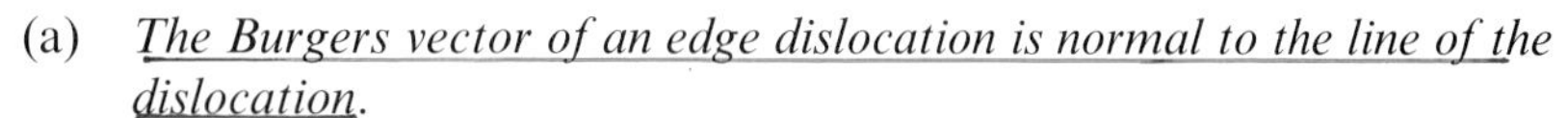
(a) *The Burgers vector of an edge dislocation is normal to the line of the dislocation.*

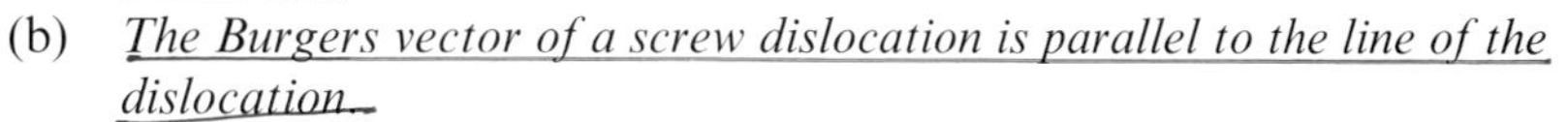
(b) *The Burgers vector of a screw dislocation is parallel to the line of the dislocation.*

In the most general case (Chapter 3) the dislocation line lies at an arbitrary angle to its Burgers vector and the dislocation line has a mixed edge and screw character. However, the Burgers vector of a single dislocation has fixed length and direction, and is independent of the position and orientation of the dislocation line.

Burgers circuits taken around other defects, such as vacancies and interstitials, do not lead to closure failures. Two rules are implied by the Burgers circuit construction used above. First, when looking along the dislocation line, which defines the positive *line sense* or direction of

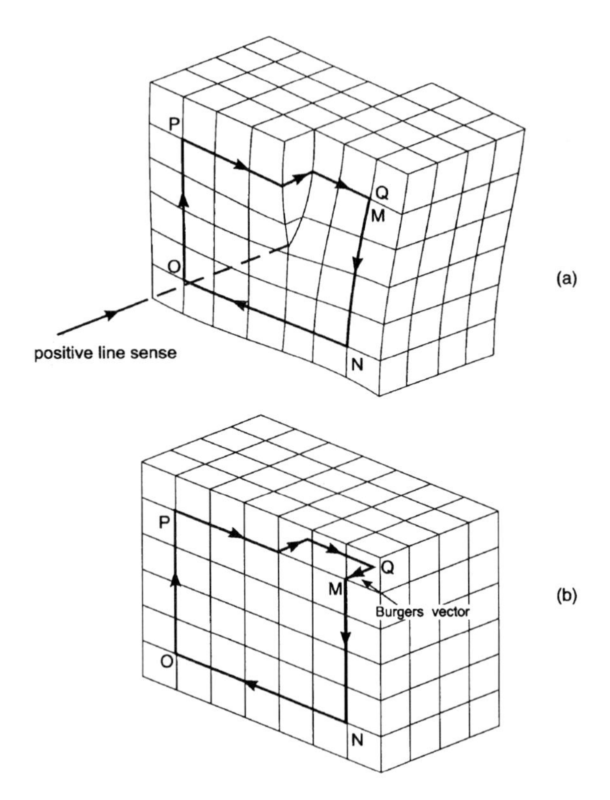

Figure 1.20 (a) Burgers circuit round a screw dislocation with positive line sense in the direction shown; (b) the same circuit in a perfect crystal; the closure failure is the Burgers vector.

the dislocation, the circuit is taken in a clockwise fashion (Figs 1.19(a), 1.20(a)). Second, the Burgers vector is taken to run from the finish to the start point of the reference circuit in the perfect crystal. This defines the right-hand/finish-start (RH/FS) convention. It is readily shown by use of sketches similar to those of Figs 1.19 and 1.20, that *reversing the line sense reverses the direction of the Burgers vector* for a given dislocation. Furthermore, *dislocations with the same line sense but opposite Burgers vectors* (or alternatively with opposite line senses and the same Burgers vector) *are physical opposites*, in that if one is a positive edge, the other is a negative edge, and if one is a right-handed screw, the other is left-handed. Dislocations which are physical opposites of each other annihilate and restore perfect crystal if brought together.

The Burgers vectors defined in the simple cubic crystals of Figs 1.19 and 1.20 are the shortest *lattice translation vectors* which join two points in the lattice. A dislocation whose Burgers vector is a lattice translation vector is known as a *perfect* or *unit dislocation*. The Burgers vector **b** is conveniently described using the indices defined in section 1.1. For example, the lattice vector from the origin to the centre of a body-

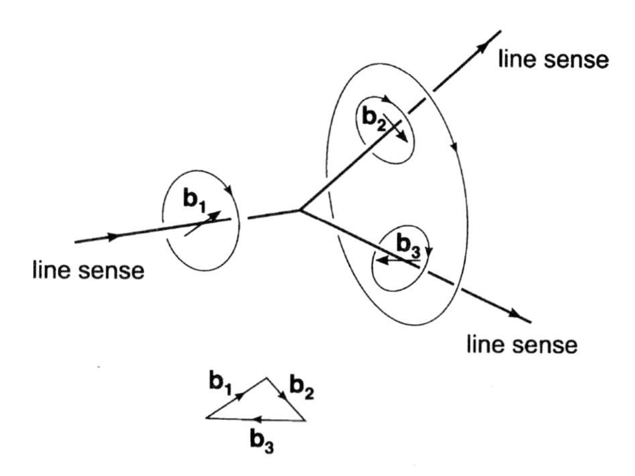

Figure 1.21 Three dislocations forming a node.

centred cubic cell is defined both in magnitude and direction by displacements of $a/2$ in the x-direction, $a/2$ in the y-direction and $a/2$ in the z-direction, and the notation used is $\mathbf{b} = \frac{1}{2}[111]$. The magnitude (or length) b of the vector is

$$b = \sqrt{\left(\frac{a^2}{4} + \frac{a^2}{4} + \frac{a^2}{4}\right)} = \frac{a\sqrt{3}}{2} \tag{1.6}$$

Similarly, if $\mathbf{b}$ is the shortest lattice translation vector in the face-centred cubic structure, i.e. $\mathbf{b} = \frac{1}{2}\langle 110\rangle$, then $b = a/\sqrt{2}$.

Dislocation lines can end at the surface of a crystal and at grain boundaries, but never inside a crystal. Thus, *dislocations must either form closed loops or branch into other dislocations*. When three or more dislocations meet at a point, or *node*, it is a necessary condition that the Burgers vector is conserved. Consider the dislocation $\mathbf{b}_1$ (Fig. 1.21) which branches into two dislocations with Burgers vectors $\mathbf{b}_2$ and $\mathbf{b}_3$. A Burgers circuit has been drawn round each according to the line senses indicated, and it follows from the diagram that

$$\mathbf{b}_1 = \mathbf{b}_2 + \mathbf{b}_3 \tag{1.7}$$

The large circuit on the right-hand side of the diagram encloses two dislocations, but since it passes through the same good material as the $\mathbf{b}_1$ circuit on the left-hand side the Burgers vector must be the same, i.e. $\mathbf{b}_1$. It is more usual to define the Burgers circuits by making a clockwise circuit around each dislocation line looking outward from the nodal point. This reverses the line sense (and hence $\mathbf{b}_1$) on the left-hand side, and then equation (1.7) becomes

$$\mathbf{b}_1 + \mathbf{b}_2 + \mathbf{b}_3 = 0 \tag{1.8}$$

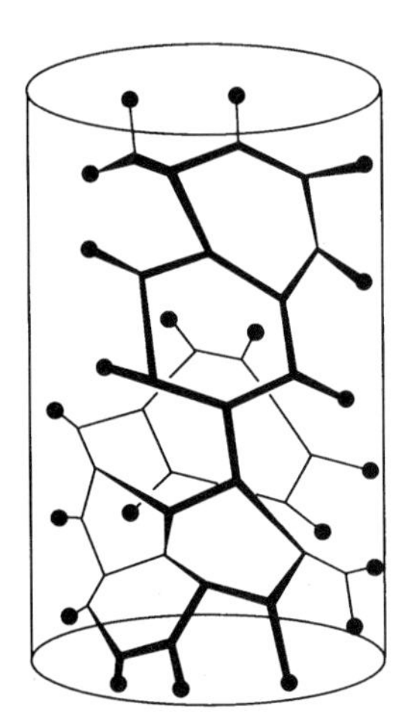

Figure 1.22 Diagrammatic illustration of the arrangement of dislocations in a well-annealed crystal; the Frank net. (From Cottrell, *The Properties of Materials at High Rates of Strain*, Inst. Mech. Eng., London, 1957.)

or, more generally, for n dislocation branches

$$\sum_{1}^{n} \mathbf{b}_i = 0 \tag{1.9}$$

The *dislocation density* ρ is defined as the total length of dislocation line per unit volume of crystal, normally quoted in units of cm^{-2} or m^{-2} ($1\,\text{m}^{-2} \equiv 10^4\,\text{cm}^{-2}$). Thus for a volume V containing line length l, $\rho = l/V$. An alternative definition, which is sometimes more convenient to use, is the number of dislocations intersecting a unit area, again measured in units of m^{-2}. If all the dislocations are parallel, the two density values are the same, but for a completely random arrangement the volume density is twice the surface density. All crystals, apart from some whiskers, contain dislocations and in well-annealed crystals the dislocations are arranged in a rather ill-defined network, the *Frank net*. In well-annealed metal crystals ρ is usually between 10^{10} and $10^{12}\,\text{m}^{-2}$, but it increases rapidly with plastic deformation, and a typical value for a heavily cold-rolled metal is about 10^{14} to $10^{15}\,\text{m}^{-2}$. The arrangement of the dislocations depends on the conditions of loading and some actual examples are presented elsewhere in the book. ρ is usually lower in nonmetallic crystals than in metal crystals, and values down to $10^5\,\text{m}^{-2}$ can be obtained in carefully grown semiconductor crystals. The average distance between dislocations in a network of density ρ is of the order of $1/\sqrt{\rho}$, i.e. 10^{-3}, 10^{-5} and 10^{-7} m for ρ equal to 10^6, 10^{10} and $10^{14}\,\text{m}^{-2}$, respectively.

Further Reading

Introduction to Materials:

Askeland, D. R. (1996) *The Science and Engineering of Materials*, Chapman and Hall.

Callister Jr, W. D. (2000) *Materials Science and Engineering*, John Wiley.

Advanced Reviews of Materials

Cahn, R. W. and Haasen, P. (1996) *Physical Metallurgy*, vols 1–3, North-Holland.

Cahn, R. W., Haasen, P. and Kramer, E. J. (1993–on) *Materials Science and Technology: A Comprehensive Treatment*, vols 1–18, Wiley-VCH.

Crystals: Atomic and Electronic Structure

Barrett, C. S. and Massalski, T. B. (1980) *Structure of Metals*, Pergamon.

Buerger, M. J. (1978) *Elementary Crystallography*, M.I.T. Press.

Cullity, B. D. (1998) *Elements of X-ray Diffraction*, Addison-Wesley.

Kelly, A., Groves, G. W. and Kidd, P. (2000) *Crystallography and Crystal Defects*, John Wiley.

Kittel, C. (1995) *Introduction to Solid State Physics*, Wiley.

Pettifor, D. (1995) *Bonding and Structure of Molecules and Solids*, Oxford University Press.

Sutton, A. P. (1993) *Electronic Structure of Materials*, Oxford University Press.
Wert, C. A. and Thomson, R. M. (1970) *Physics of Solids*, McGraw-Hill.
Windle, A. H. (1977) *A First Course in Crystallography*, Bell.

Point Defects, Stacking Faults, Twins and Grain Boundaries

Christian, J. W. (1975) *The Theory of Transformations in Metals and Alloys*, Pergamon.
Crawford, J. H. and Slifkin, L. M. (Eds.) (1972) *Point Defects in Solids*, Plenum.
Henderson, B. (1972) *Defects in Crystalline Solids*, Edward Arnold.
Mahajan, S. and Williams, D. F. (1973) 'Deformation twinning in metals and alloys', *Inter. Metall. Rev.* **18**, 43.
Peterson, N. L. and Siegel, R. W. (Eds.) (1978) *Properties of Atomic Defects in Metals*, North-Holland.
Reed-Hill, R., Hirth, J. P. and Rogers, H. C. (Eds.) (1965) *Deformation Twinning*, Gordon and Breach.
Smallman, R. E. and Harris, J. E. (Eds.) (1977) *Vacancies* '76, The Metals Soc., London.
Sutton, A. P. and Balluffi, R. W. (1995) *Interfaces in Crystalline Materials*, Oxford University Press.
Takamura, J., Doyama, M. and Kiritani, M. (Eds.) (1982) *Point Defects and Defect Interaction in Metals*, Univ. of Tokyo Press.

Dislocations

Cottrell, A. H. (1953) *Dislocations and Plastic Flow in Crystals*, Oxford University Press.
Friedel, J. (1964) *Dislocations*, Pergamon Press.
Gilman, J. J. (1969) *Micromechanics of Flow in Solids*, McGraw-Hill.
Hirsch, P. B. (Ed.) (1975) *The Physics of Metals 2. Defects*, Cambridge University Press.
Hirth, J. P. and Lothe, J. (1982) *Theory of Dislocations*, Wiley (1st edition 1968, McGraw-Hill).
Kelly, A., Groves, G. W. and Kidd, P. (2000) *Crystallography and Crystal Defects*, John Wiley.
Nabarro, F. R. N. (1967) *The Theory of Crystal Dislocations*, Oxford University Press.
Nabarro, F. R. N. (Ed.) (1979–1996) *Dislocations in Solids*, vols 1–10, North-Holland.
Read, W. T. (1953) *Dislocations in Crystals*, McGraw-Hill.
Seeger, A. (1955) 'Theory of lattice imperfections', *Handbuch der Physik*, Vol. VII, part 1, p. 383, Springer-Verlag.
Seeger, A. (1958) 'Plasticity of crystals', *Handbuch der Physik*, Vol. VII, part II, p. 1, Springer-Verlag.
Veysièrre, P. (1999) 'Dislocations and the plasticity of crystals', *Mechanics and Materials: Fundamental Linkages*, p. 271 (eds M. A. Meyers, R. W. Armstrong and H. Kirchner), John Wiley.
Weertman, J. and Weertman, J. R. (1964) *Elementary Dislocation Theory*, Macmillan.

2 Observation of Dislocations

2.1 Introduction

A wide range of techniques has been used to study the distribution, arrangement and density of dislocations and to determine their properties. The techniques can be divided into five main groups. (1) Surface methods, in which the point of emergence of a dislocation at the surface of a crystal is revealed. (2) Decoration methods, in which dislocations in bulk specimens transparent to light are decorated with precipitate particles to show up their position. (3) Transmission electron microscopy, in which the dislocations are studied at very high magnification in specimens 0.1 to 4.0 μm thick. This is the most widely applied technique. (4) X-ray diffraction, in which local differences in the scattering of X-rays are used to show up the dislocations. (5) Field ion microscopy, which reveals the position of individual atoms. Except for (5) and isolated examples in (3), these techniques do not reveal directly the arrangement of atoms at the dislocation, but rely on such features as the strain field of the dislocation (Chapter 4) to make them 'visible'.

Computer-based methods have also been employed extensively in recent years. Computer simulation has been used, for example, to assist in the analysis of images obtained by transmission electron microscopy and to predict the atomic configuration around dislocations in model crystals.

2.2 Surface Methods

If a crystal containing dislocations is subjected to an environment which removes atoms from the surface, the rate of removal of atoms around the point at which a dislocation emerges at the surface may be different from that for the surrounding matrix. The difference in the rate of removal arises from one or more of a number of properties of the dislocation: (1) lattice distortion and strain field of the dislocation; (2) geometry of planes associated with a screw dislocation so that the reverse process to crystal growth produces a surface pit; (3) concentration of impurity atoms at the dislocation which changes the chemical composition of the material near the dislocation. If the rate of removal is more rapid around the dislocation, pits are formed at these sites, Fig. 2.1, and if less rapid small hillocks are formed. Many methods are available for the slow, controlled removal of atoms from the surface. The most common and most useful are chemical and electrolytic etching. Other methods include thermal etching, in which the atoms are removed by evaporation when the crystal is heated in a low pressure

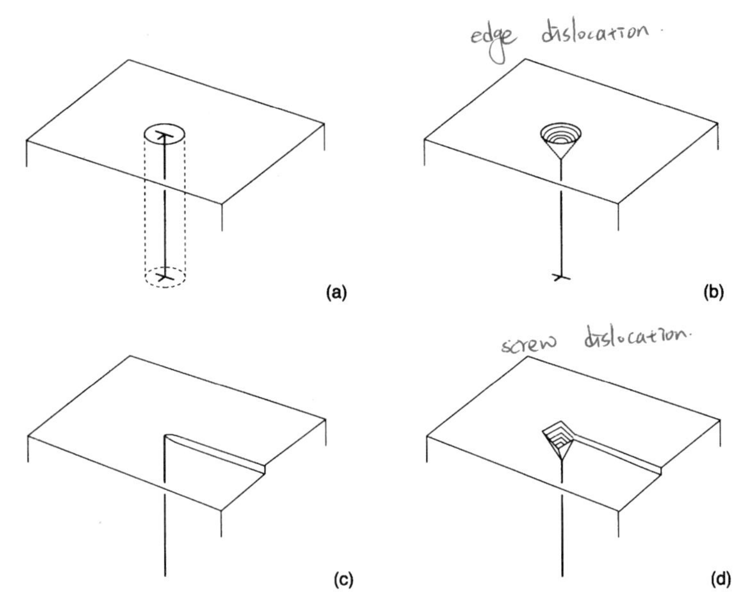

Figure 2.1 Formation of etch pits at the site where a dislocation meets the surface. (a) Edge dislocation, the cylindrical zone around the dislocation represents the region of the crystal with different physical and chemical properties from the surrounding crystal. (b) Conical-shaped pit formed at an edge dislocation due to preferential removal of atoms from the imperfect region. (c) Emergent site of a screw dislocation. (d) Spiral pit formed at a screw dislocation; the pits form by the reverse process to the crystal growth mechanism.

atmosphere at high temperature, and sputtering in which the surface atoms are removed by gas ion bombardment. The last two methods usually reveal only screw dislocations.

The specimen surface in Fig. 2.2 has been etched and preferential attack has produced black spots identified as pits. A number of methods have been used to confirm a *one-to-one correspondence* between dislocations and etch pits. The first was that due to Vogel and co-workers (1953) and is illustrated in Fig. 2.3. The photograph shows a regularly spaced row of etch pits formed at the boundary between two germanium crystals. Very precise X-ray measurements were made to determine the misorientation between the crystals and it was found that the boundary was a symmetrical pure tilt boundary (section 9.2) which can be described as an array of edge dislocations spaced one above the other as shown in Fig. 2.3(b). Such an array produces a tilt between the grains on opposite sides of the boundary. If b is the magnitude of the Burgers vector of the edge dislocations and D their distance apart, the angle of misorientation is $\theta = b/D$ (equation (9.2)). For the boundary in Fig. 2.3(b), the measured value of θ was 65 sec of arc and the predicted value of D was 1.3 μm. This agrees closely with the spacing of the etch pits shown in the photograph, confirming that in this example there is a one-to-one correspondence between etch pits and dislocations.

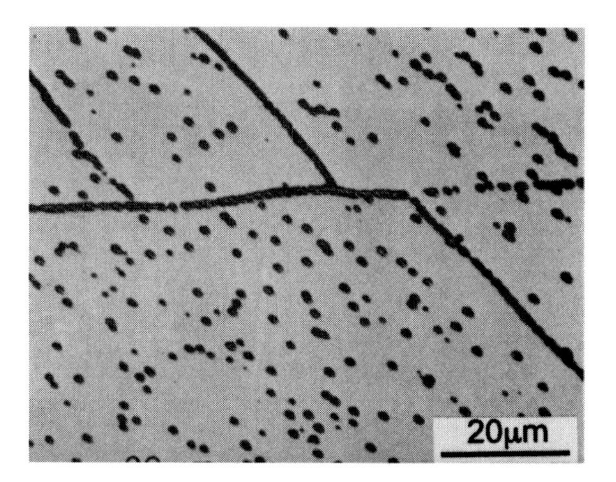

Figure 2.2 Etch pits produced on the surface of a single crystal of tungsten. (From Schadler and Low, unpublished.)

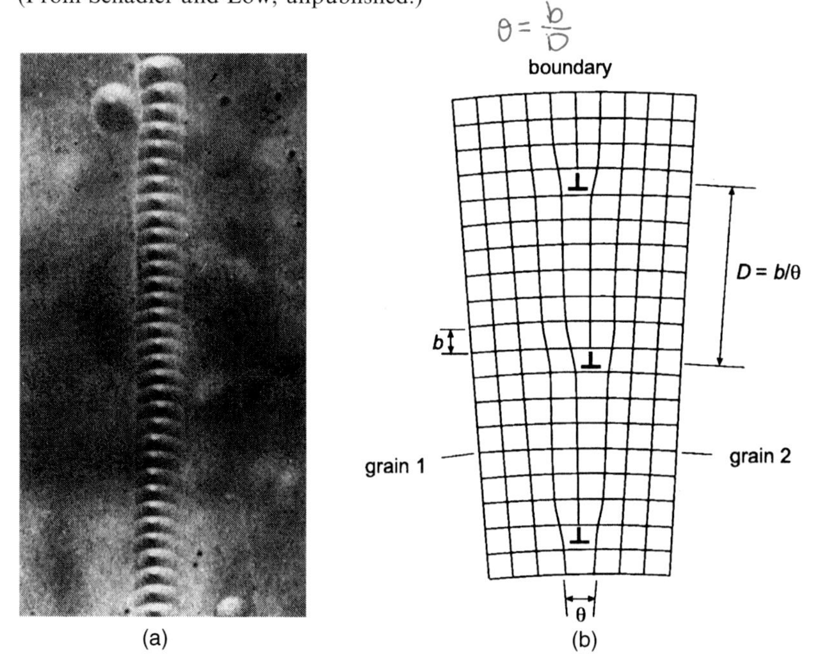

Figure 2.3 (a) A row of etch pits formed at the boundary between two germanium crystals. The etch pits are uniformly spaced. (b) Diagrammatic representation of the arrangement of dislocations in the boundary revealed by the etch pits in (a). This is a symmetrical pure tilt boundary which consists of a vertical array of edge dislocations with parallel Burgers vectors of the same sign. (After Vogel, Pfann, Corey and Thomas, *Physical Review* **90**, 489, 1953.)

Although etch pit studies are limited to the surface examination of bulk specimens the technique can be used to study the movement of dislocations as demonstrated by Gilman and Johnston (1957) in lithium

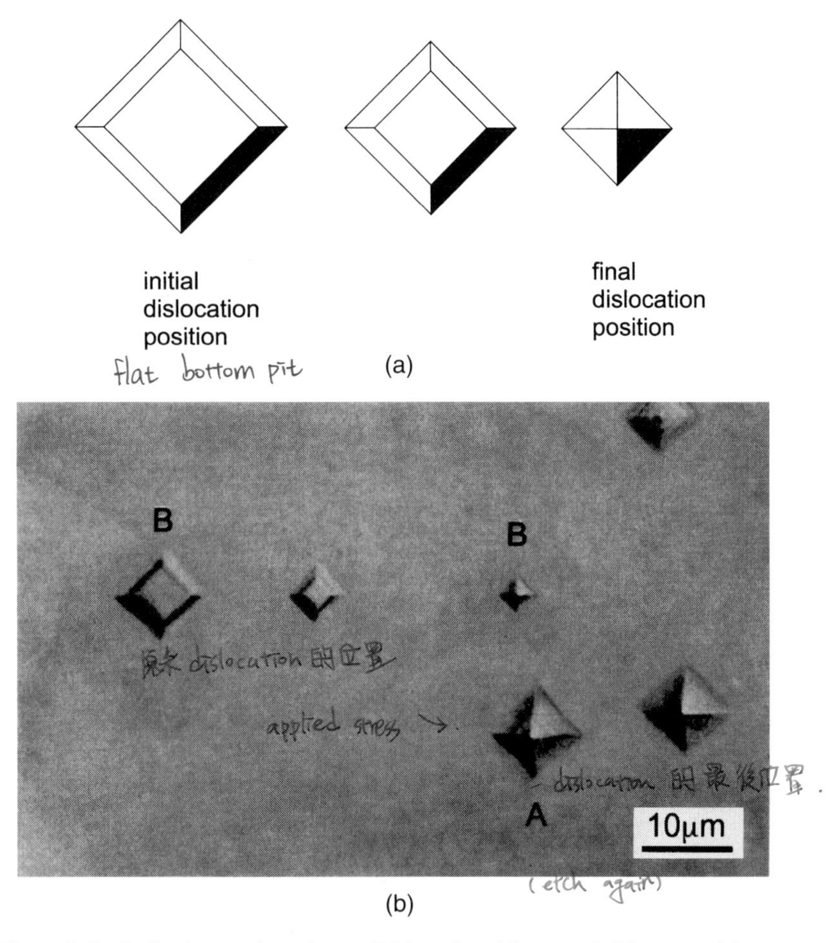

Figure 2.4 Etch pits produced on a lithium fluoride crystal. The crystal has been etched three times. The dislocation at *A* has not moved between each etching treatment and a large pyramid-shaped pit has formed. The dislocation revealed by the three pits *B* moved between etching treatments to the positions indicated by the pits. Subsequent etching of a pit after the dislocation has moved produces a flat bottom pit. (From Gilman and Johnston (1957), *Dislocations and Mechanical Properties of Crystals*, p. 116, Wiley.)

fluoride. This is illustrated in Fig. 2.4. The site of a stationary dislocation in a crystal appears as a sharp bottom pit. When a stress is applied to the crystal the dislocation moves; the distance moved depends on the applied stress and the length of time the stress is applied. If the crystal is etched again, the new position of the dislocation is revealed by a new sharp bottom pit. The etchant also attacks the old pit which develops into a flat bottom pit.

In general, surface techniques are limited to crystals with a low dislocation density, less than $10^{10}\ \mathrm{m}^{-2}$, because the etch pits have a finite size and are very difficult to resolve when they overlap each other. The distribution of dislocations in three dimensions can be obtained, using this technique, by taking successive sections through the crystal.

2.3 Decoration Methods

The dislocations in crystals which are transparent to visible light and infrared radiation are not normally visible. However, it is possible to *decorate* the dislocations by inducing precipitation along the line of the dislocation. Sites along the dislocation are favoured for precipitation by the lattice distortion there. The effect produced is similar in appearance to a row of beads along a fine thread. The position of the dislocation is revealed by the scattering of the light at the precipitates and can be observed in an optical microscope. In most applications of this method the decoration process involves the heating of the crystals before examination and this restricts the use of the method to the study of 'recovered' or high-temperature deformation structures. It is not suitable for studying structures formed by low-temperature deformation.

The procedure with the widest range of application, which has been used to study dislocations in alkali halide and semiconductor crystals, is to dope with impurity atoms. By suitable heat treatment, the precipitation of the foreign atoms can subsequently be induced along the dislocations. Amelinckx (1958) revealed dislocations in KCl by adding

Figure 2.5 A thin crystal of KCl examined in an optical microscope. Particles of silver have precipitated on the dislocations, which are in the form of a network. Only part of the network is in focus. (From Amelinckx, *Acta Metall.* **6**, 34, 1958.)

AgCl to the melt prior to crystal growth. The crystals were annealed in a reducing atmosphere of hydrogen to precipitate silver particles on the dislocations present, as shown by the optical micrograph in Fig. 2.5. By a similar method, dislocations in silicon can be decorated and then observed with infrared radiation by diffusing in metallic elements such as copper or aluminium. These decoration techniques can be used in combination with etch-pit studies to demonstrate the one-to-one correspondence between dislocations and etchpits. They have been reviewed by Amelinckx (1964).

An optical method which does not depend on decoration has been used to study dislocations in magnetic bubble-domain materials. It utilises detection by polarised light of the *birefringence* induced by the stress from the dislocation. A brief review is given by Humphreys (1980).

2.4 Electron Microscopy

General Principles

Transmission electron microscopy has been the technique most widely used for the observation of dislocations and other crystal defects, such as stacking faults, twin and grain boundaries, and voids. Static arrangements of defects are usually studied, but in some cases, miniaturised tensile stages have been used to deform specimens (less than 3 mm in length) within a microscope, thus providing direct observation of dislocation interaction and multiplication processes. Transmission electron microscopy is applicable to a wide range of materials, subject only to the conditions that specimens can be prepared in very thin section and that they remain stable when exposed to a beam of high-energy electrons within a high vacuum. The electrons in a *conventional* transmission electron microscope have an energy of typically 100 keV. Inelastic scattering of the electrons as they pass through a solid sets an upper limit to the specimen thickness of from about 100 nm for heavy, high atomic number elements to 1000 nm for light elements. Only for *high-voltage* microscopes of energy ~1 MeV can these limits for electron transparency be exceeded, and then only by about five times. Like its light counterpart, the electron microscope uses lenses to focus the beam and produce an image, but the lenses are electromagnetic. The electrons behave as de Broglie waves, with a wavelength of 3.7 pm at 100 keV, and this sets a theoretical limit to the resolution of the same order of magnitude. In practice, lens aberrations and electrical and mechanical stabilities place the limit at typically 0.2 to 0.4 nm.

Schematic ray diagrams for a conventional transmission microscope are shown in Fig. 2.6. Electrons emerge from the condenser lens onto the specimen, which is usually a disc a few mm in diameter. Within a *crystalline* sample, the scattered electrons are concentrated into discrete directions. These *diffracted* beams satisfy the geometrical requirements of *Bragg's Law*, in that each set of parallel crystal planes diffract

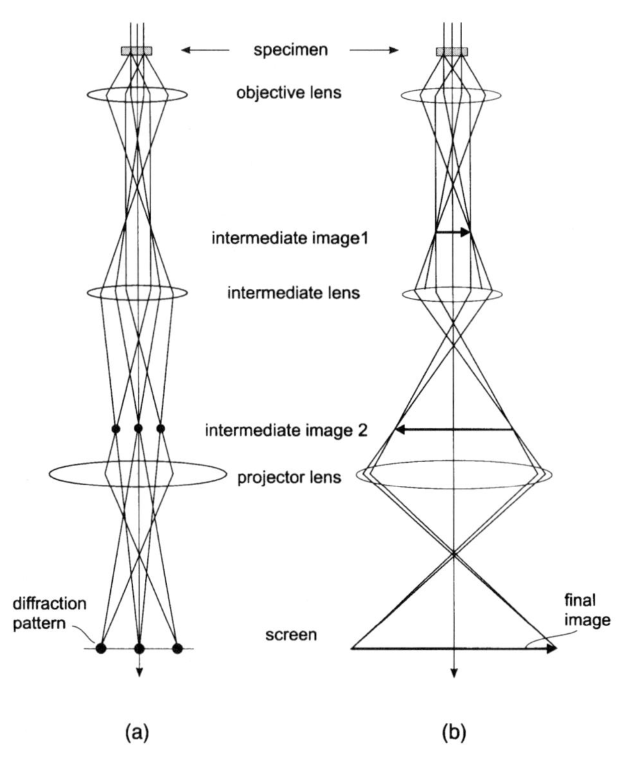

Figure 2.6 Two basic operations of the TEM imaging systems involving projecting (a) the diffraction pattern and (b) the image on to the viewing screen. The intermediate lens selects either the back focal plane or the image plane of the objective lens as its object.

electrons in a specific direction. The diffracted electron beams are brought to focus in the back focal plane of the objective lens (Fig. 2.6(a)), which is the plane of the diffraction pattern. When the microscope is operated in the *diffraction mode*, the diffraction lens is focused on the back focal plane and the subsequent lenses project a magnified diffraction pattern on the fluorescent screen. The objective lens also produces an inverted image of the specimen in the first image plane, however, and if the diffraction lens is focused on this plane, the microscope is in the *imaging mode* and produces a magnified image (Fig. 2.6(b)), the final magnification being variable between about 10^2 and 10^6 times. An aperture is placed in the objective back focal plane to permit only one beam to form the image. When the beam transmitted directly through the specimen is selected, a *bright-field* image is displayed, whereas when one of the diffracted beams is chosen, a *dark-field* image is observed. For the analysis of defects observed in the imaging mode, information on the diffraction conditions is required and is obtained in the diffraction mode. Of particular importance is the *diffraction vector* **g** perpendicular to the diffracting planes.

The image simply reveals the variation in intensity of the selected electron beam as it leaves the specimen. In the theory of this diffraction contrast, it is usually assumed for simplicity that the intensity variation across the bottom surface of the specimen may be obtained by considering the specimen to consist of narrow columns (a few nm wide) with axes parallel to the beam: the electron intensity does not vary across a narrow column and is independent of neighbouring columns. The final image is therefore an intensity map of the grid of columns. Furthermore, a two-beam condition is assumed to apply, wherein the contribution of the electrons in the directly transmitted beam and *one* diffracted beam only, is considered, i.e. only one set of crystal planes is at, or close to, the Bragg angle. These *column* and *two-beam approximations* have been shown to be valid for conventional microscopy. If the specimen is perfectly flat, uniformly thick and free of defects, the image is homogeneous with no variations in intensity. Image contrast only arises if variations in beam intensity occur from one part of the specimen to another. For example, if the specimen is bent in one region sufficiently to affect crystal planes which are close to the Bragg condition, strong variations in intensity can result. Dislocations and many other crystal defects bend crystal planes.

The effect of crystal defects on the image depends on the vector $\mathbf{u}$ by which atoms are displaced from their perfect lattice sites. In the column approximation, $\mathbf{u}$ within a column varies only with z, the coordinate along the column axis. Solutions of the equations for the electron intensities contain a factor $\mathbf{g} \cdot \mathbf{u}$, or alternatively $\mathbf{g} \cdot \mathrm{d}\mathbf{u}/\mathrm{d}z$, which is not present for a perfect crystal. Thus, diffraction conditions for which $\mathbf{g} \cdot \mathbf{u}$ (or $\mathbf{g} \cdot \mathrm{d}\mathbf{u}/\mathrm{d}z$) is zero will not lead to contrast in the image. The physical interpretation of these effects follows from the fact that $\mathrm{d}\mathbf{u}/\mathrm{d}z$ is the variation of displacement with depth and it therefore measures the bending of atomic planes. If the reflecting planes normal to $\mathbf{g}$ are not tilted by the defect, there is no change in diffracted intensity and the defect is invisible.

Dislocations

It can be seen from Fig. 1.18 that certain planes near a dislocation are bent, and the bending decreases with increasing distance from the dislocation. (Expressions for $\mathbf{u}$ are given in Chapter 4.) If these planes are bent into a strongly diffracting orientation, the intensity of the directly transmitted beam will be reduced (and that of the diffracted beam increased) in columns near the dislocation. This is shown schematically in Fig. 2.7. The dislocation will appear as a dark line in the bright-field image (or as a bright line in a dark-field image). An example of two rows of dislocations is shown by the bright-field micrograph of Fig. 2.8. The actual dislocation structure observed in a particular specimen depends very much on the material and, if it has been deformed, on the strain, strain rate and temperature it has experienced. Examples are to be found

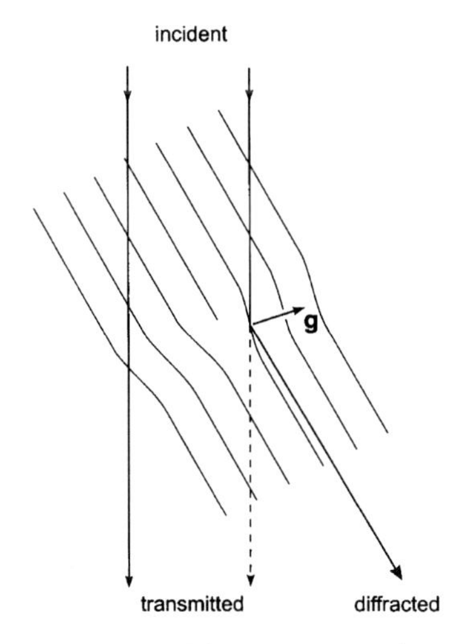

Figure 2.7 Planes near an edge dislocation bent into the orientation for diffraction.

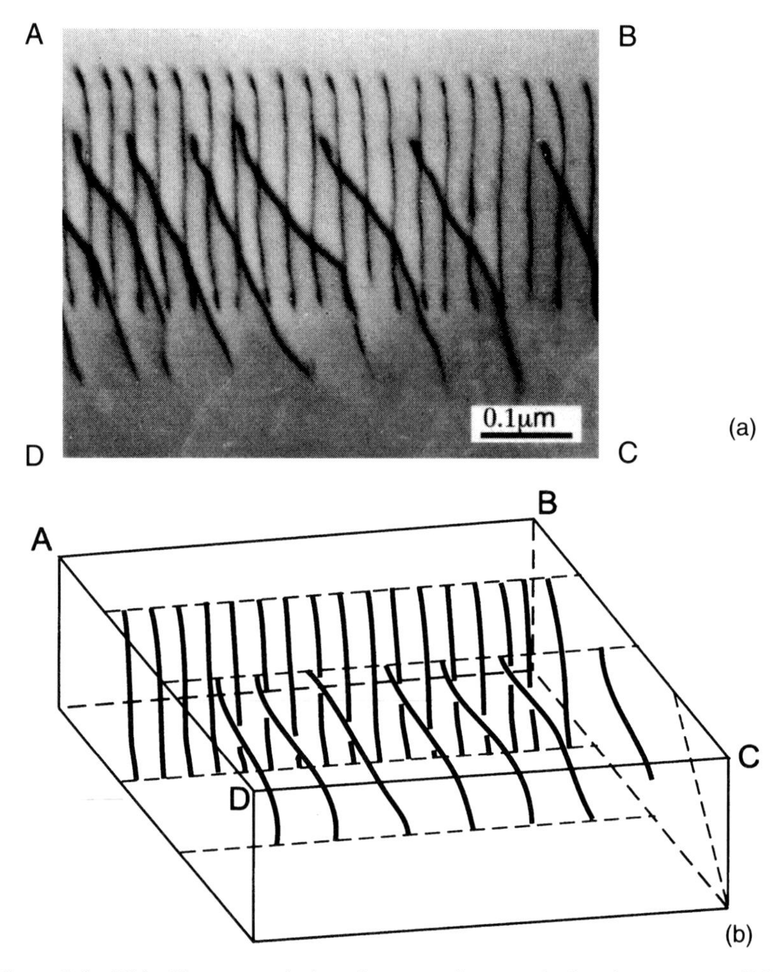

Figure 2.8 Thin film transmission electron micrograph showing two parallel rows of dislocations. Each dark line is produced by a dislocation. The dislocations extend from top to bottom of the foil which is about 200 nm thick. The line diagram illustrates the distribution of the dislocations in the foil and demonstrates that the photograph above represents a projected image of a three-dimensional array of dislocations.

in Figs 2.9, 6.15, 9.1, 9.2 and 9.9. The width of a dislocation image is determined by the variation of $d\mathbf{u}/dz$ across the specimen, i.e. by the range of the dislocation strain field, and is typically 10 to 50 nm. The actual form of the image depends on the diffraction conditions, the nature of the dislocation and its depth in the foil. It may appear as a single line (not necessarily centred on the real dislocation), a double line, a wavy line or a broken line. Also, the line may be invisible, as explained above, and this may be exploited to determine the Burgers vector **b**.

The *invisibility criterion* is $\mathbf{g} \cdot \mathbf{u} = 0$. Consider first a straight screw dislocation. In an isotropic medium, planes parallel to the line remain

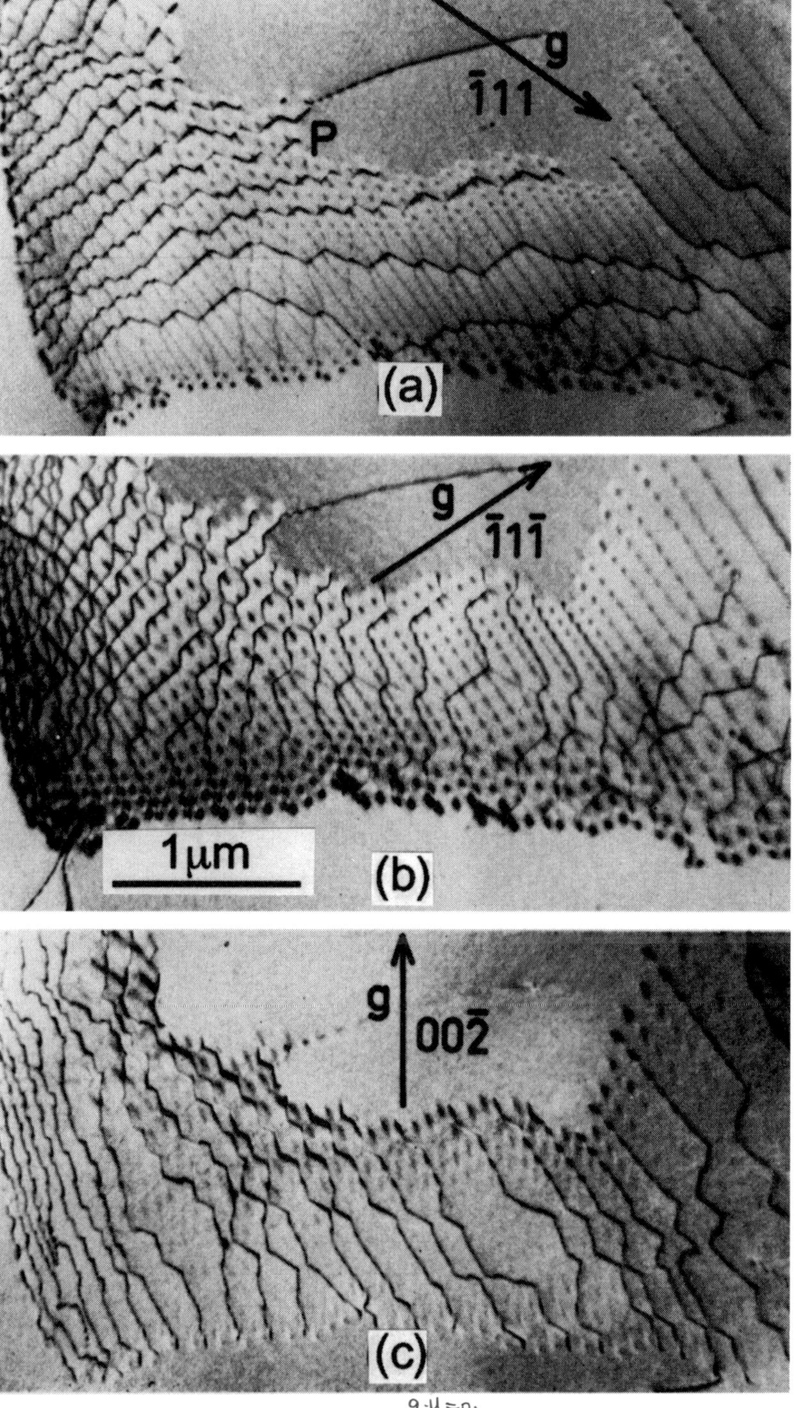

Figure 2.9 Illustration of the use of the $\mathbf{g} \cdot \mathbf{b} = 0$ method to determine the Burgers vector of dislocations. (From Lindroos, *Phil. Mag.* **24**, 709, 1971.)

flat (see Fig. 1.18), for **u** is parallel to **b** (section 4.3). Hence, when **g** is perpendicular to **b**, $\mathbf{g}\cdot\mathbf{u} = \mathbf{g}\cdot\mathbf{b} = 0$ and the invisibility criterion is satisfied. If the specimen is tilted in the microscope to find two sets of diffracting planes $\mathbf{g}_1$ and $\mathbf{g}_2$ for which the line is invisible, **b** must be perpendicular to both $\mathbf{g}_1$ and $\mathbf{g}_2$ and therefore has the direction of $\mathbf{g}_1 \times \mathbf{g}_2$. For an edge dislocation, all planes parallel to the line are bent, and **u** is non-zero in all directions perpendicular to the line. The criterion $\mathbf{g}\cdot\mathbf{u} = 0$, therefore, requires in this case that both $\mathbf{g}\cdot\mathbf{b}$ and $\mathbf{g}\cdot(\mathbf{b}\times\mathbf{t})$ be zero, where **t** is a vector along the line. It is satisfied only when **g** is parallel to the line, for only planes perpendicular to the line remain flat (see Fig. 1.18). For the mixed dislocation, which produces a combination of edge and screw displacements, there is no condition for which $\mathbf{g}\cdot\mathbf{u}$ is exactly zero. Also, in anisotropic crystals no planes remain flat around edge and screw dislocations, except in a few special cases. Thus, the invisibility method for determining **b** often relies on finding diffraction vectors which result in weak contrast rather than complete invisibility. It has proved, nevertheless, a powerful technique in dislocation studies. An example is illustrated in Fig. 2.9. Here, three different diffraction

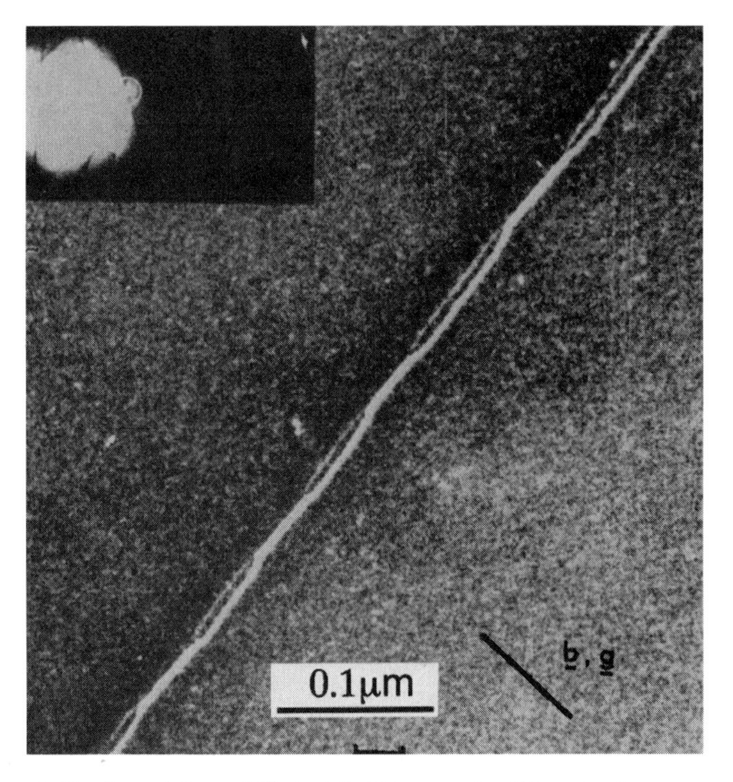

Figure 2.10 A weak-beam 220 dark-field image of a dislocation in an annealed silicon specimen showing constricted segments. The inset shows the diffracting conditions used to form the image. (From Ray and Cockayne, *Proc. Roy. Soc.* **A325**, 543, 1971.)

vectors **g** have been chosen to produce three different images of the same field of view. It contains a network of four sets of dislocation lines.

Although modern transmission electron microscopes have resolution limits of less than 1 nm, the dislocation image width is much greater than this, and fine structure of the dislocation, such as splitting into partials (Chapters 5 and 6) may not be resolved. The *weak-beam* technique can overcome this disadvantage. It utilizes the fact that when the specimen is tilted away from the Bragg orientation so that the diffracted beam from the perfect crystal is weak; strong diffraction from a dislocation can only occur when $\mathbf{g} \cdot (d\mathbf{u}/dz)$ is large. The strong bending of planes this requires only occurs close to the centre of the dislocation. The weak-beam images of dislocations are therefore much narrower (typically 1–2 nm) than those obtained under strong-beam conditions, although the experimental procedure is more demanding. An example of closely-spaced dislocations resolved in silicon is shown in Fig. 2.10. The invisibility criteria are the same as those discussed above, and the weak-beam method has been applied to many materials.

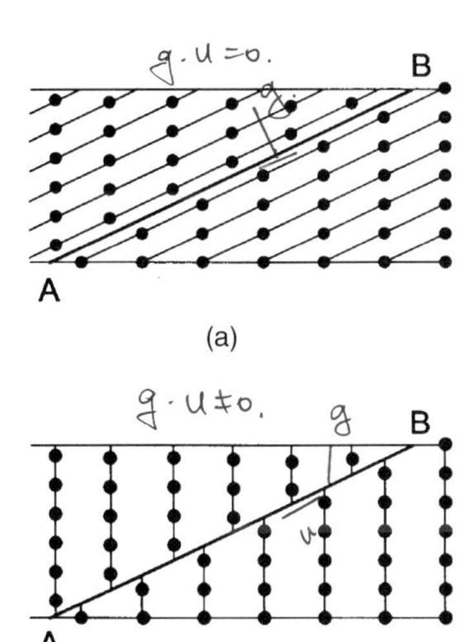

Figure 2.11 Displacements of the reflecting planes shown by lines at a stacking fault *AB*. Diffraction vector **g** is perpendicular to these planes. In (a) $\mathbf{g} \cdot \mathbf{u} = 0$ and no contrast occurs. In (b) $\mathbf{g} \cdot \mathbf{u} \neq 0$ and interference between waves from above and below the fault gives a fringe pattern. (From Howie, *Metallurgical Reviews*, **6**, 467, 1961.)

Planar Defects

Stacking faults and other coherent interfaces without long-range strain fields also form characteristic image contrast in the transmission electron microscope. A fault is produced by the displacement of the crystal above the fault plane by a *constant* vector **u** relative to the crystal below (section 1.3). For columns passing through the fault, this therefore introduces a phase factor $2\pi\,\mathbf{g} \cdot \mathbf{u}$ into the main and diffracted beam amplitudes. Thus, when $\mathbf{g} \cdot \mathbf{u}$ is zero or an integer, no contrast is present and the fault is invisible. This is demonstrated schematically in Fig. 2.11(a), where **u** lies in the plane of the fault and the diffracting planes are parallel to **u**. For other values of $\mathbf{g} \cdot \mathbf{u}$, however, the fault produces contrast, as implied by Fig. 2.11(b). When the fault is inclined to the specimen surface, contrast takes the form of light and dark fringes parallel to the line of intersection of the fault plane with the surface, as shown by the example in Fig. 2.12. The precise form of the fringe contrast depends on the diffraction conditions employed, and this enables faults and defects such as coherent twin, grain and phase boundaries to be analysed.

Lattice Imaging

Under certain conditions, it is possible to study the atomic structure of defects by exploiting the full resolution of the transmission electron microscope to image the actual columns of atoms in a thin sample. An edge dislocation lying normal to the specimen surface, for example, produces an extra half-plane in the image, as shown by the example in Fig. 9.21. Another experimental image is shown in Fig. 9.27(b). Unlike conventional microscopy, in which the image is formed by a single

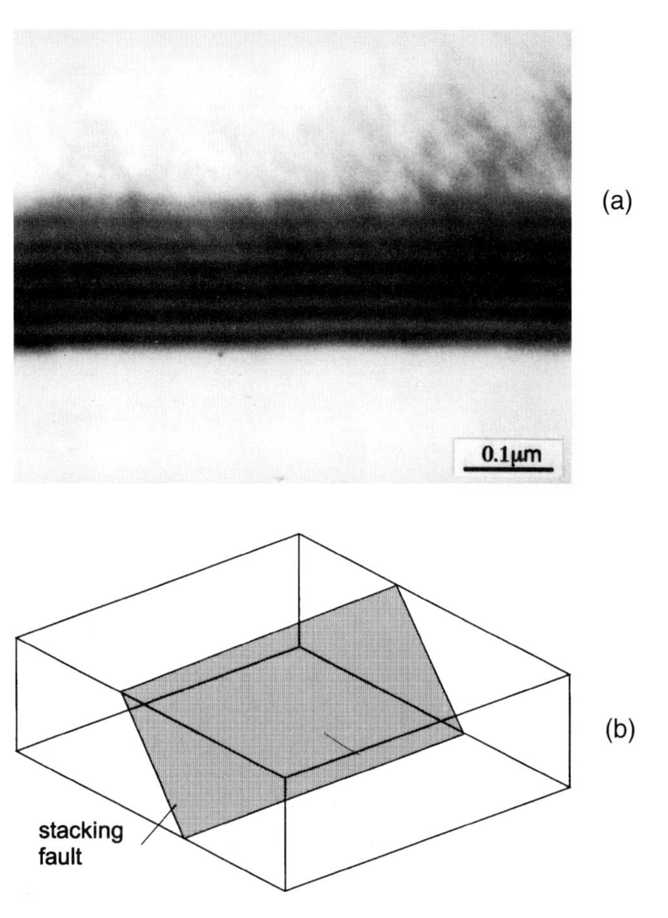

Figure 2.12 (a) Thin film transmission electron micrograph showing the fringe pattern typical of stacking faults which lie at an angle to the plane of the foil, as illustrated schematically in (b).

(direct or diffracted) beam, lattice imaging requires two or more beams to pass through the objective aperture in order that the periodicity of the object, i.e. the lattice planes, be resolved. Spacings down to 0.1 nm have been resolved by this technique. In early studies, it was assumed that one-to-one correspondence exists between the fringes and the crystal planes, and that atomic positions can be visualized directly from the image. It is now known, however, that this is only true in special circumstances. They require the specimen to be very thin ($\sim$1–10 nm) so that rediffraction of diffracted beams cannot occur. Interpretation of lattice images is difficult if this condition is not met.

Other Factors

Some factors which limit the applicability of transmission electron microscopy have already been noted. To attain the specimen thickness required for electron transparency, it is usually necessary to first cut the samples by mechanical or spark-erosion machining to a thickness of about 10–100 μm, and then thin the region in the middle by techniques such as electrolytic polishing, chemical polishing or ion-beam sputtering. Other methods, such as mechanical cleavage for layer crystals or deposition from solution for some organic crystals, have been used less commonly. Some materials, such as organic solids, are unstable under the effect of the electron beam and can only be viewed for short times before decomposing.

For transmission electron microscopy to be of value, it is important that the dislocations seen in the thin foil sample are typical of those of the bulk material. This is sometimes not the case. For example, inadvertent deformation during thinning can introduce additional dislocations. Dislocations close to the surface, on the other hand, experience forces attracting them to the surface (section 4.8), and this effect can substantially reduce the dislocation density in thin specimens. As explained earlier, high-voltage microscopes allow thicker samples to be used so that these problems are reduced. However, this advantage has to be offset against the greater cost of the instrument. Also, if an atom in a crystal acquires a kinetic energy greater than its threshold *displacement energy*, which is typically 25 eV, it can escape from its lattice site and create a vacancy–interstitial pair. Materials with atomic weight up to about 170 suffer such displacement damage from bombardment with electrons with energy of 1 MeV or greater, and their structure is therefore changed by examination in the high-voltage microscope. This feature can, of course, be used with advantage for radiation-damage studies.

2.5 X-ray Diffraction Topography

Direct observation of dislocations with X-rays is achieved by a method somewhat similar to electron diffraction, but with a greatly reduced resolution. This rather specialized technique produces dislocation image widths of 1 μm or greater. Consequently, it is applicable only to the study of crystals with low dislocation densities $\lesssim 10^{10}\,\mathrm{m}^{-2}$, but has the advantage that the penetration of X-rays is greater than electrons so that much thicker specimens can be used. The specimen, usually a large single crystal, is oriented with respect to the X-ray beam so that a set of lattice planes is set at the Bragg angle for strong reflection. The reflected beam is examined photographically. As in electron diffraction, any local bending of the lattice associated with a dislocation results in a change in the reflection conditions and the X-rays are scattered differently in this region. The difference in the intensity of the diffracted X-rays can be recorded photographically. A photograph of dislocations in a silicon crystal revealed by X-ray diffraction topography is shown in Fig. 2.13.

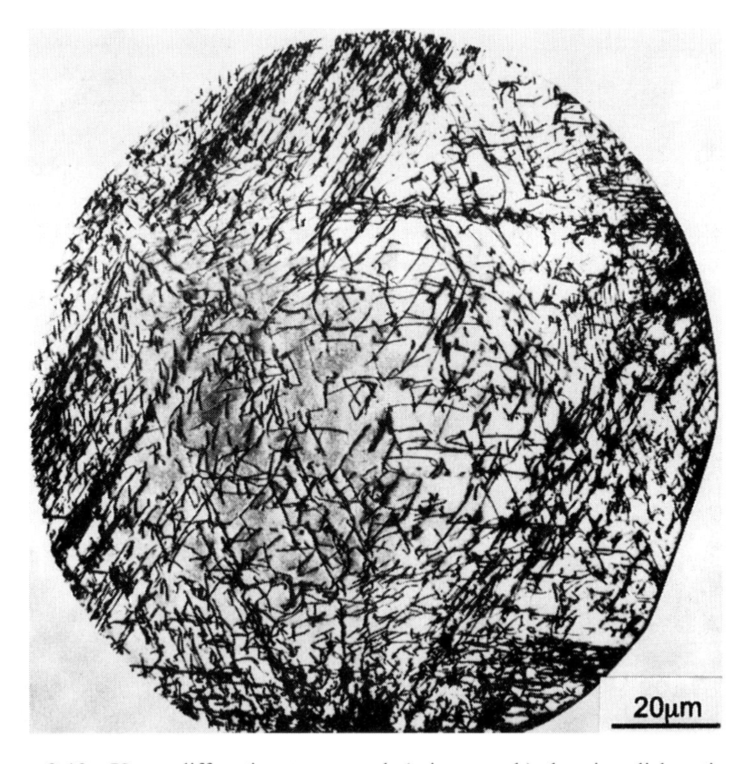

Figure 2.13 X-ray diffraction topograph (micrograph) showing dislocations in a single crystal of silicon. No magnification occurs in recording the topograph, but by using very fine grain sized photographic emulsions subsequent magnification up to about ×500 is possible. (From Jenkinson and Lang (1962), *Direct Observation of Imperfections in Crystals*, p. 471, Interscience.)

2.6 Field Ion Microscopy

The maximum resolution of the electron microscope does not, in general, allow the examination of the positions of individual atoms and, in particular, point defects cannot be detected unless they form in clusters. This limitation is overcome by the field ion microscope which has a resolution of 0.2 to 0.3 nm. The specimen is a fine wire which is electropolished at one end to a sharp hemispherical tip. Within the microscope chamber, the tip is sharpened further to a radius between 5 and 100 nm by *field evaporation*, a process in which a high electric field ($\sim$100 V nm^{-1}) removes atoms from the tip surface.

An image of the atom positions on the tip surface is obtained by *field ionisation*. The specimen is contained at cryogenic temperatures in a chamber evacuated to a high vacuum and into which a low-pressure imaging gas of helium or neon is admitted. The specimen is held at a positive potential of several kV with respect to a fluorescent screen, thereby producing a high field in the range 10–100 V nm^{-1} at the specimen tip. The field polarizes gas atoms near the tip and attracts them towards it. After repeated collisions with the cold surface, the gas atoms slow down sufficiently for them to occasionally become ionized as they pass through the strong field at the protruding surface atoms. The electrons enter the specimen and the positive gas ions are repelled by the field to the screen. Since they travel approximately radially, there is a geometrical magnification of about d/r, where d is the distance from the tip to the image and r is the tip radius. Magnifications of the order of 10^6 times are obtained. Only about 1 in 10 of the surface atoms protrude sufficiently to ionize the image gas, and they appear as bright spots in the image. These atoms are at the edge of the crystal planes where they intersect the tip surface. The image therefore consists of sets of concentric rings of bright spots, each set corresponding to the surface ledges of planes of a particular crystallographic index. A vacancy is revealed in the image as a missing spot, and more complicated defects produce greater disruption to the ring pattern. An example of a grain boundary in tungsten is shown in Fig. 2.14. It is seen that the disruption is strongly localised along the boundary, rather like the simple bubble-raft model of Fig. 1.14.

Although the image only shows details of the positions of atoms on a surface, field ion microscopy may be used to reveal three-dimensional structures. This is achieved by alternating the formation of field ionisation

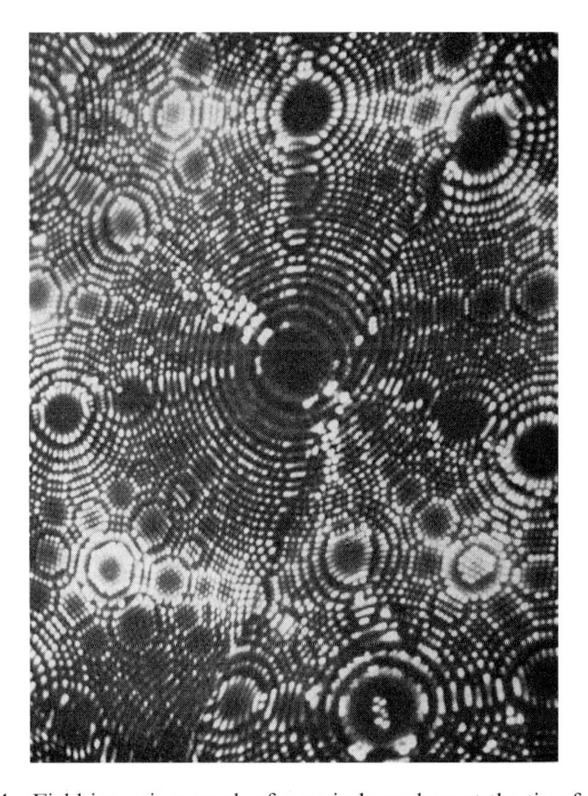

Figure 2.14 Field ion micrograph of a grain boundary at the tip of tungsten needle. Each bright spot represents a tungsten atom. (From *Scientific American*, Sept. 1967.)

images with the repeated removal of surface atoms by field evaporation. It has been used to investigate structures associated with radiation damage, dislocations and grain boundaries. Additionally, alloy and impurity effects have been studied by an *atom probe* technique. In this, a hole at the image screen permits a surface atom removed by field evaporation to enter a mass spectrometer, where it is analysed. This allows structural and chemical features to be investigated at the atomic level.

2.7 Computer Simulation

The power of computers has been exploited in two particular areas concerned with the atomic structure and morphology of dislocations. In one, it is used to assist a well-established experimental technique, namely transmission electron microscopy. In the other, it is employed to model atom behaviour in crystals and provide information not obtainable by experimental investigation.

Image Simulation

It was explained in section 2.4 that the transmission electron microscope image of a dislocation depends not only on the diffraction conditions but also on the form of the dislocation and the nature of the specimen. For example, the $\mathbf{g} \cdot \mathbf{b} = 0$ invisibility criterion only applies generally to isotropic crystals, which have the same elastic properties in all directions. Except in a few cases, real crystals are elastically anisotropic to varying degrees, and when the anisotropy is strong the criterion cannot be applied to analyse dislocations. Another example arises when the dislocation is a loop, as occurs in irradiation damage or quenching. If the loop is too small (less than about 10 nm in diameter) to be resolved distinctly, conventional analysis is not possible, although the image does contain information on loop plane and Burgers vector required by the investigator. In these situations, defect identification is best undertaken by matching real micrographs to computer-generated ones. The latter are obtained for all possible forms the unknown defect might have using the same diffraction conditions as those employed experimentally. It is possible to identify the defect uniquely by suitable choice of conditions.

An example of real and simulated images due to small vacancy loops in ruthenium is shown in Fig. 2.15. Each simulated micrograph is produced by first computing the electron intensity using the column approximation (section 2.4), with the displacement field of the trial defect obtained from elasticity theory. Typically, several thousand columns are required. The resulting grid of intensities is then used to generate a corresponding grid of dots, the size (or 'greyness') of each dot being given by the intensity. The simulated image is therefore formed in a manner analogous to photograph reproduction in newspapers.

Another example is provided in Fig. 2.16. In Fig. 2.16(a), atomic positions around a stacking-fault tetrahedron (section 5.7) formed from

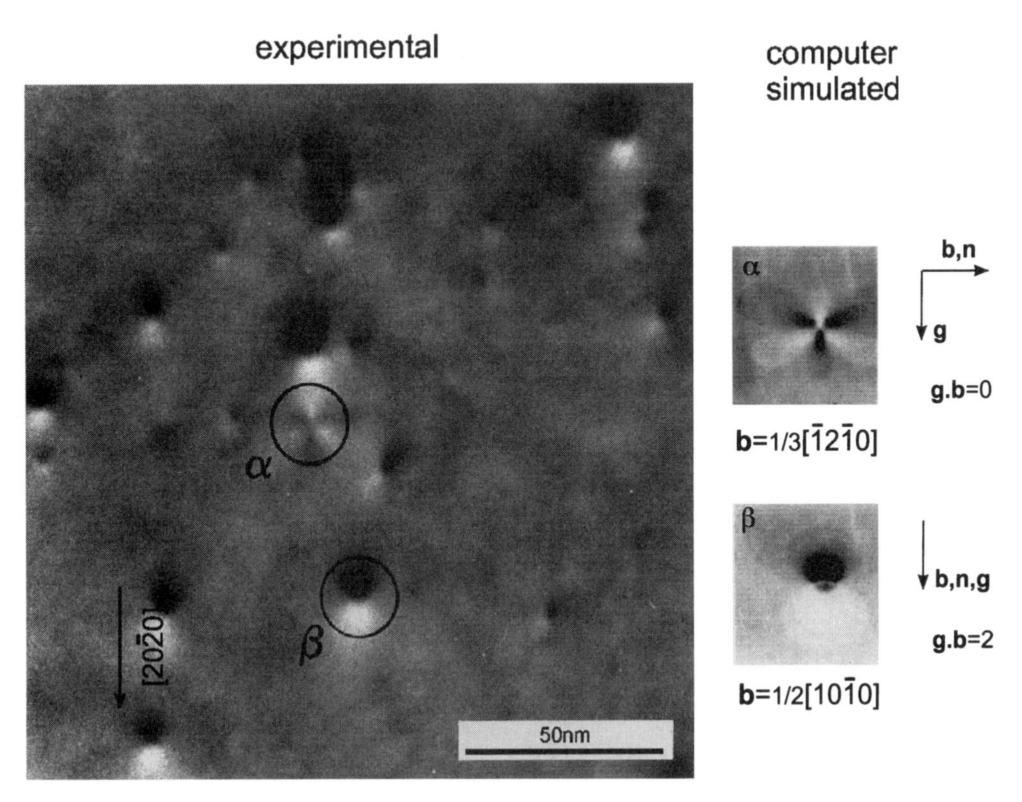

Figure 2.15 Real and computer-simulated images of vacancy loops produced by heavy-ion bombardment of ruthenium. The diffraction vector is **g**, and from image matching, loops α and β have the Burgers vector **b** and loop normals **n** shown. (Courtesy of W. Phythian.)

45 vacancies in copper have been simulated (see below). The atomic coordinates have been used to predict the weak-beam image in a transmission electron microscope (Fig. 2.16(b)). It matches the experimental image (Fig. 2.16(c)) from proton-irradiated copper very well.

Defect Simulation

Generally, the experimental techniques described here do not permit the atom positions at the centre or *core* of a dislocation to be studied. This is unfortunate, for it will be seen in later chapters that the atomic structure of the core determines characteristics such as the slip plane of the dislocation, the ease with which it slips, its effect on electrical behaviour and its interaction with other defects. A theoretical approach now widely used to investigate atomic configuration is defect simulation by computer.

In this method, computer programs are used first to generate the atomic coordinates for a perfect crystal of specified structure, orientation and size: the number of atoms is usually in the range from 10^3 to 10^6. A defect is then introduced. In the case of a dislocation, this is achieved

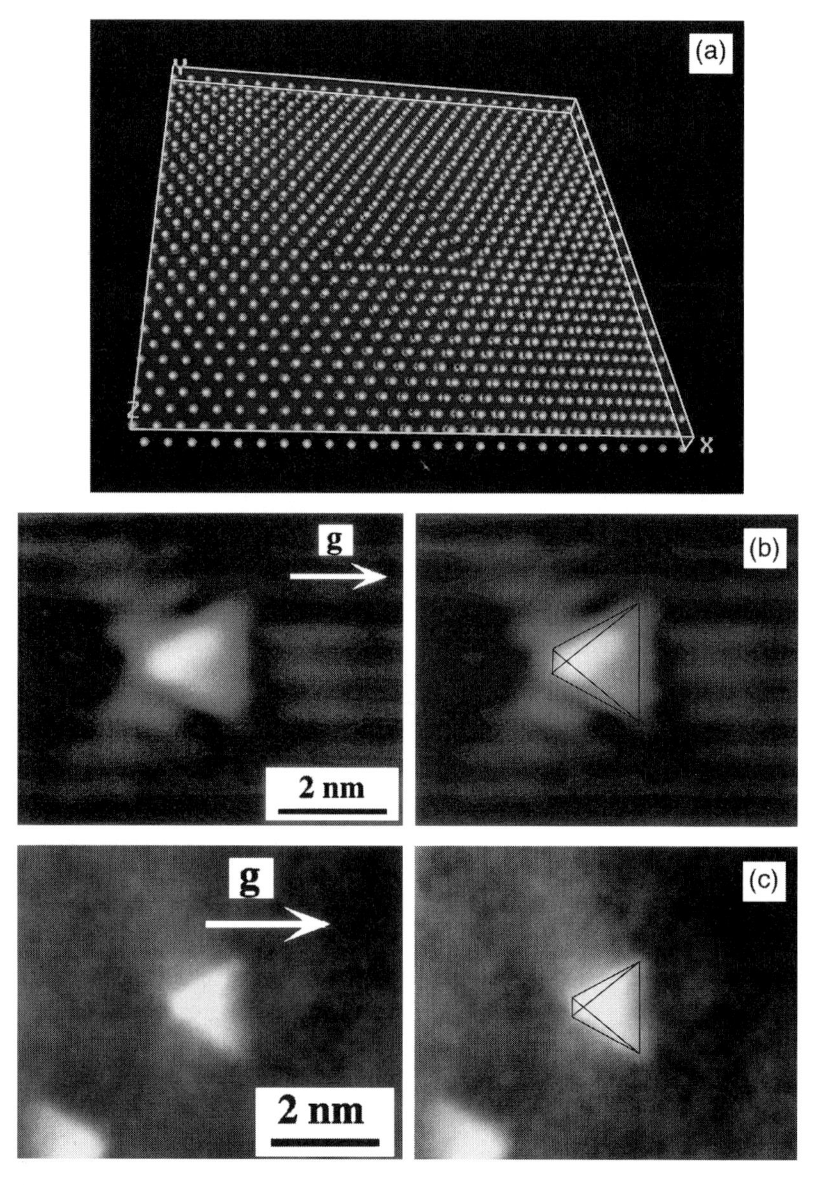

Figure 2.16 Stacking-fault tetrahedron in irradiated copper. (a) Atoms in two {111} planes through a tetrahedron in a computer simulation. (b) Experimental and (c) simulated weak-beam images, with a wire frame on the right showing the orientation of the defect. (From Schaublin, Dai, Osetsky and Victoria (1998), *Inter. Congr. on Electron Microscopy*, vol. 1, p. 173, Institute of Physics Publishing.)

by changing the atom coordinates according to the displacements predicted by elasticity theory (Chapter 4). The equilibrium atomic configuration is then computed by minimising the crystal energy with respect to the atom coordinates, i.e. the crystal is allowed to 'relax'. Calculation of the energy requires a routine to compute the interaction energy of every atom with its neighbours, and the validity of this interatomic potential determines the validity of the final configuration. Potentials are available for the simulation of many crystal structures. In most cases, they have been developed to fit experimental properties of real crystals, such as lattice parameter, elastic constants and stacking-fault energy.

Computer simulation has been applied to the modelling of dislocations, of which Figs 2.16(a), 5.8, 6.7 and 6.19 are examples, boundaries, such as that in Fig. 9.27, point defect structures, see Fig. 6.9, and surfaces. Ways in which information on the atomic structure of dislocations obtained by simulation can be presented graphically are discussed in section 10.3.

Further Reading

Amelinckx, S. (1964) 'The direct observation of dislocations', *Solid State Physics* Supplement 6.

Amelinckx, S., Gever, R. and van Landuyt, J. (Eds.) (1978) *Diffraction and Imaging Techniques in Materials Science*, 2nd edition, North-Holland.

Anstis, G. R. and Hutchinson, J. L. (1992) 'High-resolution imaging of dislocations', *Dislocations in Solids*, vol. 9, chap. 44 (ed. F. R. N. Nabarro), North-Holland.

Edington, J. W. (1974 on) *Practical Electron Microscopy in Materials Science* (5 volumes), Macmillan, Philips Technical Library.

Glauert, A. M. (Ed.) (1974 on) *Practical Methods in Electron Microscopy* (9 volumes), North-Holland.

Goodhew, P. J., Humphreys, J. and Beanland, R. (2001) *Electron Microscopy and Analysis*, Taylor and Francis.

Hammond, C. (1997) *The Basics of Crystallography and Diffraction*, Oxford University Press.

Hirsch, P. B., Howie, A., Nicholson, R. B. and Pashley, D. W. (1965) *Electron Microscopy of Thin Crystals*, Butterworth.

Hren, H. J., Goldstein, J. I. and Joy, D. C. (Eds.) (1979) *Introduction to Analytical Electron Microscopy*, Plenum Press.

Humphreys, C. J. (1980) 'Imaging of dislocations', *Dislocations in Solids*, vol. 5, p. 1 (ed. F. R. N. Nabarro), North-Holland.

Jenkins, M. L. and Kirk, M. A. (2001) *Characterization of Radiation Damage by Transmission Electron Microscopy*, Inst. of Physics Publishing.

Loretto, M. H. and Smallman, R. E. (1975) *Defect Analysis in Electron Microscopy*, Chapman & Hall.

Pashley, D. W. (1965) 'The direct observation of imperfections in crystals', *Reports on Prog. in Phys.* **28**, 291.

Williams, D. B. and Carter, C. B. (1996) *Transmission Electron Microscopy: A Textbook for Materials Science*, Plenum Press.

3 Movement of Dislocations

3.1 Concept of Slip

There are two basic types of dislocation movement. *Glide* or *conservative motion* occurs when the *dislocation moves in the surface which contains both its line and Burgers vector*: a dislocation able to move in this way is *glissile*, one which cannot is *sessile*. *Climb* or *non-conservative motion* occurs when the *dislocation moves out of the glide surface, and thus normal to the Burgers vector*. Glide of many dislocations results in *slip*, which is the most common manifestation of plastic deformation in crystalline solids. It can be envisaged as sliding or successive displacement of one plane of atoms over another on so-called *slip planes*. Discrete blocks of crystal between two slip planes remain undistorted as illustrated in Fig. 3.1. Further deformation occurs either by more movement on existing slip planes or by the formation of new slip planes.

The slip planes and slip directions in a crystal have specific crystallographic forms. The slip planes are normally the planes with the highest density of atoms, i.e. those which are most widely spaced, and the direction of slip is the direction in the slip plane corresponding to one of the shortest lattice translation vectors. Often, this direction is one in which the atoms are most closely spaced. It was seen in section 1.2 that the shortest lattice vectors are $\frac{1}{2}\langle 111\rangle$, $\frac{1}{2}\langle 110\rangle$ and $\frac{1}{3}\langle 11\bar{2}0\rangle$ in the body-centred cubic, face-centred cubic and close-packed hexagonal systems, respectively. Thus, in close-packed hexagonal crystals, slip often occurs on the (0001) basal plane in directions of the type $\langle\bar{1}2\bar{1}0\rangle$ and in face-centred cubic metals on $\{111\}$ planes in $\langle 110\rangle$ directions. In body-centred cubic metals the slip direction is the $\langle 111\rangle$ close-packed direction, but the slip plane is not well defined on a macroscopic scale. Microscopic evidence suggests that slip occurs on $\{112\}$ and $\{110\}$ planes and that $\{110\}$ slip is preferred at low temperatures. A slip plane and a slip direction in the plane constitute a *slip system*. Face-centred cubic crystals have four $\{111\}$ planes with three $\langle 110\rangle$ directions in each, and therefore have twelve $\{111\}$ $\langle 110\rangle$ slip systems. Slip results in the formation of steps on the surface of the crystal.

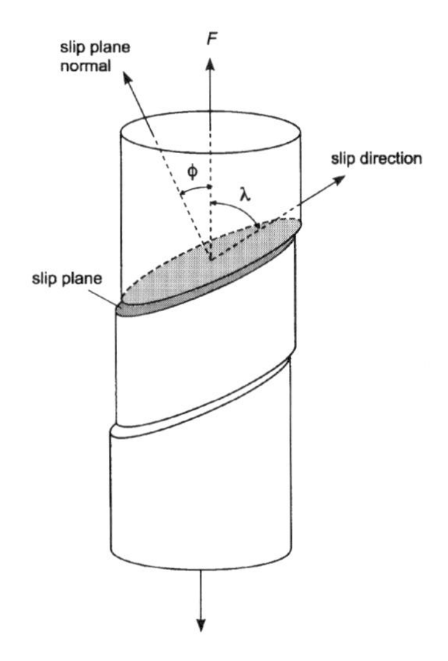

Figure 3.1 Illustration of the geometry of slip in crystalline materials. Note that $(\phi + \lambda) \neq 90°$ in general.

These are readily detected if the surface is carefully polished before plastic deformation. Figure 3.2 is an example of slip in a 3.25 per cent silicon iron crystal; the line diagram illustrates the appearance of a section through the crystal normal to the surface.

A characteristic shear stress is required for slip. Consider the crystal illustrated in Fig. 3.1 which is being deformed in tension by an applied force F along the axis of the cylindrical crystal. If the cross-sectional area is A the tensile stress parallel to F is $\sigma = F/A$. The force has a component $F\cos\lambda$ in the slip direction, where λ is the angle between F and the

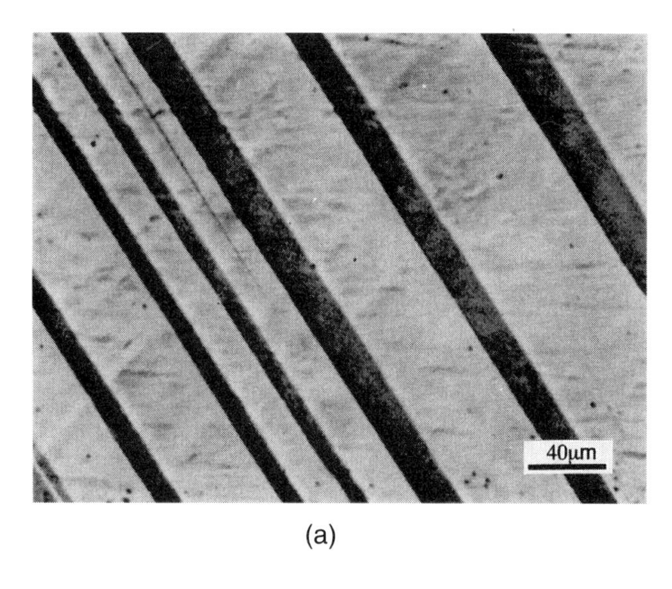

(a)

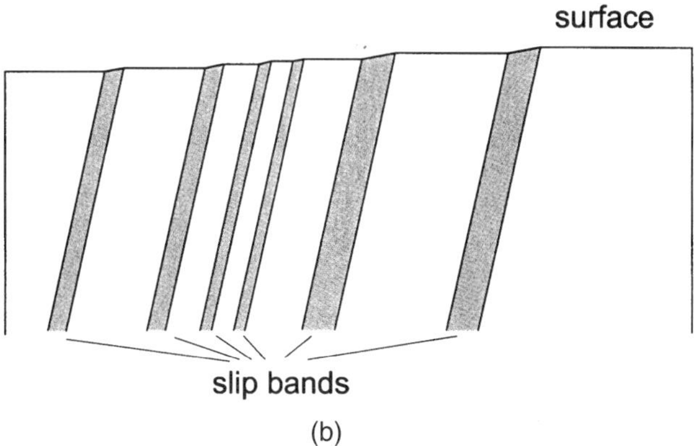

(b)

Figure 3.2 (a) Straight slip bands on a single crystal of 3.25 per cent silicon iron. (From Hull, *Proc. Roy. Soc.* **A274**, 5, 1963.) (b) Sketch of a section across the slip bands normal to surface shown in (a). Each band is made up of a large number of slip steps on closely spaced parallel slip planes.

slip direction. This force acts over the slip surface which has an area $A/\cos\phi$, where ϕ is the angle between F and the normal to the slip plane. Thus the *shear stress* τ, *resolved on the slip plane in the slip direction*, is

$$\tau = \frac{F}{A}\cos\phi\cos\lambda \tag{3.1}$$

The symbol τ will be used to denote the shear stress resolved in this way throughout this book. If F_c is the tensile force required to start slip, the corresponding value of the shear stress τ_c is called the *critical resolved shear stress for slip*. It has been found in some crystals which deform on a single slip system that τ_c is independent of the orientation of the crystal. The quantity $\cos\phi\cos\lambda$ is known as the *Schmid factor*.

3.2 Dislocations and Slip

In Section 1.4 it was shown that the theoretical shear stress for slip was many times greater than the experimentally observed stress, i.e. τ_c. The low value can be accounted for by the movement of dislocations. Consider the edge dislocation represented in Fig. 1.18. This could be formed in a different way to that described in Chapter 1, as follows: cut a slot along *AEFD* in the crystal shown in Fig. 3.3 and displace the top surface of the cut *AEFD* one lattice spacing over the bottom surface in the direction *AB*. An extra half-plane *EFGH* and a dislocation line *FE* are formed and the distortion produced is identical to that of Fig. 1.18. Although it is emphasised that dislocations are not formed in this way in practice, this approach demonstrates that the dislocation can be defined as the *boundary between the slipped and unslipped parts of the crystal*. Apart from the immediate region around the dislocation core *FE*, the atoms across *AEFD* are in perfect registry. The distortion due to the dislocations in Figs 1.18 and 3.3 has been described by giving all points on one side of an imagined cut – *ABCD* in Fig. 1.15 and *AEFD* in Fig. 3.3 – the same displacement relative to points on the other side; this displacement is the Burgers vector. These defects are *Volterra* dislocations, named after the Italian mathematician who first considered such distortions. More general forms arising from variable displacements are possible in principle, but are not treated here.

Only a relatively small applied stress is required to move the dislocation along the plane *ABCD* in the way demonstrated in Fig. 3.4. This can be understood from the following argument. Well away from the dislocation, the atom spacings are close to the perfect crystal values, and a shear stress as high as the theoretical value of equation (1.5) would be required to slide them all past each other. Near the dislocation line itself, some atom spacings are far from the ideal values, and small relative changes in position of only a few atoms are required for the dislocation to move. For example, a small shift of atom 1 relative to atoms 2 and 3 in Fig. 3.4(a) effectively moves the extra half-plane from x

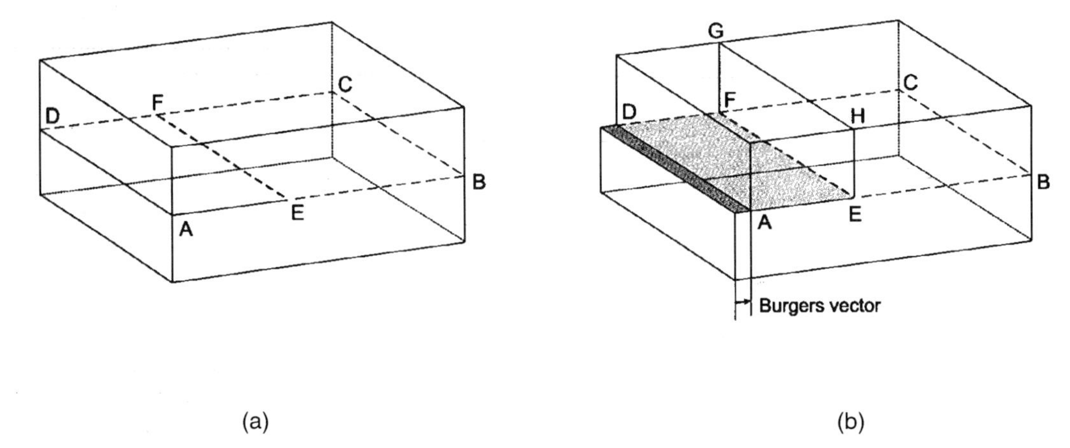

Figure 3.3 Formation of a pure edge dislocation *FE*.

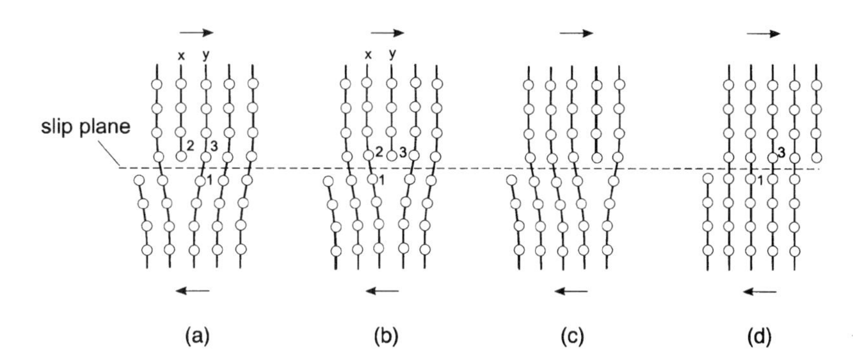

Figure 3.4 Movement of an edge dislocation: the arrows indicate the applied shear stress.

to y (Fig. 3.4(b)), and this process is repeated as the dislocation continues to glide (Figs 3.4(c), (d)). The applied stress required to overcome the lattice resistance to the movement of the dislocation is the *Peierls–Nabarro stress* (see Chapter 10) and is much smaller than the theoretical shear stress of a perfect lattice.

Figure 3.4 demonstrates why the Burgers vector is the most important parameter of a dislocation. Two neighbouring atoms (say 1 and 3) on sites adjacent across the slip plane are displaced relative to each other by the Burgers vector when the dislocation glides past. Thus, *the slip direction* (see Fig. 3.1) *is necessarily always parallel to the Burgers vector of the dislocation responsible for slip*. The movement of one dislocation across the slip plane to the surface of the crystal produces a surface slip step equal to the Burgers vector. Each surface step produced by a slip band in Fig. 3.2 must have been produced by the glide of many thousands of dislocations. The plastic shear strain in the slip direction resulting from dislocation movement is derived in section 3.9.

3.3 The Slip Plane

In Fig. 3.3 the edge dislocation has moved in the plane $ABCD$ which is the *slip plane*. This is uniquely defined as *the plane which contains both the line and the Burgers vector of the dislocation*. The glide of an edge dislocation is limited, therefore, to a specific plane. The movement of a screw dislocation, for example from AA' to BB' in Fig. 3.5, can also be envisaged to take place in a slip plane, i.e. $LMNO$, and a slip step is formed. However, the line of the screw dislocation and the Burgers vector do not define a unique plane and the glide of the dislocation is not restricted to a specific plane. It will be noted that the displacement of atoms and hence the slip step associated with the movement of a screw dislocation is parallel to the dislocation line, for that is the direction of its Burgers vector. This can be demonstrated further by considering a plan view of the atoms above and below a slip plane containing a screw dislocation (Fig. 3.6). Movement of the screw dislocation produces a displacement b parallel to the dislocation line.

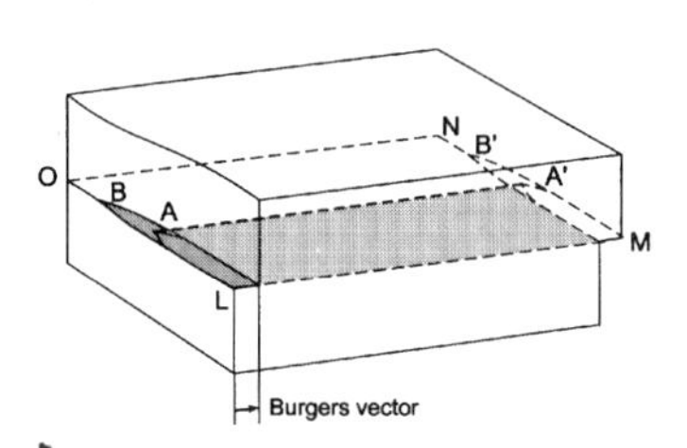

Figure 3.5 Formation and then movement of a pure screw dislocation AA' to BB' by slip.

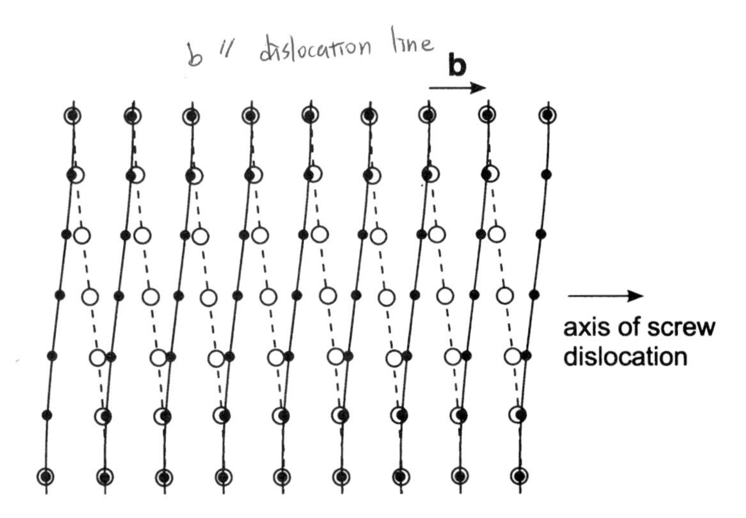

Figure 3.6 Arrangement of atoms around a screw dislocation. Open circles above plane of diagram, filled circles below (for right-handed screw).

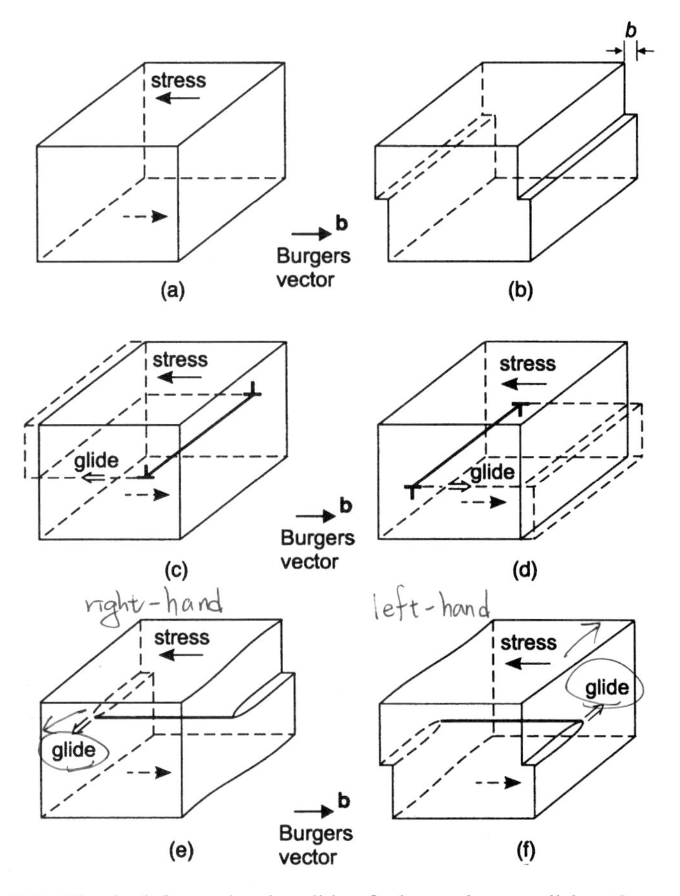

Figure 3.7 Plastic deformation by glide of edge and screw dislocations under the applied shear stress shown.

The direction in which a dislocation glides under stress can be determined by physical reasoning. Consider material under an applied shear stress (Fig. 3.7(a)) so that it deforms plastically by glide in the manner indicated in Fig. 3.7(b). From the description of section 3.2, a dislocation responsible for this deformation must have its Burgers vector in the direction shown. It is seen from Figs 3.7(c) and (d) that a positive edge dislocation glides to the left in order that the extra half-plane produces a step on the left-hand face as indicated, whereas a negative edge dislocation glides to the right. A right-handed screw glides towards the front in order to extend the surface step in the required manner (Fig. 3.7(e)), whereas a left-handed screw glides towards the back (Fig. 3.7(f)). These observations demonstrate that (a) *dislocations of opposite sign glide in opposite directions under the same stress*, as expected of physical opposites, and (b) *for dislocation glide a shear stress must act on the slip plane in the direction of the Burgers vector*, irrespective of the direction of the dislocation line.

In the examples illustrated in Figs 3.3, 3.5 and 3.7 it has been assumed that the moving dislocations remain straight. However, dislocations are generally bent and irregular, particularly after plastic deformation, as can be seen in the electron micrographs presented throughout this book. A more general shape of a dislocation is shown in Fig. 3.8(a). The

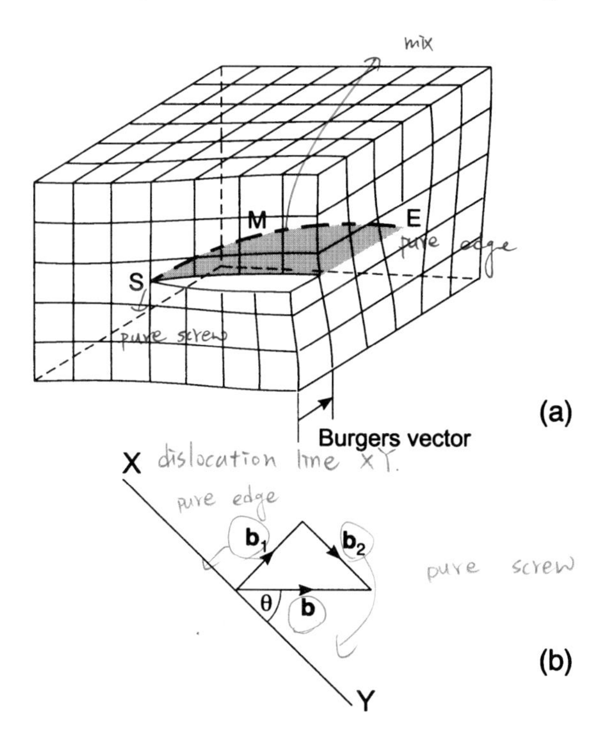

Figure 3.8 Mixed dislocations. (a) The curved dislocation *SME* is pure edge at *E* and pure screw at *S*. (b) Burgers vector **b** of dislocation *XY* is resolved into a pure edge component $\mathbf{b}_1$ and a pure screw component $\mathbf{b}_2$.

boundary separating the slipped and unslipped regions of the crystal is curved, i.e. the dislocation is curved, but the Burgers vector is the same all along its length. It follows that at point E the dislocation line is *normal* to the vector and is therefore *pure edge* and at S is *parallel* to the vector and is *pure screw*. The remainder of the dislocation (M) has a *mixed edge and screw* character. The Burgers vector $\mathbf{b}$ of a mixed dislocation, XY in Fig. 3.8(b), can be resolved into two components by regarding the dislocation as two coincident dislocations; a pure edge with vector $\mathbf{b}_1$ of length $b \sin\theta$ at right angles to XY, and a pure screw with vector $\mathbf{b}_2$ of length $b \cos\theta$ parallel to XY:

$$\mathbf{b}_1 + \mathbf{b}_2 = \mathbf{b} \tag{3.2}$$

3.4 Cross Slip

In general, screw dislocations tend to move in certain crystallographic planes (see, for example, dissociation of perfect dislocations, section 5.3). Thus, in face-centred cubic metals the screw dislocations move in $\{111\}$ type planes, but can switch from one $\{111\}$ type plane to another if it contains the direction of $\mathbf{b}$. This process, known as *cross slip*, is illustrated in Fig. 3.9. In Fig. 3.9(a) a dislocation line, Burgers vector $\mathbf{b} = \frac{1}{2}[\bar{1}01]$, is gliding to the left in the (111) plane under the action of an applied shear stress. The only other $\{111\}$ plane containing this slip vector is $(1\bar{1}1)$. Suppose that as the loop expands the local stress field which is producing dislocation motion changes so that motion is preferred on $(1\bar{1}1)$ instead of (111). Only a pure screw segment is free to move in both $(1\bar{1}1)$ and (111) planes and so cross slip occurs at S

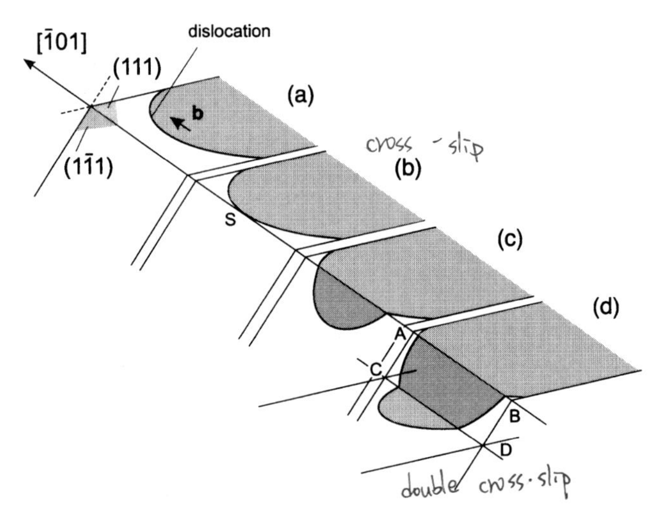

Figure 3.9 Sequence of events (a), (b), (c) in cross slip in a face-centred cubic metal. The $[\bar{1}01]$ direction is common to the (111) and $(1\bar{1}1)$ close-packed planes. A screw dislocation at S is free to glide in either of these planes. Cross slip produces a non-planar slip surface. Double cross slip is shown in (d).

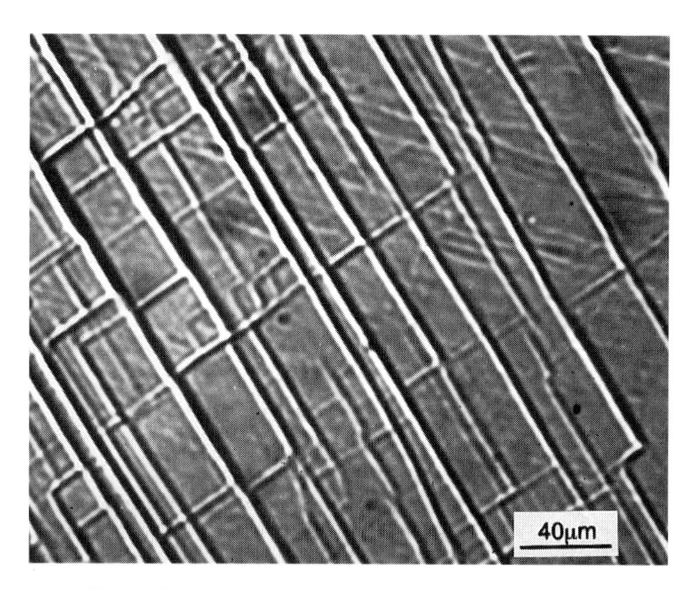

Figure 3.10 Cross slip on the polished surface of a single crystal of 3.25 per cent silicon iron.

(Fig. 3.9(b)). Glide of the dislocation then occurs on the $(1\bar{1}1)$ plane (Fig. 3.9(c)). *Double cross slip* is illustrated in Fig. 3.9(d).

The cross slip of moving dislocations is readily seen by transmission electron microscopy because a moving dislocation leaves a track which slowly fades. In body-centred cubic metals the slip plane is less well defined and the screw dislocation with $\mathbf{b} = \frac{1}{2}\langle 111 \rangle$ can glide on three $\{110\}$ planes and three $\{112\}$ planes. Slip often wanders from one plane to another producing wavy slip lines on prepolished surfaces. An example of the result of cross slip is shown in Fig. 3.10.

3.5 Velocity of Dislocations

Dislocations move by glide at velocities which depend on the applied shear stress, purity of crystal, temperature and type of dislocation. A direct method of measuring dislocation velocity was developed by Johnston and Gilman using etch pits to reveal the position of dislocations at different stages of deformation as illustrated in Fig. 2.4. A crystal containing freshly introduced dislocations, usually produced by lightly deforming the surface, is subjected to a constant stress pulse for a given time. From the positions of the dislocations before and after the stress pulse, the distance each dislocation has moved, and hence the average dislocation velocity, can be determined. By repeating the experiment for different times and stress levels the velocity can be determined as a function of stress as shown in Fig. 3.11(a) for lithium fluoride. The dislocation velocity was measured over 12 orders of magnitude and was a very sensitive function of the resolved shear stress. In the range of velocities between 10^{-9} and $10^{-3}\,\mathrm{m\,s^{-1}}$ the logarithm of the velocity varies linearly with the logarithm of the applied stress, thus

$$v \propto \left(\frac{\tau}{\tau_0}\right)^n \tag{3.3a}$$

where v is velocity, τ is the applied shear stress resolved in the slip plane, τ_0 is the shear stress for $v = 1\,\mathrm{m\,s^{-1}}$, n is a constant and was found experimentally to be $\sim$25 for lithium fluoride. It must be emphasised that equation (3.3a) is purely empirical and implies no physical interpretation of the mechanism of dislocation motion.

The velocity of edge and screw components was measured independently and in the low velocity range edge dislocations moved 50 times faster than screw dislocations. There is a critical stress, which represents the onset of plastic deformation, required to start the dislocations moving. The effect of temperature on dislocation velocity is illustrated in

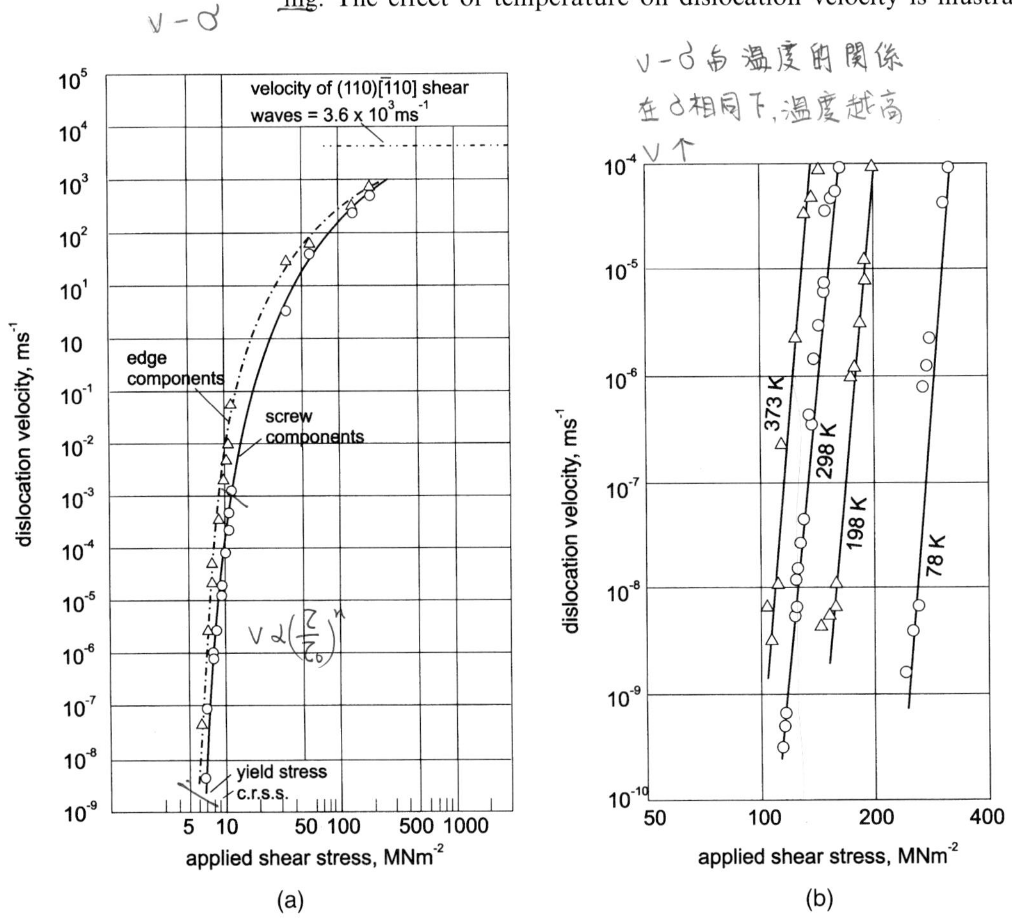

Figure 3.11 (a) Stress dependence of the velocity of edge and screw dislocations in lithium fluoride. (From Johnston and Gilman, *J. Appl. Phys.* **30**, 129, 1959.) (b) Stress dependence of the velocity of edge dislocations in 3.25 per cent silicon iron at four temperatures. (After Stein and Low, *J. Appl. Phys.* **31**, 362, 1960.)

Fig. 3.11(b) for results obtained by Stein and Low who studied 3.25 per cent silicon iron by the same method. The dislocation velocities were only measured below 10^{-4} m s^{-1} and therefore the curves do not show the bending over found in lithium fluoride. The curves are of the same form as equation (3.3a). At 293 K, $n \sim 35$, and at 78 K, $n \sim 44$. τ_0 increased with decreasing temperature.

The stress dependence of dislocation velocity varies significantly from one material to another, as illustrated in Fig. 3.12. For a given material, the velocity of transverse shear-wave propagation is the limiting velocity for uniform dislocation motion. However, damping forces increasingly oppose motion when the velocity increases above about 10^{-3} of this limit, and thus n in equation (3.3a) decreases rapidly in this range. Studies on face-centred cubic and hexagonal close-packed crystals have shown that at the critical resolved shear stress for macroscopic slip, dislocation velocity is approximately 1 m s^{-1} ($\sim 10^{-3}$ of shear wave velocity) and satisfies

$$v = A\tau^m \tag{3.3b}$$

where A is a material constant. m is approximately 1 at 300 K in pure crystals, and increases to the range 2–5 with alloying and to 4–12 at 77 K. This form of stress-dependence is used in the dislocation dynamics computer simulations described in sections 10.8 and 10.10. The increase in v with decreasing temperature is due to the fact that at high dislocation velocities in metals, the dominant damping forces arise from the

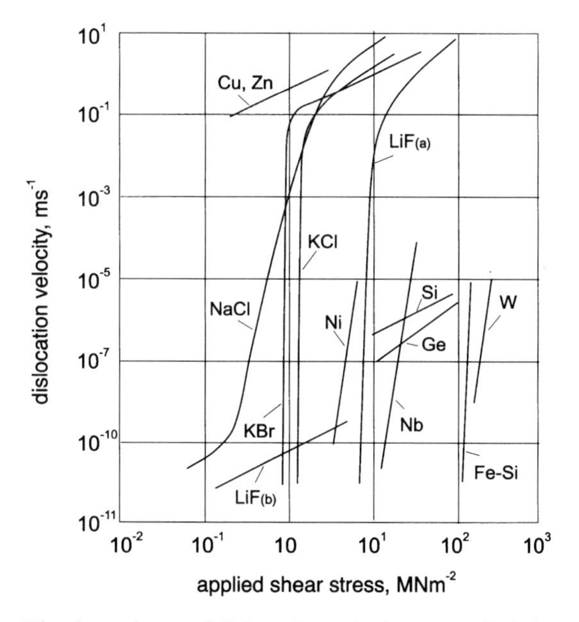

Figure 3.12 The dependence of dislocation velocity on applied shear stress. The data are for 20 °C except Ge (450 °C) and Si (850 °C). (After Haasen, *Physical Metallurgy*, Cambridge University Press.)

scattering of lattice vibrations (*phonons*), and these are less numerous at low temperatures. Additional effects such as thermoelastic dissipation and radiative emission of phonons can be important in other materials.

3.6 Climb

At low temperatures where diffusion is difficult, and in the absence of a non-equilibrium concentration of point defects, the movement of dislocations is restricted almost entirely to glide. However, at higher temperatures an edge dislocation can move out of its slip plane by a process called *climb*. Consider the diagram of an edge dislocation in Fig. 3.13. If

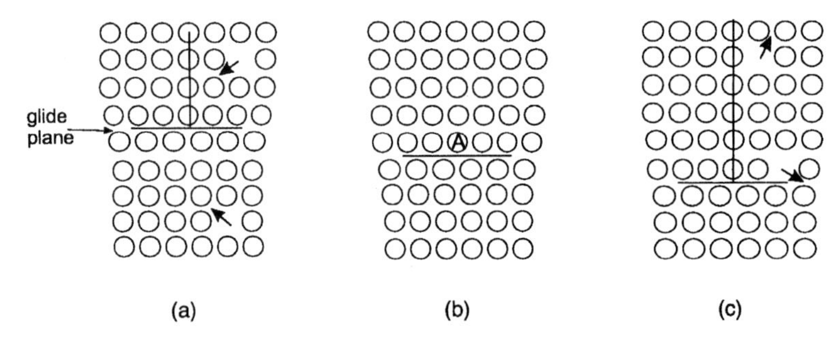

Figure 3.13 Positive and negative climb of an edge dislocation. In (b) the dislocation is centred on the row of atoms A normal to the plane of the diagram. If the vacancies in the lattice diffuse to the dislocation at A the dislocation will climb in a positive sense as in (a). If vacancies are generated at the dislocation line and then diffuse away the dislocation will climb in the negative sense as in (c).

the row of atoms A normal to the plane of the diagram is removed, the dislocation line moves up one atom spacing out of its original slip plane; this is called *positive climb*. Similarly, if a row of atoms is introduced below the extra half-plane the dislocation line moves down one atom spacing, *negative climb*. Positive climb can occur by either diffusion of vacancies to A or the formation of an interstitial atom at A and its diffusion away. Negative climb can occur either by an interstitial atom diffusing to A or the formation of a vacancy at A and its diffusion away.

More generally, if a small segment $\mathbf{l}$ of line undergoes a small displacement $\mathbf{s}$ (Fig. 3.14), the local change in volume is

$$dV = \mathbf{b} \cdot \mathbf{l} \times \mathbf{s} = \mathbf{b} \times \mathbf{l} \cdot \mathbf{s} \tag{3.4}$$

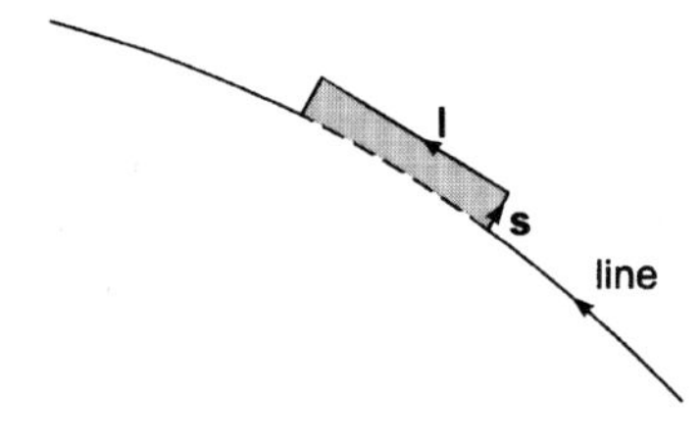

Figure 3.14 Displacement of a segment of line $\mathbf{l}$ by $\mathbf{s}$. The area of the shaded element is $|\mathbf{l} \times \mathbf{s}|$.

for during the movement the two sides of the area element $\mathbf{l} \times \mathbf{s}$ are displaced by $\mathbf{b}$ relative to each other. The glide plane of the element is by definition perpendicular to $\mathbf{b} \times \mathbf{l}$, and so when either $\mathbf{s}$ is perpendicular to $\mathbf{b} \times \mathbf{l}$ or $\mathbf{b} \times \mathbf{l} = 0$, which means the element is pure screw, dV is zero. This is the condition for glide (conservative motion) discussed previously. For other cases, volume is not conserved ($dV \neq 0$) and the motion is climb, the number of point defects required being dV/Ω, where Ω is the volume per atom. The mass transport involved occurs

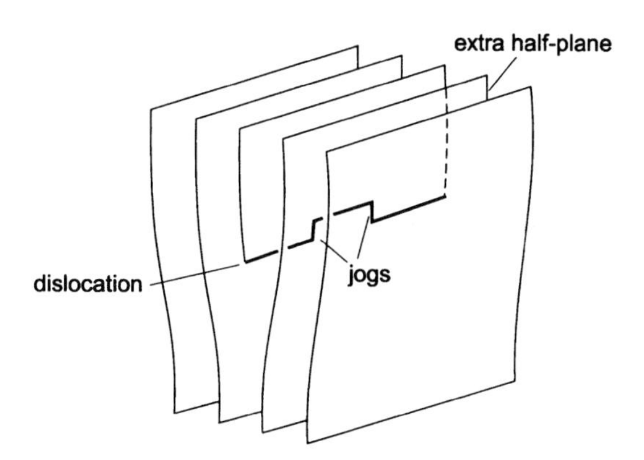

Figure 3.15 Single jogs on an edge dislocation.

by diffusion and therefore climb requires thermal activation. The most common climb processes involve the diffusion of vacancies either towards or away from the dislocation.

It has been implied above that a complete row of atoms is removed simultaneously, whereas in practice individual vacancies or small clusters of vacancies diffuse to the dislocation. The effect of this is illustrated in Fig. 3.15 which shows climb of a short section of a dislocation line resulting in the formation of two steps called *jogs*. Both positive and negative climb proceeds by the nucleation and motion of jogs. Conversely, jogs are *sources* and *sinks* for vacancies.

Jogs are steps on the dislocation which move it from one atomic slip plane to another. Steps which displace it on the same slip plane are called *kinks*. The two are distinguished in Fig. 3.16. Jogs and kinks are short elements of dislocation with the same Burgers vector as the line on which they lie, and the usual rules apply for their conservative and non-conservative movement. Thus, the kink, having the same slip plane as the dislocation line, does not impede glide of the line. (In fact, it may assist it – section 10.3.) Similarly, the jog on an edge dislocation (Fig. 3.16(c)) does not affect glide. The jog on a screw dislocation (Fig. 3.16(d)) has edge character, however, and can only glide along the line; movement at right angles to the Burgers vector requires climb. This impedes glide of the screw and results in point defect production during slip (Chapter 7). Since E_f^i is much larger than E_f^v in metals (section 1.3), it is very difficult for interstitial-producing jogs to move perpendicular to the line and vacancy production predominates.

The jogs described have a height of one lattice spacing and a characteristic energy $E_j \sim 1\,\text{eV}$ (0.16 aJ) resulting from the increase in dislocation line length (section 7.2). They are produced extensively during plastic deformation by the intersection of dislocations (see Chapter 7), and exist even in well-annealed crystals, for there is a thermodynamic equilibrium number of jogs (*thermal jogs*) per unit length of dislocation given by

$$n_j = n_0 \exp(-E_j/kT) \tag{3.5}$$

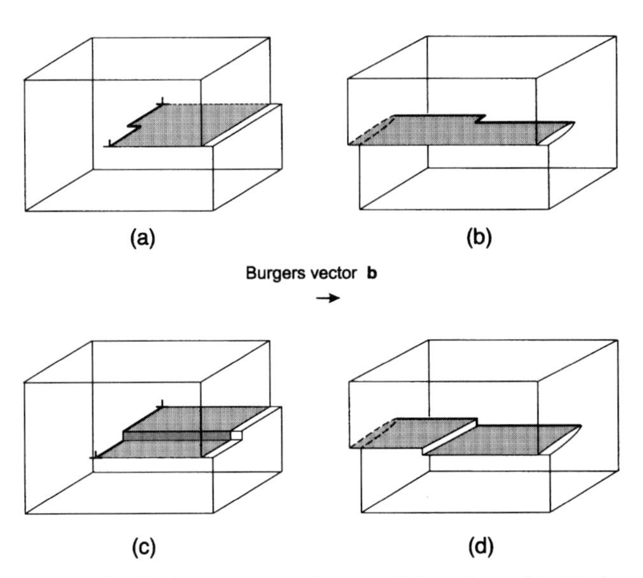

Figure 3.16 (a), (b) Kinks in edge and screw dislocations. (c), (d) Jogs in edge and screw dislocations. The slip planes are shaded.

where n_0 is the number of atom sites per unit length of dislocation, i.e. n_j/n_0 is of the order of 10^{-5} at $T = 1000\,\mathrm{K}$ and 10^{-17} at $T = 300\,\mathrm{K}$.

The climb of dislocations by jog formation and migration is analogous to crystal growth by surface step transport. There are two possible mechanisms. In one, a pre-existing single jog (as in Fig. 3.16(c)), or a jog formed at a site without the need for thermal nucleation, migrates along the line by vacancy emission or absorption. This involves no change in dislocation length and the activation energy for jog diffusion is

$$(E_f^v + E_m^v) = E_d$$

where E_m^v is the vacancy migration energy and E_d the activation energy for self-diffusion. In the other, thermal jogs are nucleated on an otherwise straight line and the jog migration energy is $(E_d + E_j)$. In most situations the first process dominates. The effective activation energy in some circumstances can be less than one half the value of E_d measured in the bulk, for the crystal is distorted at atom sites close to the dislocation line itself and E_d is smaller there. The process of mass transport occurring along the line, rather than to or from it, is known as *pipe diffusion*, and an example is discussed in section 3.8. Although jog diffusion in thermal equilibrium occurs at a rate proportional to $\exp(-E_d/kT)$, or a rate with E_d modified as discussed above, the actual rate of climb of the dislocation depends also on the forces acting on it. Several factors are involved, as discussed in section 4.7.

Pure screw dislocations have no extra half-plane and in principle cannot climb. However, a small edge component or a jog on a screw dislocation will provide a site for the start of climb. Two examples will serve to illustrate the climb process in both edge and predominantly screw dislocations.

3.7 Experimental Observation of Climb

Prismatic Dislocation Loops

The shaded plane bounded by the curved line *SME* in Fig. 3.8 contains the Burgers vector **b** and is therefore the slip plane of the dislocation

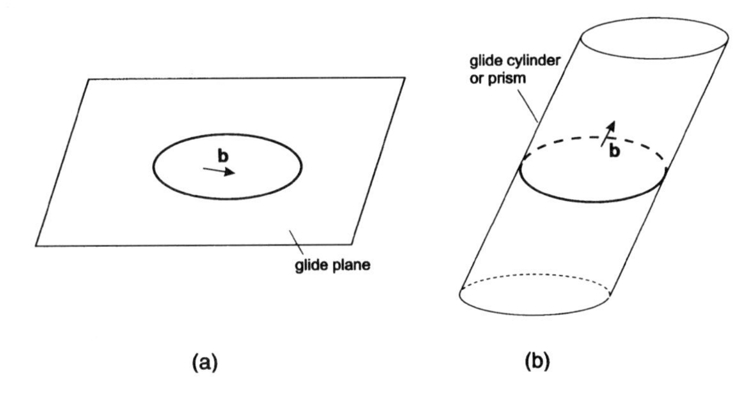

Figure 3.17 Glide plane and glide cylinder for dislocation loops with Burgers vector (a) in the loop plane and (b) inclined to the loop plane.

(section 3.3). If a dislocation forms a loop in a plane and **b** lies in that plane (Fig. 3.17(a)), the plane is the slip plane and the loop can expand or shrink by glide, depending on the forces acting on it (section 4.5). When the Burgers vector is not in the plane of the loop, the slip surface defined by the dislocation line and its Burgers vector is a cylindrical surface (Fig. 3.17(b)). The dislocation is called a *prismatic dislocation*. It follows that the dislocation can only move conservatively, i.e. by glide, along the cylindrical surface and if the loop expands or shrinks *climb* must be occurring.

Numerous examples of prismatic loops have been observed using transmission electron microscopy. They can be formed in the following way. The supersaturation (excess concentration) of vacancies resulting either from rapid quenching from a high temperature (see section 1.3) or from atomic displacements produced by irradiation with energetic particles may precipitate out in the form of a disc on a close-packed plane. If the disc is large enough, it is energetically favourable for it to collapse to produce a dislocation loop (Fig. 3.18). The Burgers vector of the loop is normal to the plane of the loop, so that an edge dislocation has been formed. In the presence of an excess concentration of vacancies the loop will expand by positive climb. Alternatively, at high temperature when the equilibrium concentration of vacancies in the lattice is large (section 1.3) or if there is a nearby sink for vacancies the loop will emit vacancies and shrink by negative climb. Figure 3.19 shows an example of the latter. The dislocation loops were formed in a thin sheet of aluminium

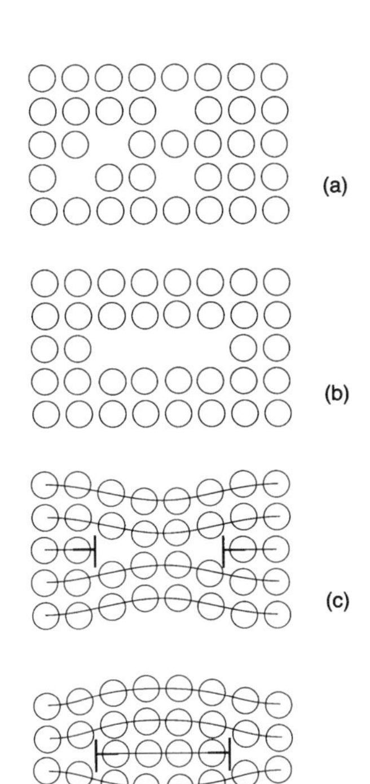

Figure 3.18 Formation of a prismatic dislocation loop. (a) Represents a crystal with a large non-equilibrium concentration of vacancies. In (b) the vacancies have collected on a close-packed plane and in (c) the disc has collapsed to form an edge dislocation loop. (d) Loop formed by a platelet of self-interstitial atoms.

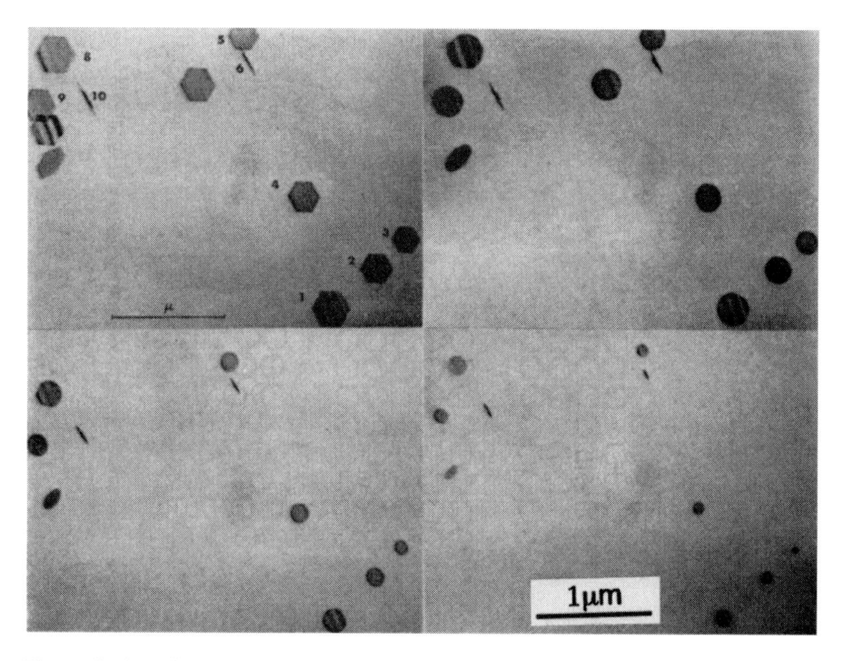

Figure 3.19 Electron transmission photographic sequence showing the shrinkage of $\frac{1}{3}\langle 111\rangle$ stacking fault dislocation loops in aluminium by negative climb. The foil was annealed at 102 °C for 0, 213, 793 and 1301 min, respectively. (From Tartour and Washburn, *Phil. Mag.* **18**, 1257, 1968.)

by quenching. The sheet was thinned to about 100 nm and examined by transmission electron microscopy. The surfaces of the sheet are very effective sinks for vacancies and when the foil was heated in the microscope to allow thermal activation to assist in the formation and diffusion of vacancies, the stacking-fault and line-tension forces (section 4.7) acting on the dislocations induced the loops to shrink and disappear.

In a similar fashion, platelets of self-interstitial atoms can form dislocations in irradiated materials (Fig. 3.18(d)) and can grow by absorption of self-interstitials created by subsequent radiation damage.

Helical Dislocations

Dislocations in the form of a long spiral have been observed in crystals which have been thermally treated to produce climb conditions. Thus, in Fig. 3.20 the helical dislocation in CaF_2 was formed by heating the crystal to a high temperature. A mechanism for the formation of helical dislocations is as follows. The dislocation AB, in Fig. 3.21, is pinned or locked at A and B, and is partly edge and partly screw in character. Motion of the dislocation in the plane ABA' corresponds to glide since this plane contains the line and the Burgers vector. Motion at right angles to this plane corresponds to climb. An excess of vacancies at a suitable temperature will cause the dislocation to climb. The configura-

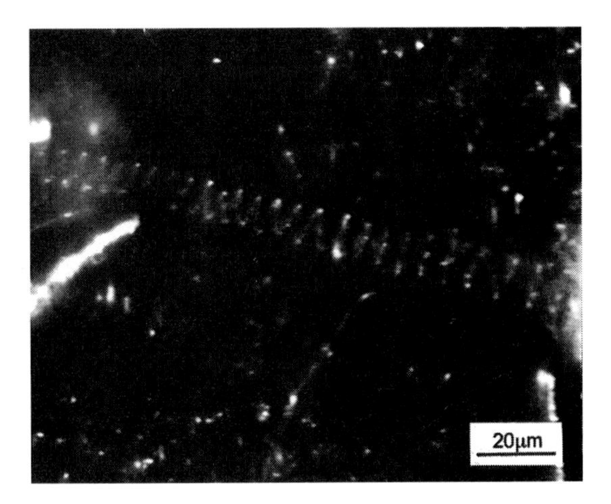

Figure 3.20 Spiral or helical dislocation in CaF_2 (fluorite) revealed by the decoration technique. (After Bontinck and Amelinckx, *Phil. Mag.* **2**, 94, 1957.)

tion of the dislocation after a certain amount of climb is shown in Fig. 3.21(b). The dislocation now lies on the surface of a cylinder whose axis is parallel to the Burgers vector (i.e. prismatic dislocation) and it can glide on the surface of the cylinder. Further climb displaces each part of the dislocation in a direction normal to the surface of the cylinder. The dislocation will change into a double spiral by spiralling around node points A and B, as shown in Fig. 3.21(c). The radius of the spiral will be smallest at the nodes, since, for a given number of vacancies, the angle of rotation will be greater there. If $A'B$ is small compared with AA', combination of prismatic glide and climb will result in the formation of a uniformly spaced spiral. The Burgers vector of the helical dislocation illustrated in Fig. 3.21 must therefore lie along the axis of the helix. The helix consists essentially of a screw dislocation parallel to the axis of the helix and a set of prismatic loops.

3.8 Conservative Climb

From the description of the basic types of dislocation movement at the beginning of this chapter, the heading of this section appears to be a contradiction. However, if a prismatic loop (Fig. 3.18) moves in its plane without shrinking or expanding at temperatures too low for bulk diffusion of point defects, it follows that climb is occurring without net loss or gain of vacancies or self-interstitial atoms. An example is seen in Fig. 3.22. Loop translation occurs by the transfer of vacancies around the loop by *pipe diffusion*, the vacancies producing positive climb at one side of the loop and negative climb at the other side. The process is called *conservative climb*.

3.9 Plastic Strain due to Dislocation Movement

It is implicit in the description of dislocation behaviour presented in preceding sections that dislocation movement, usually under the influence of an externally-applied load, results in *plastic strain*. This is in

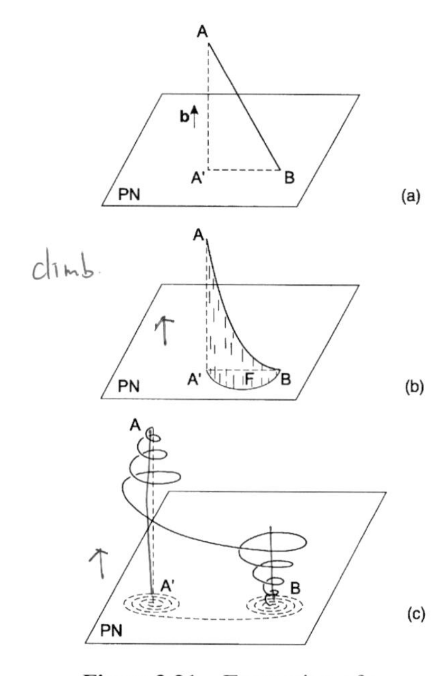

Figure 3.21 Formation of a helical dislocation by climb of a straight dislocation with a screw component. (a) A straight dislocation line AB and its projection in the plane PN which is normal to the Burgers vector **b** and passes through B. A' is the projection of A on to plane PN. (b) Change produced in the dislocation AB by climb. AB is now curved and lies on a cylinder whose axis is parallel to **b**. The dislocation can glide on this cylinder. The area F is proportional to the amount of material added or lost in climb. (c) Helical dislocation produced after further climb. The projection of this dislocation on to PN is the double spiral shown in the diagram. (After Amelinckx, Bontinck, Dekeyser and Seitz, *Phil. Mag.* **2**, 355, 1957.)

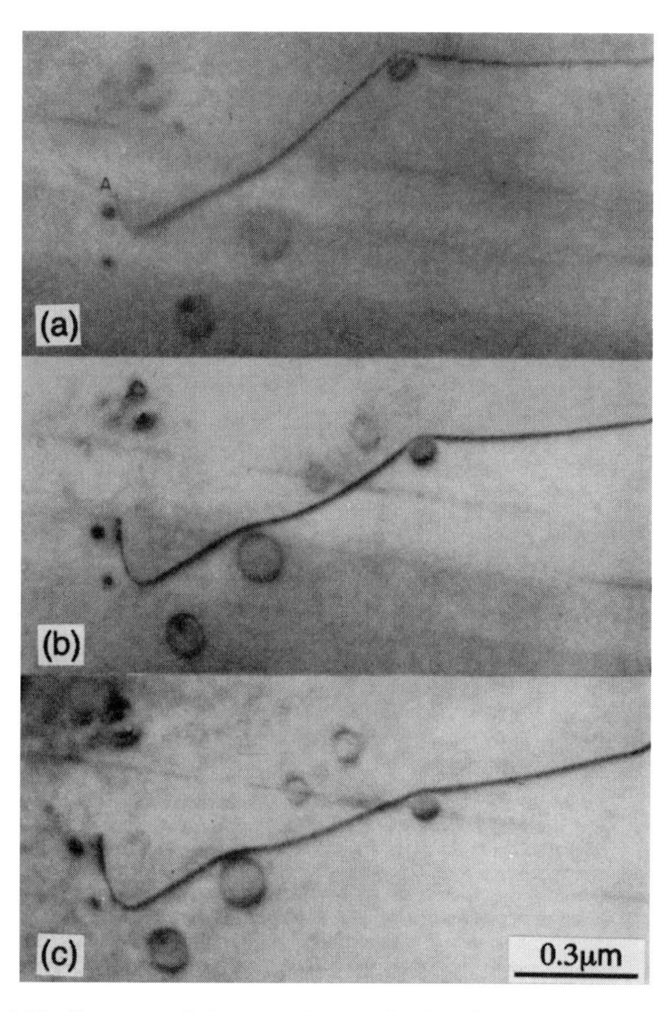

Figure 3.22 Sequence of electron micrographs showing the conservative climb motion of a dislocation loop, with a Burgers vector normal to the plane of the loop, due to its interaction with an edge dislocation. (From Kroupa and Price, *Phil. Mag.* **6**, 243, 1961.)

addition to the *elastic strain*, which is simply related to the external stress by Hooke's law (Chapter 4). The relation between plastic strain and the applied stress is more complicated and depends on factors such as temperature, applied strain rate and, in particular, the microstructure of the material: these aspects are discussed further in Chapter 10. There is, however, a simple relationship between the plastic strain and the dislocation density defined as in section 1.4. It is based on the fact that when a dislocation moves, two atoms on sites adjacent across the plane of motion are displaced relative to each other by the Burgers vector **b** (section 3.2).

The relationship for *slip* is derived first. Consider a crystal of volume hld containing, for simplicity, straight edge dislocations (Fig. 3.23(a)).

Under a high enough applied shear stress acting on the slip plane in the direction of **b**, as shown, the dislocations will glide, positive ones to the right, negative ones to the left. The top surface of the sample is therefore displaced plastically relative to the bottom surface as demonstrated in Fig. 3.23(b). If a dislocation moves completely across the slip plane through the distance d, it contributes b to the total displacement D. Since b is small in comparison to d and h, the contribution made by a dislocation which moves a distance x_i may be taken as the fraction (x_i/d) of b. Thus, if the number of dislocations which move is N, the total displacement is

$$D = \frac{b}{d}\sum_{i=1}^{N} x_i \tag{3.6}$$

and the macroscopic *plastic shear strain* ε is given by

$$\varepsilon = \frac{D}{h} = \frac{b}{hd}\sum_{i=1}^{N} x_i \tag{3.7}$$

This can be simplified by noting that the *average* distance moved by a dislocation $\bar{x}$ is

$$\bar{x} = \frac{1}{N}\sum_{i=1}^{N} x_i \tag{3.8}$$

and that the density of mobile dislocations ρ_m is (Nl/hld). Hence

$$\varepsilon = b\rho_m\bar{x} \tag{3.9}$$

The *strain rate* is therefore

$$\dot{\varepsilon} = \frac{\mathrm{d}\varepsilon}{\mathrm{d}t} = b\rho_m\bar{v} \tag{3.10}$$

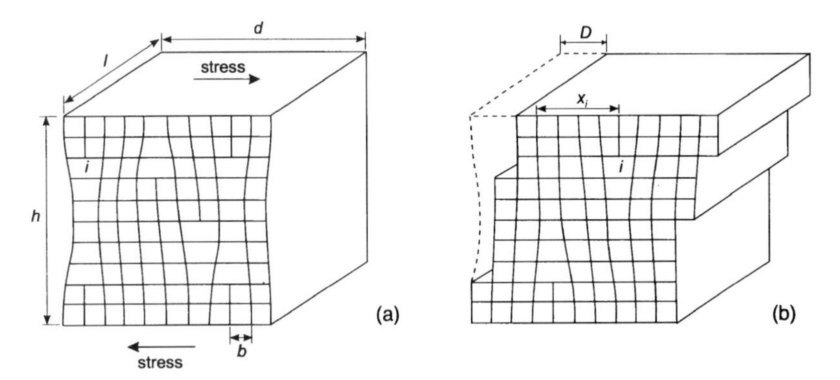

Figure 3.23 (a) Edge dislocations in a crystal subjected to an external shear stress resolved for slip. (b) Plastic displacement D produced by glide of the dislocations. Dislocation i has moved a distance x_i, as shown.

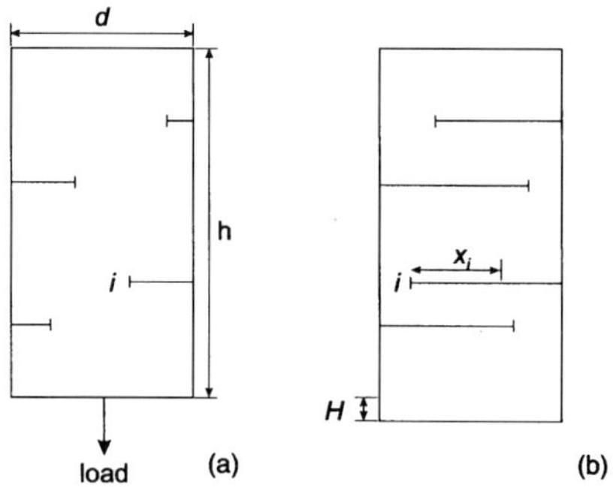

Figure 3.24 (a) Edge dislocations, denoted by their extra half planes, in a crystal under a tensile load. (b) The dislocations have climbed, producing a plastic elongation H.

where $\bar{v}$ is the average dislocation velocity. The same relationships hold for screw and mixed dislocations. Also, since the average area of slip plane swept by a dislocation $\bar{A}$ equals $l\bar{x}$, an alternative to equation (3.9) is

$$\varepsilon = bn\bar{A} \tag{3.11}$$

where n is the number of lines per unit volume. It is emphasised that ρ_m appearing in the above equations is the *mobile* dislocation density, for dislocations which do not move do not contribute to the plastic strain.

Climb under an external tensile load is shown schematically in Fig. 3.24. When an edge dislocation climbs, an extra plane of thickness b is inserted into, or removed from, the crystal in the area over which the line moves. As in the analysis for slip, if a dislocation moves through distance x_i, it contributes $b(x_i/d)$ to the plastic displacement H of the external surface. It is therefore easy to show that the total plastic *tensile* strain (H/h) parallel to the Burgers vector and the strain rate are given by equations (3.9) and (3.10) respectively. The same relations also hold for mixed dislocations, except that b is then the magnitude of the edge component of the Burgers vector.

Further Reading

The books and paper listed under 'Dislocations' at the end of Chapter 1 provide excellent sources of further information.

Balluffi, R. W. and Granato, A. V. (1979) 'Dislocations, vacancies and interstitials', *Dislocations in Solids* (ed. F. R. N. Nabarro), vol. 4, p. 1, North-Holland.

Gilman, J. J. (1969) *Micromechanics of Flow in Solids*, McGraw-Hill.

Gillis, P. P., Gilman, J. J. and Taylor, J. W. (1969) 'Stress dependences of dislocation velocities', *Phil. Mag.* **20**, 279.

Imura, T. (1972) 'Dynamic studies of plastic deformation by means of high voltage electron microscopy', *Electron Microscopy and Strength of Materials*, p. 104, University of California Press.

Indenbom, V. L. and Lothe, J. (Eds.) (1992) *Elastic Strain Fields and Dislocation Mobility*, Elsevier.

Johnston, W. G. (1962) 'Yield points and delay times in single crystals', *J. Appl. Phys.* **33**, 2716.

Johnston, W. G. and Gilman, J. J. (1959) 'Dislocation velocities, dislocation densities and plastic flow in lithium fluoride crystals', *J. Appl. Phys.* **30**, 129.

Rosenfield, A. R., Hahn, G. T., Bement, A. L. and Jaffee, R. I. (Eds.) (1968) *Dislocation Dynamics*, McGraw-Hill.

Schmid, E. and Boas, W. (1950) *Plasticity of Crystals*, Hughes & Co. Ltd.

Stein, D. F. and Low, J. W. (1960) 'Mobility of edge dislocations in silicon iron crystals', *J. Appl. Phys.* **31**, 362.

Taylor, G. I. (1934) 'Mechanism of plastic deformation in crystals', *Proc. Roy. Soc.* A **145**, 362.

Vreeland, T. Jr. (1968) 'Dislocation velocity measurements', *Techniques of Metals Research*. (ed. R. F. Bunshaw), vol. 2, p. 341, Wiley.

4 Elastic Properties of Dislocations

4.1 Introduction

The atoms in a crystal containing a dislocation are displaced from their perfect lattice sites, and the resulting distortion produces a stress field in the crystal around the dislocation. The dislocation is therefore a source of *internal stress* in the crystal. For example, consider the edge dislocation in Fig. 1.18(b). The region above the slip plane contains the extra half-plane forced between the normal lattice planes, and is in compression: the region below is in tension. The stresses and strains in the bulk of the crystal are sufficiently small for conventional elasticity theory to be applied to obtain them. This approach only ceases to be valid at positions very close to the centre of the dislocation. Although most crystalline solids are elastically *anisotropic*, i.e. their elastic properties are different in different crystallographic directions, it is much simpler to use *isotropic elasticity* theory. This still results in a good approximation in most cases. From a knowledge of the elastic field, the energy of the dislocation, the force it exerts on other dislocations, its energy of interaction with point defects, and other important characteristics, can be obtained. The elastic field produced by a dislocation is not affected by the application of stress from *external* sources: the total stress on an element within the body is the superposition of the internal and external stresses.

4.2 Elements of Elasticity Theory

The *displacement* of a point in a strained body from its position in the unstrained state is represented by the vector

$$\mathbf{u} = [u_x, u_y, u_z] \tag{4.1}$$

The components u_x, u_y, u_z represent projections of $\mathbf{u}$ on the x, y, z axes, as shown in Fig. 4.1. In *linear* elasticity, the nine components of *strain* are given in terms of the first derivatives of the displacement components thus:

$$\begin{aligned} e_{xx} &= \frac{\partial u_x}{\partial x} \\ e_{yy} &= \frac{\partial u_y}{\partial y} \\ e_{zz} &= \frac{\partial u_z}{\partial z} \end{aligned} \tag{4.2}$$

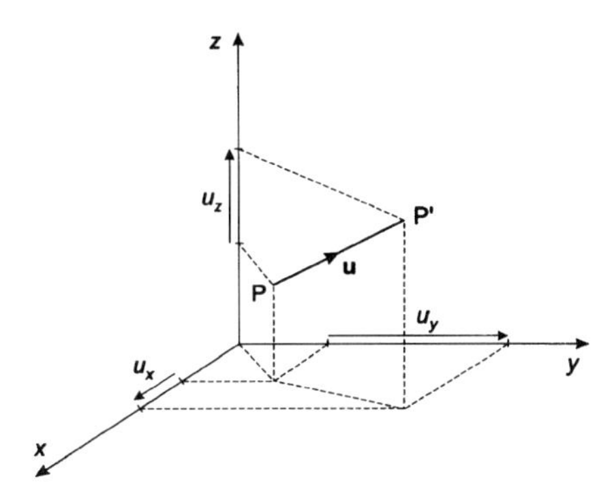

Figure 4.1 Displacement of P to P' by displacement vector **u**.

and

$$
\begin{aligned}
e_{yz} = e_{zy} &= \frac{1}{2}\left(\frac{\partial u_y}{\partial z} + \frac{\partial u_z}{\partial y}\right) \\
e_{zx} = e_{xz} &= \frac{1}{2}\left(\frac{\partial u_z}{\partial x} + \frac{\partial u_x}{\partial z}\right) \\
e_{xy} = e_{yx} &= \frac{1}{2}\left(\frac{\partial u_x}{\partial y} + \frac{\partial u_y}{\partial x}\right)
\end{aligned}
\tag{4.3}
$$

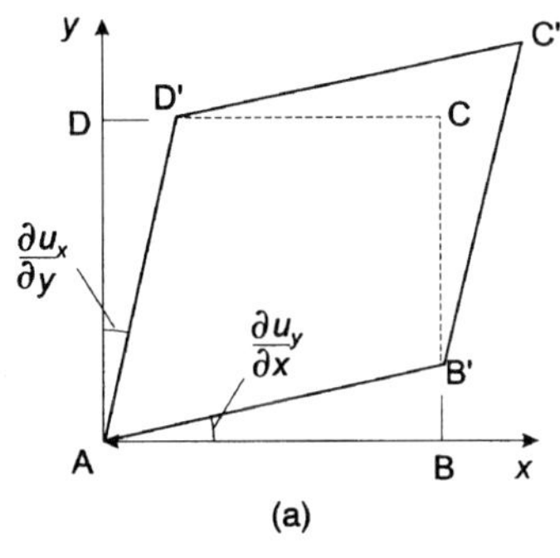

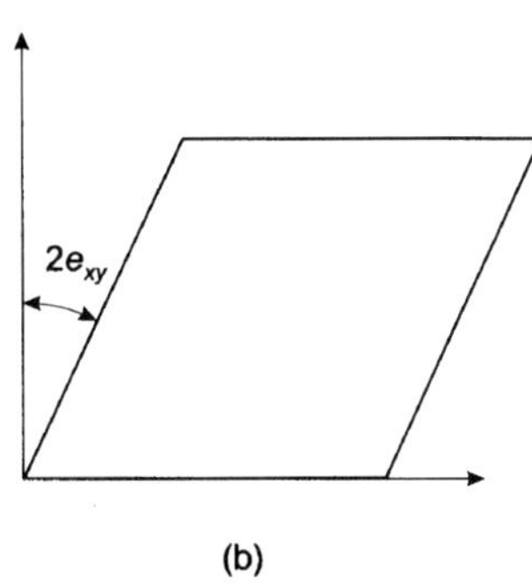

Figure 4.2 (a) Pure shear and (b) simple shear of an area element in the xy plane.

The magnitude of these components is $\ll 1$. Partial differentials are used because in general each displacement component is a function of position (x, y, z). The three strains defined in (4.2) are the *normal* strains. They represent the fractional change in length of elements parallel to the x, y and z axes respectively. The six components defined in (4.3) are the *shear* strains, and they also have simple physical meaning. This is demonstrated by e_{xy} in Fig. 4.2(a), in which a small area element $ABCD$ in the xy plane has been strained to the shape $AB'C'D'$ without change of area. The angle between the sides AB and AD initially parallel to x and y respectively has decreased by $2e_{xy}$. By rotating, but not deforming, the element as in Fig. 4.2(b), it is seen that the element has undergone a simple shear. The simple shear strain often used in engineering practice is $2e_{xy}$, as indicated.

The volume V of a small volume element is changed by strain to $(V + \Delta V) = V(1 + e_{xx})(1 + e_{yy})(1 + e_{zz})$. The fractional change in volume Δ, known as the *dilatation*, is therefore

$$
\Delta = \Delta V/V = (e_{xx} + e_{yy} + e_{zz}) \tag{4.4}
$$

Δ is independent of the orientation of the axes x, y, z.

In elasticity theory, an element of volume experiences forces via stresses applied to its surface by the surrounding material. *Stress* is the force per unit area of surface. A complete description of the stresses acting therefore requires not only specification of the magnitude and direction of the force but also of the orientation of the surface, for as the

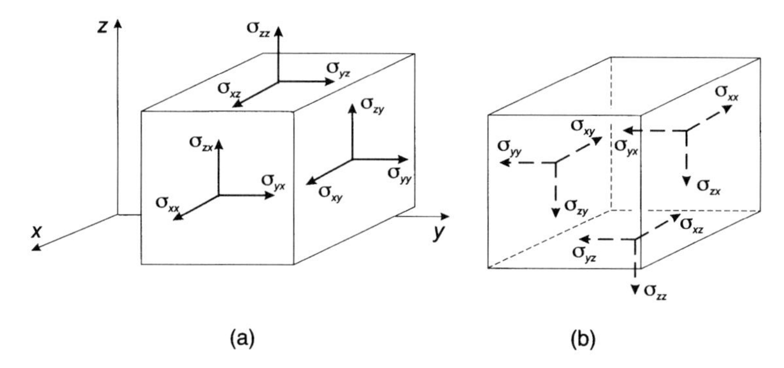

Figure 4.3 Components of stress acting on (a) the top and front faces and (b) the bottom and back faces of an elemental cube.

orientation changes so, in general, does the force. Consequently, nine components must be defined to specify the state of stress. They are shown with reference to an elemental cube aligned with the x, y, z axes in Fig. 4.3(a). The component σ_{ij}, where i and j can be x, y or z, is defined as the force per unit area exerted in the $+i$ direction on a face with outward normal in the $+j$ direction by the material *outside* upon the material *inside*. For a face with outward normal in the $-j$ direction, i.e. the bottom and back faces shown in Fig. 4.3(b), σ_{ij} is the force per unit area exerted in the $-i$ direction. For example, σ_{yz} acts in the positive y direction on the top face and the negative y direction on the bottom face.

The six components with $i \neq j$ are the *shear* stresses. (As explained in section 3.1, it is customary in dislocation studies to represent the shear stress acting on the slip plane in the slip direction of a crystal by the symbol τ.) By considering moments of forces taken about x, y and z axes placed through the centre of the cube, it can be shown that rotational equilibrium of the element, i.e. net couple = 0, requires

$$\sigma_{yz} = \sigma_{zy} \quad \sigma_{zx} = \sigma_{xz} \quad \sigma_{xy} = \sigma_{yx} \tag{4.5}$$

Thus, the order in which subscripts i and j is written is immaterial. The three remaining components σ_{xx}, σ_{yy}, σ_{zz} are the *normal* components. From the definition given above, a *positive* normal stress results in *tension* and a *negative* one in *compression*. The effective pressure acting on a volume element is therefore

$$p = -\frac{1}{3}(\sigma_{xx} + \sigma_{yy} + \sigma_{zz}) \tag{4.6}$$

For some problems, it is more convenient to use cylindrical polar coordinates (r, θ, z). The stresses are still defined as above, and are shown in Fig. 4.4. The notation is easier to follow if the second subscript j is considered as referring to the face of the element having a constant value of the coordinate j.

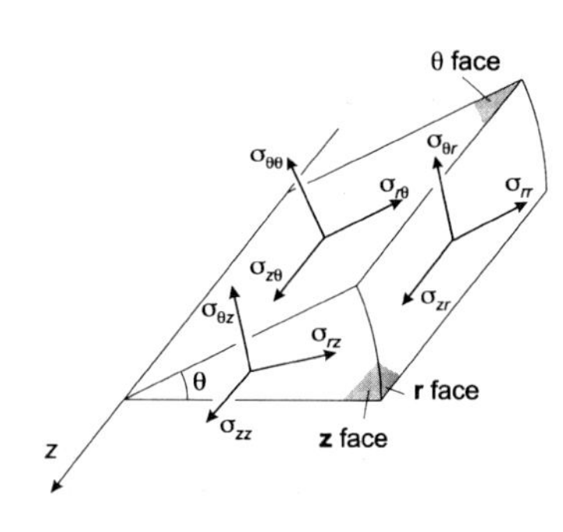

Figure 4.4 Components of stress in cylindrical polar coordinates.

The relationship between stress and strain in linear elasticity is *Hooke's Law*, in which each stress component is linearly proportional to each strain. For isotropic solids, only two proportionality constants are required:

$$\begin{aligned}
\sigma_{xx} &= 2Ge_{xx} + \lambda(e_{xx} + e_{yy} + e_{zz}) \\
\sigma_{yy} &= 2Ge_{yy} + \lambda(e_{xx} + e_{yy} + e_{zz}) \\
\sigma_{zz} &= 2Ge_{zz} + \lambda(e_{xx} + e_{yy} + e_{zz}) \\
\sigma_{xy} &= 2Ge_{xy} \quad \sigma_{yz} = 2Ge_{yz} \quad \sigma_{zx} = 2Ge_{zx}
\end{aligned} \tag{4.7}$$

λ and G are the Lamé constants, but G is more commonly known as the *shear modulus*. Other elastic constants are frequently used, the most useful being *Young's modulus*, E, *Poisson's ratio*, ν, and *bulk modulus*, K. Under uniaxial, normal loading in the longitudinal direction, E is the ratio of longitudinal stress to longitudinal strain and ν is minus the ratio of lateral strain to longitudinal strain. K is defined to be $-p/\Delta$. Since only two material parameters are required in Hooke's law, these constants are interrelated. For example,

$$E = 2G(1 + \nu) \quad \nu = \lambda/2(\lambda + G) \quad K = E/3(1 - 2\nu) \tag{4.8}$$

Typical values of E and ν for metallic and ceramic solids lie in the ranges 40–600 GN m^{-2} and 0.2–0.45 respectively.

The internal energy of a body is increased by strain. The *strain energy* per unit volume is one-half the product of stress times strain for each component. Thus, for an element of volume dV, the elastic strain energy is

$$\mathrm{d}E_{\mathrm{el}} = \frac{1}{2}\mathrm{d}V \sum_{i=x,y,z} \sum_{j=x,y,z} \sigma_{ij}e_{ij} \tag{4.9}$$

and similarly for polar coordinates.

4.3 Stress Field of a Straight Dislocation

Screw Dislocation

The elastic distortion around an infinitely-long, straight dislocation can be represented in terms of a cylinder of elastic material. Consider the screw dislocation AB shown in Fig. 4.5(a); the elastic cylinder in Fig. 4.5(b) has been deformed to produce a similar distortion. A radial slit $LMNO$ was cut in the cylinder parallel to the z-axis and the free surfaces displaced rigidly with respect to each other by the distance b, the magnitude of the Burgers vector of the screw dislocation, in the z-direction.

The elastic field in the dislocated cylinder can be found by direct inspection. First, it is noted that there are no displacements in the x and y directions:

$$u_x = u_y = 0 \tag{4.10}$$

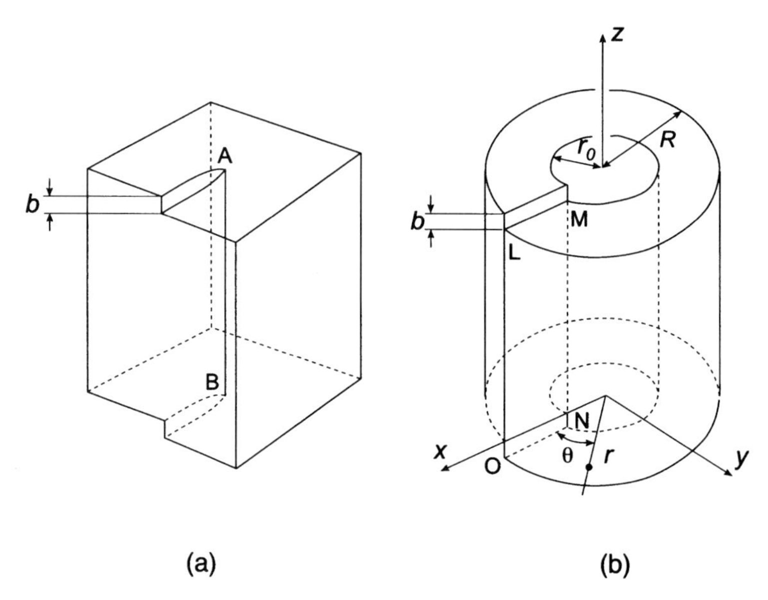

Figure 4.5 (a) Screw dislocation AB formed in a crystal. (b) Elastic distortion of a cylindrical ring simulating the distortion produced by the screw dislocation in (a).

Secondly, the displacement in the z-direction increases uniformly from zero to b as θ increases from 0 to 2π:

$$u_z = \frac{b\theta}{2\pi} = \frac{b}{2\pi}\tan^{-1}(y/x) \tag{4.11}$$

It is then readily found from equations (4.2) and (4.3) that

$$\begin{aligned} &e_{xx} = e_{yy} = e_{zz} = e_{xy} = e_{yx} = 0 \\ &e_{xz} = e_{zx} = -\frac{b}{4\pi}\frac{y}{(x^2+y^2)} = -\frac{b}{4\pi}\frac{\sin\theta}{r} \\ &e_{yz} = e_{zy} = \frac{b}{4\pi}\frac{x}{(x^2+y^2)} = \frac{b}{4\pi}\frac{\cos\theta}{r} \end{aligned} \tag{4.12}$$

From equations (4.7) and (4.12), the components of stress are

$$\begin{aligned} &\sigma_{xx} = \sigma_{yy} = \sigma_{zz} = \sigma_{xy} = \sigma_{yx} = 0 \\ &\sigma_{xz} = \sigma_{zx} = -\frac{Gb}{2\pi}\frac{y}{(x^2+y^2)} = -\frac{Gb}{2\pi}\frac{\sin\theta}{r} \\ &\sigma_{yz} = \sigma_{zy} = \frac{Gb}{2\pi}\frac{x}{(x^2+y^2)} = \frac{Gb}{2\pi}\frac{\cos\theta}{r} \end{aligned} \tag{4.13}$$

The components in cylindrical polar coordinates (Fig. 4.4) take a simpler form. Using the relations

$$\begin{aligned}\sigma_{rz} &= \sigma_{xz}\cos\theta + \sigma_{yz}\sin\theta \\ \sigma_{\theta z} &= -\sigma_{xz}\sin\theta + \sigma_{yz}\cos\theta\end{aligned} \tag{4.14}$$

and similarly for the shear strains, the only non-zero components are found to be

$$\begin{aligned}e_{\theta z} &= e_{z\theta} = \frac{b}{4\pi r} \\ \sigma_{\theta z} &= \sigma_{z\theta} = \frac{Gb}{2\pi r}\end{aligned} \tag{4.15}$$

The elastic distortion contains no tensile or compressive components and consists of pure shear: $\sigma_{z\theta}$ acts parallel to the z-axis in radial planes of constant θ and $\sigma_{\theta z}$ acts in the fashion of a torque on planes normal to the axis (Fig. 4.4). The field exhibits complete radial symmetry and the cut $LMNO$ can be made on any radial plane $\theta = \text{constant}$. For a dislocation of *opposite* sign, i.e. a left-handed screw, the signs of all the field components are *reversed*.

The stresses and strains are proportional to $1/r$ and therefore diverge to infinity as $r \to 0$. Solids cannot withstand infinite stresses, and for this reason the cylinder in Fig. 4.5 is shown as hollow with a hole of radius r_0. Real crystals are not hollow, of course, and so as the centre of a dislocation in a crystal is approached, elasticity theory ceases to be valid and a non-linear, atomistic model must be used (see section 10.3). The region within which the linear-elastic solution breaks down is called the *core* of the dislocation. From equation (4.15) it is seen that the stress reaches the theoretical limit (equation (1.5)) and the strain exceeds about 10% when $r \approx b$. A reasonable value for the *dislocation core radius* r_0 therefore lies in the range b to $4b$, i.e. $r_0 \lesssim 1$ nm in most cases.

Edge Dislocation

The stress field is more complex than that of a screw but can be represented in an isotropic cylinder in a similar way. Considering the edge dislocation in Fig. 4.6(a), the same elastic strain field can be produced in the cylinder by a rigid displacement of the faces of the slit by a distance b in the x-direction (Fig. 4.6(b)). The displacement and strains in the z-direction are zero and the deformation is called plane strain. Derivation of the field components is beyond the scope of the present treatment, however. The stresses are found to be

$$\begin{aligned}\sigma_{xx} &= -Dy\frac{(3x^2 + y^2)}{(x^2 + y^2)^2} \\ \sigma_{yy} &= Dy\frac{(x^2 - y^2)}{(x^2 + y^2)^2}\end{aligned}$$

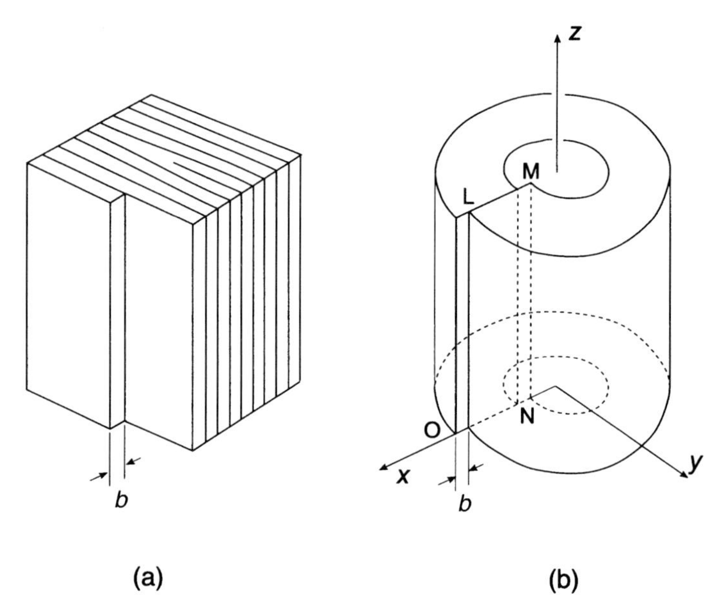

Figure 4.6 (a) Edge dislocation formed in a crystal. (b) Elastic distortion of a cylindrical ring simulating the distortion produced by the edge dislocation in (a).

$$\begin{aligned}
\sigma_{xy} &= \sigma_{yx} = Dx\frac{(x^2 - y^2)}{(x^2 + y^2)^2} \\
\sigma_{zz} &= \nu(\sigma_{xx} + \sigma_{yy}) \\
\sigma_{xz} &= \sigma_{zx} = \sigma_{yz} = \sigma_{zy} = 0
\end{aligned} \tag{4.16}$$

where

$$D = \frac{Gb}{2\pi(1-\nu)}$$

The stress field has, therefore, both dilational and shear components. The largest normal stress is σ_{xx} which acts parallel to the slip vector. Since the slip plane can be defined as $y = 0$, the maximum compressive stress (σ_{xx} negative) acts immediately above the slip plane and the maximum tensile stress (σ_{xx} positive) acts immediately below the slip plane. The effective pressure (equation (4.6)) on a volume element is

$$p = \frac{2}{3}(1+\nu)D\frac{y}{(x^2 + y^2)} \tag{4.17}$$

It is compressive above the slip plane and tensile below. These observations are implied qualitatively by the type of distortion illustrated in Figs 1.18 and 4.6(a).

As in the case of the screw, the signs of the components are reversed for a dislocation of opposite sign, i.e. a negative edge dislocation with extra half-plane along the negative y-axis. Again, the elastic solution has an inverse dependence on distance from the line axis and breaks down when x and y tend to zero. It is valid only outside a core of radius r_0.

The elastic field produced by a *mixed* dislocation (Fig. 3.8(b)) having *edge and screw character* is obtained from the above equations by adding the fields of the edge and screw constituents, which have Burgers vectors given by equation (3.2). The two sets are independent of each other in isotropic elasticity.

4.4 Strain Energy of a Dislocation

The existence of distortion around a dislocation implies that a crystal containing a dislocation is not in its lowest energy state. The extra energy is the *strain energy*. The total strain energy may be divided into two parts

$$E_{\text{total}} = E_{\text{core}} + E_{\text{elastic strain}} \tag{4.18}$$

The elastic part, stored outside the core, may be determined by integration of the energy of each small element of volume. This is a simple calculation for the screw dislocation, because from the symmetry the appropriate volume element is a cylindrical shell of radius r and thickness dr. From equation (4.9), the elastic energy stored in this volume *per unit length* of dislocation is

$$\begin{aligned} \mathrm{d}E_{\text{el}}(\text{screw}) &= \frac{1}{2}2\pi r\,\mathrm{d}r(\sigma_{\theta z}e_{\theta z} + \sigma_{z\theta}e_{z\theta}) \\ &= 4\pi r\,\mathrm{d}r\,Ge_{\theta z}^2 \end{aligned} \tag{4.19}$$

Thus, from equation (4.15), the total elastic energy stored in the cylinder (Fig. 4.5) per unit length of dislocation is

$$E_{\text{el}}(\text{screw}) = \frac{Gb^2}{4\pi}\int_{r_0}^{R}\frac{\mathrm{d}r}{r} = \frac{Gb^2}{4\pi}\ln\left(\frac{R}{r_0}\right) \tag{4.20}$$

where R is the outer radius.

The above approach is much more complicated for other dislocations having less symmetric fields. It is generally easier to consider E_{el} as the work done in displacing the faces of the cut $LMNO$ by b (Figs 4.5 and 4.6) against the resisting internal stresses. For an infinitesimal element of area dA of $LMNO$, the work done is

$$\begin{aligned} \mathrm{d}E_{\text{el}}(\text{screw}) &= \frac{1}{2}\sigma_{zy}b\mathrm{d}A \\ \mathrm{d}E_{\text{el}}(\text{edge}) &= \frac{1}{2}\sigma_{xy}b\mathrm{d}A \end{aligned} \tag{4.21}$$

with the stresses evaluated on $y = 0$. The factor $\frac{1}{2}$ enters because the stresses build up from zero to the final values given by equations (4.13) and (4.16) during the displacement process. The element of area is a strip of width dx parallel to the z-axis, and so the total strain energy per unit length of dislocation is

$$\begin{aligned} E_{el}(\text{screw}) &= \frac{Gb^2}{4\pi}\int_{r_0}^{R}\frac{dx}{x} = \frac{Gb^2}{4\pi}\ln\left(\frac{R}{r_0}\right) \\ E_{el}(\text{edge}) &= \frac{Gb^2}{4\pi(1-\nu)}\int_{r_0}^{R}\frac{dx}{x} = \frac{Gb^2}{4\pi(1-\nu)}\ln\left(\frac{R}{r_0}\right) \end{aligned} \qquad (4.22)$$

The screw result is the same as equation (4.20). Strictly, equations (4.22) neglect small contributions from the work done on the core surface $r = r_0$ of the cylinder, but they are adequate for most requirements.

Equations (4.22) demonstrate that E_{el} depends on the core radius r_0 and the crystal radius R, but only logarithmically. E_{el} (edge) is greater than E_{el} (screw) by $1/(1-\nu) \approx 3/2$. Taking $R = 1$ mm, $r_0 = 1$ nm, $G = 40\,\text{GN}\,\text{m}^{-2}$ and $b = 0.25$ nm, the elastic strain energy of an edge dislocation will be about $4\,\text{nJ}\,\text{m}^{-1}$ or about 1 aJ (6 eV) for each atom plane threaded by the dislocation. In crystals containing many dislocations, the dislocations tend to form in configurations in which the superimposed long-range elastic fields cancel. The energy per dislocation is thereby reduced and an appropriate value of R is approximately half the average spacing of the dislocations arranged at random.

Estimates of the energy of the core of the dislocation are necessarily very approximate. However, the estimates that have been made suggest that the core energy will be of the order of 1 eV for each atom plane threaded by the dislocation, and is thus only a small fraction of the elastic energy. However, in contrast to the elastic energy, the energy of the core will vary as the dislocation moves through the crystal and this gives rise to the lattice resistance to dislocation motion discussed in section 10.3.

The validity of elasticity theory for treating dislocation energy outside a core region has been demonstrated by computer simulation (section 2.7). Figure 4.7 shows data for an atomic model of alpha iron containing a straight edge dislocation with Burgers vector $\frac{1}{2}[111]$ and line direction $[11\bar{2}]$. E_{total} is the strain energy within a cylinder of radius R with the dislocation along its axis. The energy varies logarithmically with R, as predicted by equation (4.22), outside a core of radius 0.7 nm, which is about $2.6b$. The core energy is about $7\,\text{eV}\,\text{nm}^{-1}$.

It was mentioned in the preceding section that the elastic field of a *mixed* dislocation (see Fig. 3.8(b)) is the superposition of the fields of its edge and screw parts. As there is no interaction between them, the total elastic energy is simply the sum of the edge and screw energies with b replaced by $b\sin\theta$ and $b\cos\theta$ respectively:

$$\begin{aligned} E_{el}(\text{mixed}) &= \left[\frac{Gb^2\sin^2\theta}{4\pi(1-\nu)} + \frac{Gb^2\cos^2\theta}{4\pi}\right]\ln\left(\frac{R}{r_0}\right) \\ &= \frac{Gb^2(1-\nu\cos^2\theta)}{4\pi(1-\nu)}\ln\left(\frac{R}{r_0}\right) \end{aligned} \qquad (4.23)$$

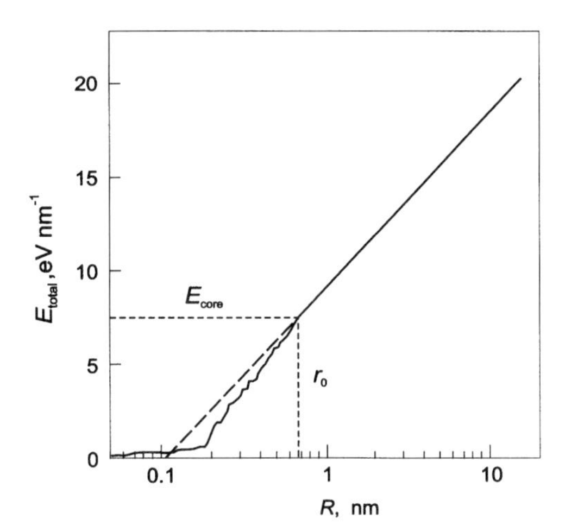

Figure 4.7 The strain energy within a cylinder of radius R that contains a straight edge dislocation along its axis. The data was obtained by computer simulation for a model of iron. (Courtesy Yu. N. Osetsky.)

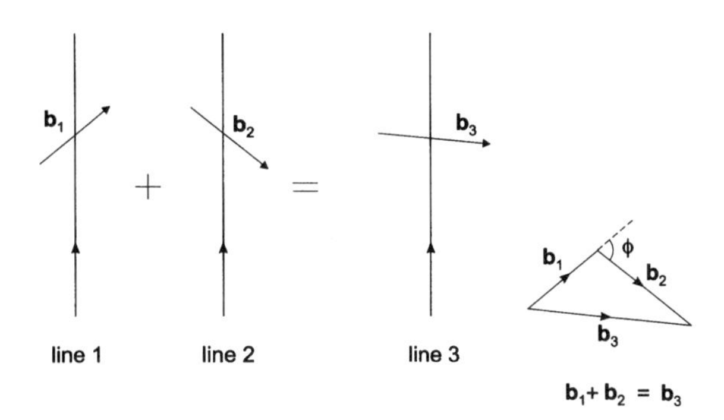

Figure 4.8 Reaction of two dislocations to form a third.

which falls between the energy of an edge and a screw dislocation.

From the expressions for edge, screw and mixed dislocations it is clear that the energy per unit length is relatively insensitive to the character of the dislocation and also to the values of R and r_0. Taking realistic values for R and r_0 all the equations can be written approximately as

$$E_{el} = \alpha G b^2 \tag{4.24}$$

where $\alpha \approx 0.5-1.0$. This leads to a very simple rule (*Frank's rule*) for determining whether or not it is energetically feasible for two dislocations to react and combine to form another. Consider the two dislocations in Fig. 4.8 with Burgers vectors $\mathbf{b}_1$ and $\mathbf{b}_2$ given by the Burgers circuit construction (section 1.4). Allow them to combine to form a new dislocation with Burgers vector $\mathbf{b}_3$ as indicated. From equation (4.24), the elastic energy per unit length of the dislocations is proportional to b_1^2, b_2^2 and b_3^2 respectively. Thus, if $(b_1^2 + b_2^2) > b_3^2$, the reaction is favourable for it results in a reduction in energy. If $(b_1^2 + b_2^2) < b_3^2$, the reaction is unfavourable and the dislocation with Burgers vector b_3 is liable to dissociate into the other two. If $(b_1^2 + b_2^2) = b_3^2$, there is no energy change. These three conditions correspond to the angle ϕ in Fig. 4.8 satisfying $\pi/2 < \phi \leq \pi$, $0 \leq \phi < \pi/2$ and $\phi = \pi/2$ respectively. In this argument the assumption is made that there is no additional interaction energy involved, i.e. that before and after the reaction the reacting dislocations are separated sufficiently so that the interaction energy is small. If this is not so, the reactions are still favourable and unfavourable, but the energy changes are smaller than implied above. Frank's rule is used to consider the feasibility of various dislocation reactions in Chapters 5 and 6.

4.5 Forces on Dislocations

When a sufficiently high stress is applied to a crystal containing dislocations, the dislocations move and produce plastic deformation either by slip as described in section 3.3 or, at sufficiently high temperatures, by climb (section 3.6). The load producing the applied stress therefore does work on the crystal when a dislocation moves, and so the dislocation responds to the stress as though it experiences a force equal to the work done divided by the distance it moves. The force defined in this way is a virtual, rather than real, force, but the force concept is useful for treating the mechanics of dislocation behaviour. The *glide* force is considered in this section and the climb force in section 4.7.

Consider a dislocation moving in a slip plane under the influence of a uniform resolved shear stress τ (Fig. 4.9). When an element $\mathrm{d}l$ of the dislocation line of Burgers vector **b** moves forward a distance $\mathrm{d}s$ the crystal planes above and below the slip plane will be displaced relative to each other by b. The average shear displacement of the crystal surface produced by glide of $\mathrm{d}l$ is

$$\left(\frac{\mathrm{d}s\,\mathrm{d}l}{A}\right)b \tag{4.25}$$

where A is the area of the slip plane. The external force due to τ acting over this area is $A\tau$, so that the work done when the element of slip occurs is

$$\mathrm{d}W = A\tau\left(\frac{\mathrm{d}s\,\mathrm{d}l}{A}\right)b \tag{4.26}$$

The glide force F on a unit length of dislocation is defined as the work done when unit length of dislocation moves unit distance. Therefore

$$F = \frac{\mathrm{d}W}{\mathrm{d}s\,\mathrm{d}l} = \frac{\mathrm{d}W}{\mathrm{d}A} = \tau b \tag{4.27}$$

The stress τ *is the shear stress in the glide plane resolved in the direction of* **b** and the glide force F acts normal to the dislocation at every point

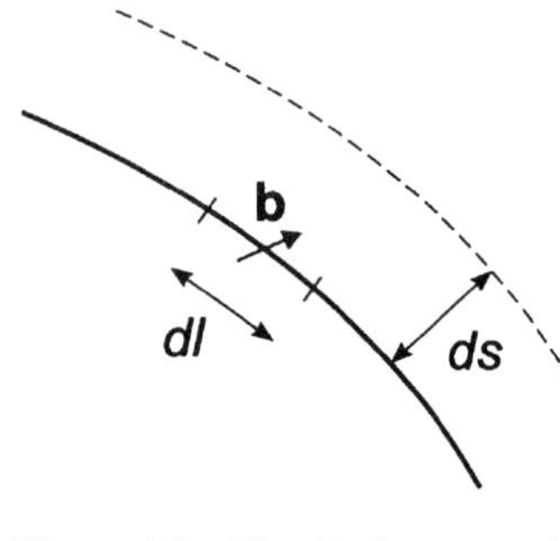

Figure 4.9 The displacement $\mathrm{d}s$ used to determine the glide force on an element $\mathrm{d}l$ in its glide plane.

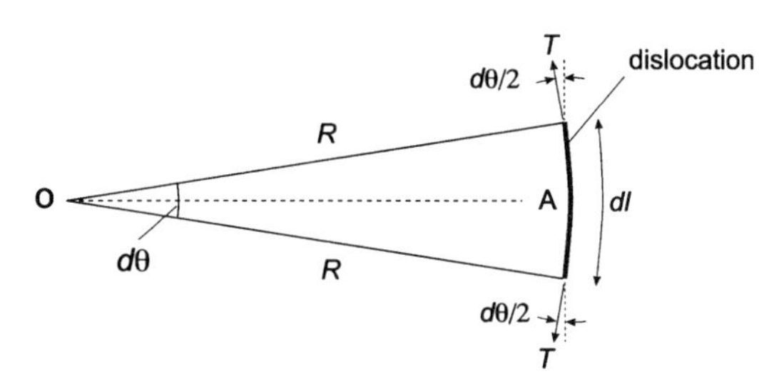

Figure 4.10 Curved element of dislocation under line tension forces T.

along its length, irrespective of the line direction. The positive sense of the force is given by the physical reasoning of section 3.3.

In addition to the force due to an externally applied stress, a dislocation has a *line tension* which is analogous to the surface tension of a soap bubble or a liquid. This arises because, as outlined in the previous section, the strain energy of a dislocation is proportional to its length and an increase in length results in an increase in energy. The line tension has units of energy per unit length. From the approximation used in equation (4.24), the line tension, which may be defined as *the increase in energy per unit increase in the length of a dislocation line*, will be

$$T = \alpha G b^2 \tag{4.28}$$

Consider the curved dislocation in Fig. 4.10. The line tension will produce forces tending to straighten the line and so reduce the total energy. The direction of the net force is perpendicular to the dislocation and towards the centre of curvature. The line will only remain curved if there is a shear stress which produces a force on the dislocation line in the opposite sense. The shear stress τ_0 needed to maintain a radius of curvature R is found in the following way. The angle subtended at the centre of curvature is $\mathrm{d}\theta = \mathrm{d}l/R$, assumed to be $\ll 1$. The outward force along OA due to the applied stress acting on the elementary piece of dislocation is $\tau_0 b \mathrm{d}l$ from equation (4.27), and the opposing inward force along OA due to the line tension T at the ends of the element is $2T\sin(\mathrm{d}\theta/2)$, which is equal to $T\mathrm{d}\theta$ for small values of $\mathrm{d}\theta$. The line will be in equilibrium in this curved position when

$$\begin{aligned} T\mathrm{d}\theta &= \tau_0 b\, \mathrm{d}l \\ \text{i.e.} \quad \tau_0 &= \frac{T}{bR} \end{aligned} \tag{4.29}$$

Substituting for T from equation (4.28)

$$\tau_0 = \frac{\alpha G b}{R} \tag{4.30}$$

This gives an expression for the stress required to bend a dislocation to a radius R and is used many times in subsequent chapters. A particularly direct application is in the understanding of the Frank–Read dislocation multiplication source described in Chapter 8.

Equation (4.30) assumes from equation (4.24) that edge, screw and mixed segments have the same energy per unit length, and the curved dislocation of Fig. 4.10 is therefore the arc of a circle. This is only strictly valid if Poisson's ratio ν equals zero. In all other cases, the line experiences a torque tending to rotate it towards the screw orientation where its energy per unit length is lower. The true line tension of a mixed segment is

$$T = E_{el}(\theta) + \frac{d^2 E_{el}(\theta)}{d\theta^2} \tag{4.31}$$

where $E_{el}(\theta)$ is given by equation (4.23). T for a screw segment is four times that of an edge when $\nu = 1/3$. Thus, for a line bowing under a uniform stress, the radius of curvature at any point is still given by equation (4.29), but the overall line shape is approximately elliptical with major axis parallel to the Burgers vector; the axial ratio is approximately $1/(1-\nu)$. For most calculations, however, equation (4.30) is an adequate approximation.

4.6 Forces between Dislocations

A simple semi-qualitative argument will illustrate the significance of the concept of a force between dislocations. Consider two parallel edge dislocations lying in the same slip plane. They can either have the same sign as in Fig. 4.11(a) or opposite sign as in Fig. 4.11(b). When the dislocations are separated by a large distance the total elastic energy per unit length of the dislocations in both situations will be, from equation (4.24)

$$\alpha G b^2 + \alpha G b^2 \tag{4.32}$$

When the dislocations in Fig. 4.11(a) are very close together the arrangement can be considered approximately as a single dislocation with a Burgers vector magnitude $2b$ and the elastic energy will be given by

$$\alpha G (2b)^2 \tag{4.33}$$

which is twice the energy of the dislocations when they are separated by a large distance. Thus the dislocations will tend to repel each other to reduce their total elastic energy. When dislocations of opposite sign (Fig. 4.11(b)) are close together, the effective magnitude of their Burgers vectors will be zero, and the corresponding long-range elastic energy zero also. Thus dislocations of opposite sign will attract each other to reduce their total elastic energy. The positive and negative edge dislocations in Fig. 4.11(b) will combine and annihilate each other. These conclusions regarding repulsion and attraction also follow for dislocations of mixed orientation from Frank's rule by putting $\phi = 0$ or π in Fig. 4.8. Similar effects occur when the two dislocations do not lie in the same slip plane (Fig. 4.11(c)), but the conditions for attraction and repulsion are usually more complicated, as discussed below.

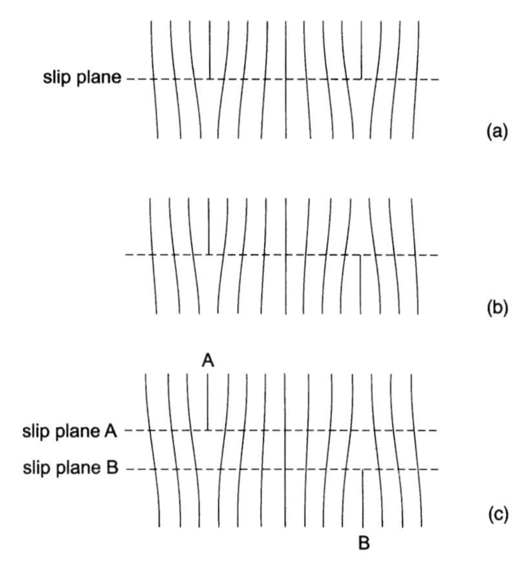

Figure 4.11 Arrangement of edge dislocations with parallel Burgers vectors lying in parallel slip planes. (a) Like dislocations on the same slip plane, (b) unlike dislocations on the same slip plane, and (c) unlike dislocations on slip planes separated by a few atomic spacings.

The basis of the method used to obtain the force between two dislocations is the determination of the additional work done in introducing the second dislocation into a crystal which already contains the first. Consider two dislocations lying parallel to the z-axis (Fig. 4.12). The total energy of the system consists of (a) the self-energy of dislocation I, (b) the self-energy of dislocation II, and (c) the elastic interaction energy between I and II. The *interaction energy* E_{int} is the work done in displacing the faces of the cut which creates II in the presence of the stress field of I. The displacements across the cut are b_x, b_y, b_z, the components of the Burgers vector **b** of II. By visualising the cut parallel to either the x or y axes, two alternative expressions for E_{int} per unit length of II are

$$\begin{aligned} E_{\text{int}} &= +\int_x^\infty (b_x\sigma_{xy} + b_y\sigma_{yy} + b_z\sigma_{zy})\mathrm{d}x \\ E_{\text{int}} &= -\int_y^\infty (b_x\sigma_{xx} + b_y\sigma_{yx} + b_z\sigma_{zx})\mathrm{d}y \end{aligned} \tag{4.34}$$

where the stress components are those due to I. (The signs of the right-hand side of these equations arise because if the displacements of **b** are taken to occur on the face of a cut with outward normal in the positive y and x directions, respectively, they are in the direction of positive x, y, z for the first case (x-axis cut) and negative x, y, z for the second (y-axis cut).)

The interaction force on II is obtained simply by differentiation of these expressions, i.e. $F_x = -\partial E_{\text{int}}/\partial x$ and $F_y = -\partial E_{\text{int}}/\partial y$. For the two parallel edge dislocations with parallel Burgers vectors shown in Fig. 4.12, $b_y = b_z = 0$ and $b_x = b$, and the components of the force per unit length acting on II are therefore

$$F_x = \sigma_{xy} b \quad F_y = -\sigma_{xx} b \tag{4.35}$$

where σ_{xy} and σ_{xx} are the stresses of I evaluated at position (x, y) of II. The forces are reversed if II is a negative edge i.e. the dislocations have opposite sign. Equal and opposite forces act on I. F_x is the force in the glide direction and F_y the force perpendicular to the glide plane. Substituting from equation (4.16) gives

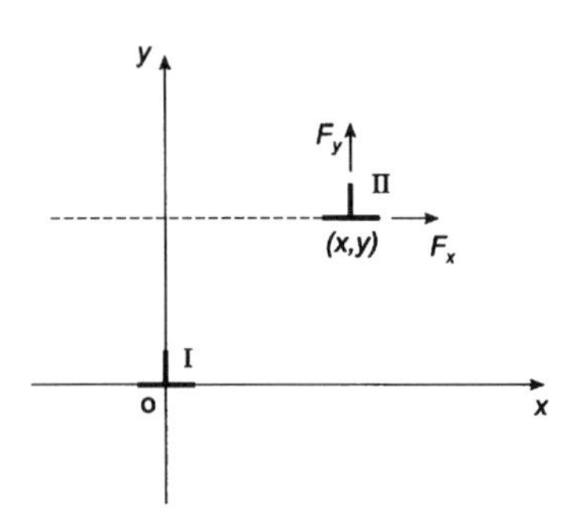

Figure 4.12 Forces considered for the interaction between two edge dislocations.

$$\begin{aligned} F_x &= \frac{Gb^2}{2\pi(1-\nu)} \frac{x(x^2 - y^2)}{(x^2 + y^2)^2} \\ F_y &= \frac{Gb^2}{2\pi(1-\nu)} \frac{y(3x^2 + y^2)}{(x^2 + y^2)^2} \end{aligned} \tag{4.36}$$

Since an edge dislocation can move by slip only in the plane contained by the dislocation line and its Burgers vector, the component of force which is most important in determining the behaviour of the dislocations in Fig. 4.12 is F_x. For dislocations of the same sign, inspection of the variation of F_x with x reveals the following:

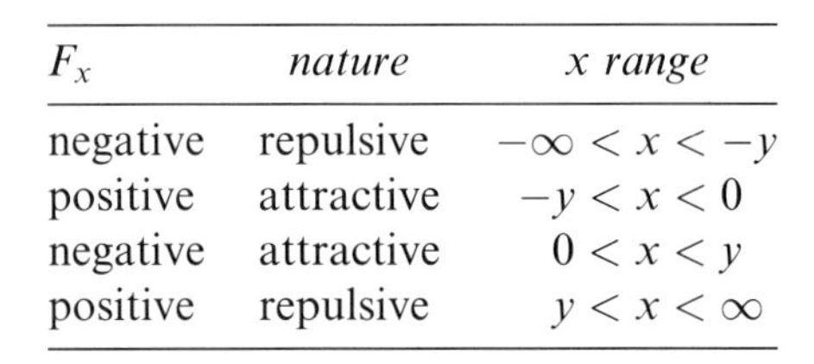

F_x	*nature*	*x range*
negative	repulsive	$-\infty < x < -y$
positive	attractive	$-y < x < 0$
negative	attractive	$0 < x < y$
positive	repulsive	$y < x < \infty$

The sign and nature of F_x is reversed if I and II are edge dislocations of opposite sign. F_x is plotted against x, expressed in units of y, in Fig. 4.13. It is zero when $x = 0, \pm y, \pm\infty$, but of these, the positions of stable equilibrium are seen to be $x = 0, \pm\infty$ for edges of the same sign and $\pm y$ if they have the opposite sign.

It follows that an array of edge dislocations of the same sign is most stable when the dislocations lie vertically above one another as in Fig. 4.14(a). This is the arrangement of dislocations in a small angle pure tilt boundary described in Chapter 9. Furthermore, edge dislocations of opposite sign gliding past each other on parallel slip planes tend to form stable *dipole* pairs as in Fig. 4.14(b) at low applied stresses (section 10.8).

Comparison of the glide force F_x in equation (4.35) with F in equation (4.27) shows that since σ_{xy} is the shear stress in the glide plane of

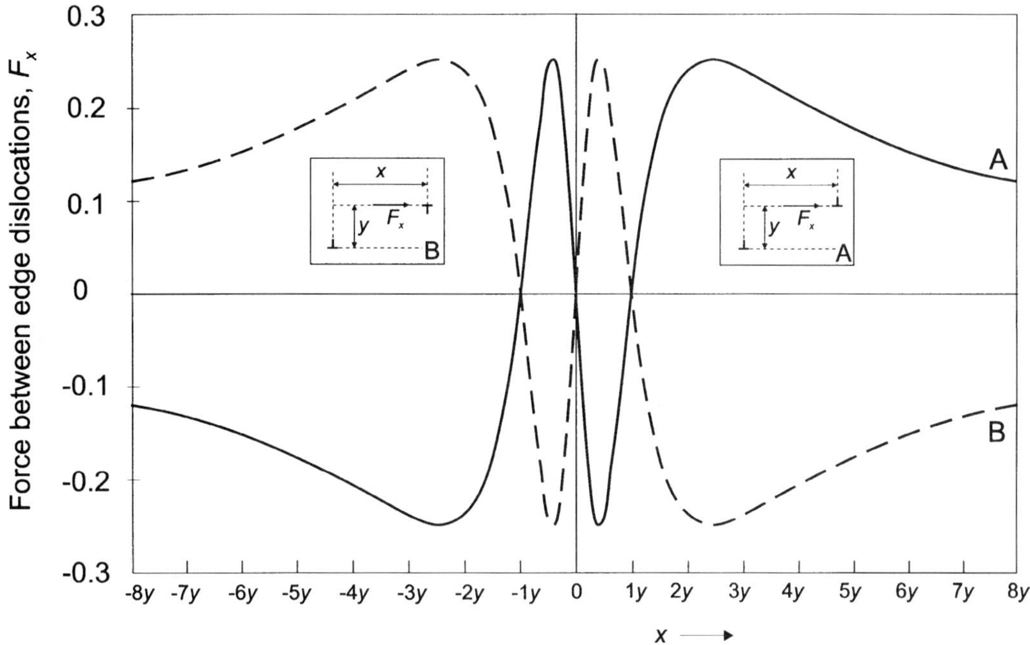

Figure 4.13 Glide force per unit length between parallel edge dislocations with parallel Burgers vectors from equation (4.36). Unit of force F_x is $Gb^2/2\pi(1-\nu)y$. The full curve A is for like dislocations and the broken curve B for unlike dislocations.

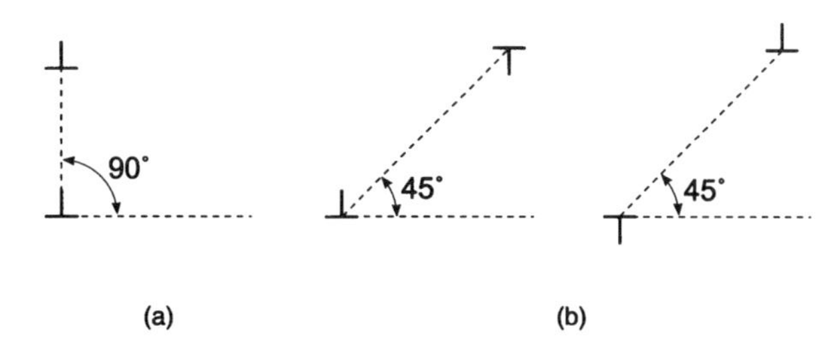

Figure 4.14 Stable positions for two edge dislocations of (a) the same sign and (b) opposite sign.

dislocation II acting in the direction of its Burgers vector, *equation* (4.27) *holds for both external and internal sources of stress.*

Consider two parallel screw dislocations, one lying along the z-axis. The radial and tangential components of force on the other are

$$F_r = \sigma_{z\theta}b \quad F_\theta = \sigma_{zr}b \tag{4.37}$$

and substituting from equations (4.15)

$$F_r = Gb^2/2\pi r \quad F_\theta = 0 \tag{4.38}$$

The force is much simpler in form than that between two edge dislocations because of the radial symmetry of the screw field. F_r is repulsive for screws of the same sign and attractive for screws of opposite sign. It is readily shown from either equations (4.35) or equations (4.37) that no forces act between a pair of parallel dislocations consisting of a pure edge and a pure screw, as expected from the lack of mixing of their stress fields (see section 4.3).

4.7 Climb Forces

The force component F_y in equation (4.35) is a *climb force* per unit length resulting from the normal stress σ_{xx} of dislocation I attempting to squeeze the extra half-plane of II from the crystal. This can only occur physically if intrinsic point defects can be emitted or absorbed at the dislocation core of II (see section 3.6). As in the case of glide forces, climb forces can arise from external and internal sources of stress. The former are important in *creep*, and the latter provided the example of conservative climb in section 3.8. Line tension can also produce climb forces, but in this case the force acts to reduce the line length in the extra half-plane: shrinkage of prismatic loops as in Fig. 3.19 is an example. However, since the creation and annihilation of point defects are involved in climb, *chemical forces* due to defect concentration changes must be taken into account in addition to these *mechanical forces.*

It was seen in section 3.6 that when an element **l** of dislocation is displaced through **s**, the local volume change is $\mathbf{b} \times \mathbf{l} \cdot \mathbf{s}$. Consider a segment length l of a positive edge dislocation climbing upwards through distance s in response to a mechanical climb force F per unit length. The work done is Fls and the number of vacancies absorbed is bls/Ω, where Ω is the volume per atom. The vacancy formation energy is therefore changed by $F\Omega/b$. As a result of this chemical potential, the equilibrium vacancy concentration at temperature T in the presence of the dislocation is reduced to

$$\begin{aligned} c &= \exp[-(E_f^v + F\Omega/b)/kT] \\ &= c_0 \exp(-F\Omega/bkT) \end{aligned} \tag{4.39}$$

where c_0 is the equilibrium concentration in a stress-free crystal (equation (1.3)). For negative climb involving vacancy emission ($F < 0$) the sign of the chemical potential is changed so that $c > c_0$. Thus, the vacancy concentration deviates from c_0, building up a chemical force per unit length on the line

$$f = \frac{bkT}{\Omega} \ln(c/c_0) \tag{4.40}$$

until f balances F in equilibrium. Conversely, in the presence of a supersaturation c/c_0 of vacancies, the dislocation climbs up under the chemical force f until compensated by, say, external stresses or line tension. The latter is used in the analysis for a dislocation climb source in section 8.7. The nature of these forces is illustrated schematically in Fig. 4.15. By substituting reasonable values of T and Ω in equation (4.40), it is easy

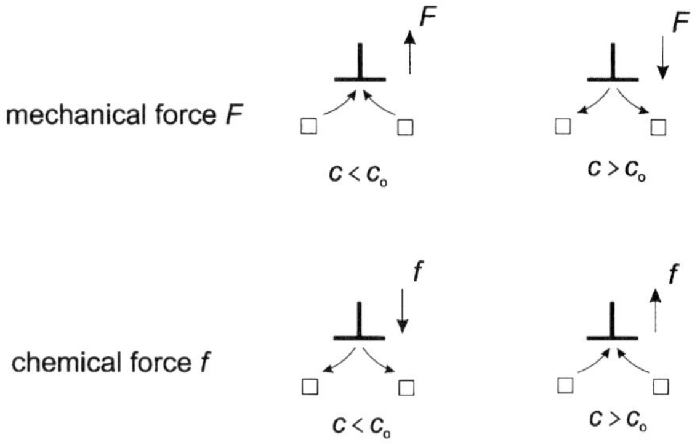

mechanical force F

$c < c_o$ $c > c_o$

chemical force f

$c < c_o$ $c > c_o$

Figure 4.15 Mechanical and chemical forces for climb of an edge dislocation. Vacancies (shown as □) have a local concentration c in comparison with the equilibrium concentration c_0 in a dislocation-free crystal.

to show that even moderate supersaturations of vacancies can produce forces much greater than those arising from external stresses.

The rate of climb of a dislocation in practice depends on (a) the direction and magnitude of the mechanical and chemical forces, F and f, (b) the mobility of jogs (section 3.6) and (c) the rate of migration of vacancies through the lattice to or from the dislocation.

4.8 Image Forces

A dislocation near a surface experiences forces not encountered in the bulk of a crystal. The dislocation is attracted towards a free surface because the material is effectively more compliant there and the dislocation energy is lower: conversely, it is repelled by a rigid surface layer. To treat this mathematically, extra terms must be added to the infinite-body stress components given in section 4.3 in order that the required surface conditions are satisfied. When evaluated at the dislocation line, as in equations (4.35) and (4.37), they result in a force. The analysis for infinite, straight dislocation lines parallel to the surface is relatively straight-forward.

Consider screw and edge dislocations parallel to, and distance d from, a surface $x = 0$ (Fig. 4.16); the edge dislocation has Burgers vector **b** in the x direction. For a *free* surface, the tractions σ_{xx}, σ_{yx} and σ_{zx} must be zero on the plane $x = 0$. Consideration of equation (4.13) shows that these boundary conditions are met for the screw if the infinite-body result is modified by adding to it the stress field of an imaginary screw dislocation of opposite sign at $x = -d$ (Fig. 4.16(a)). The required solution for the stress in the body ($x > 0$) is therefore

$$\sigma_{zx} = \frac{-Ay}{(x_-^2 + y^2)} + \frac{Ay}{(x_+^2 + y^2)} \tag{4.41}$$

$$\sigma_{zy} = \frac{Ax_-}{(x_-^2 + y^2)} - \frac{Ax_+}{(x_+^2 + y^2)} \tag{4.42}$$

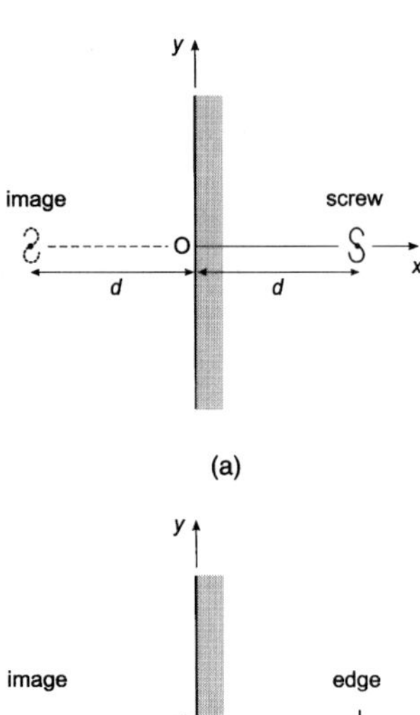

Figure 4.16 (a) Screw and (b) edge dislocations a distance d from a surface $x = 0$. The image dislocations are in space a distance d from the surface.

where $x_- = (x - d)$, $x_+ = (x + d)$ and $A = Gb/2\pi$. The force per unit length in the x-direction $F_x(= \sigma_{zy} b)$ induced by the surface is obtained from the second term in σ_{zy} evaluated at $x = d$, $y = 0$. It is

$$F_x = -Gb^2/4\pi d \tag{4.43}$$

and is simply the force due to the *image* dislocation at $x = -d$. For the edge dislocation (Fig. 4.16(b)), superposing the field of an imaginary edge dislocation of opposite sign at $x = -d$ annuls the stress σ_{xx} on $x = 0$, but not σ_{yx}. When the extra terms are included to fully match the boundary conditions, the shear stress in the body is found to be

$$\sigma_{yx} = \frac{Dx_-(x_-^2 - y^2)}{(x_-^2 + y^2)^2} - \frac{Dx_+(x_+^2 - y^2)}{(x_+^2 + y^2)^2} - \frac{2Dd[x_- x_+^3 - 6xx_+ y^2 + y^4]}{(x_+^2 + y^2)^3} \tag{4.44}$$

where $D = A/(1 - \nu)$. The first term is the stress in the absence of the surface, the second is the stress appropriate to an image dislocation at $x = -d$, and the third is that required to make $\sigma_{yx} = 0$ when $x = 0$. The force per unit length $F_x(= \sigma_{yx} b)$ arising from the surface is given by putting $x = d$, $y = 0$ in the second and third terms. The latter contributes zero, so that the force is

$$F_x = -Gb^2/4\pi(1 - \nu)d \tag{4.45}$$

and is again equivalent to the force due to the image dislocation.

The *image forces* decrease slowly with increasing d and are capable of removing dislocations from near-surface regions. They are important, for example, in specimens for transmission electron microscopy (section 2.4) when the slip planes are orientated at large angles ($\simeq 90°$) to the surface. It should be noted that a second dislocation near the surface would experience a force due to its own image *and* the surface terms in the field of the first. The interaction of dipoles, loops and curved dislocations with surfaces is therefore complicated, and only given approximately by images.

Further Reading

Bacon, D. J., Barnett, D. M. and Scattergood, R. O. (1979) 'Anisotropic continuum theory of lattice defects', *Prog. in Mater. Sci.* **23**, 51.

Cottrell, A. H. (1953) *Dislocations and Plastic Flow in Crystals*, Oxford University Press.

Friedel, J. (1964) *Dislocations*, Pergamon Press.

Hirth, J. P. and Lothe, J. (1982) *Theory of Dislocations*, Wiley.

Lardner, R. W. (1974) *Mathematical Theory of Dislocations and Fracture*, Univ. of Toronto Press.

Mura, T. (1982) *Micromechanics of Defects in Solids*, Martinus Nijhoff.

Nabarro, F. R. N. (1967) *The Theory of Crystal Dislocations*, Oxford University Press.

Nye, J. F. (1967) *Physical Properties of Crystals*, Oxford University Press.

Seeger, A. (1955) 'Theory of lattice imperfections', *Handbuch der Physik*, Vol. VII, part 1, p. 383, Springer-Verlag.
Steeds, J. W. (1973) *Anisotropic Elasticity Theory of Dislocations*, Oxford.
Teodosiu, C. (1982) *Elastic Models of Crystal Defects*, Springer.
Timoshenko, S. P. and Goodier, J. N. (1970) *Theory of Elasticity*, McGraw-Hill.
Weertman, J. and Weertman, J. R. (1964) *Elementary Dislocation Theory*, Macmillan.

5 Dislocations in Face-centred Cubic Metals

5.1 Perfect Dislocations

Many common metals such as copper, silver, gold, aluminium, nickel and their alloys, have a face-centred cubic crystal structure (Fig. 1.7). The pure metals are soft, with critical resolved shear stress values for single crystals $\simeq 0.1-1\,\mathrm{MN\,m^{-2}}$. They are ductile but can be hardened considerably by plastic deformation and alloying. The deformation behaviour is closely related to the atomic structure of the core of dislocations, which is more complex than that described in Chapters 1 and 3.

The shortest lattice vectors, and therefore the most likely Burgers vectors for dislocations in the face-centred cubic structure, are of the type $\frac{1}{2}\langle 110\rangle$ and $\langle 001\rangle$. Since the energy of a dislocation is proportional to the square of the magnitude of its Burgers vector b^2 (section 4.4), the energy of $\frac{1}{2}\langle 110\rangle$ dislocations will be only half that of $\langle 001\rangle$, i.e. $2a^2/4$ compared with a^2. Thus, $\langle 001\rangle$ dislocations are much less favoured energetically and, in fact, are only rarely observed. Since $\frac{1}{2}\langle 110\rangle$ is a translation vector for the lattice, glide of a dislocation with this Burgers vector leaves behind a perfect crystal and the dislocation is a *perfect dislocation*. Figure 5.1 represents a $\frac{1}{2}[110]$ edge dislocation in a face-centred cubic lattice. The (110) planes perpendicular to **b** are illustrated and have a two-fold stacking sequence *ABAB*... (section 1.2). The 'extra half-plane' consists of two (110) half-planes in the *ABAB*... sequence. Movement of this unit dislocation by glide retains continuity of the *A* planes and the *B* planes across the glide plane, except at the dislocation core where the extra half-planes terminate.

5.2 Partial Dislocations – the Shockley Partial

Examination of Fig. 5.1 suggests that the two extra (110) planes need not necessarily be immediately adjacent to each other. If separated, each would appear as a single dislocation with a Burgers vector shorter than $\frac{1}{2}\langle 110\rangle$. The perfect dislocation would split into two *partial dislocations*, as explained in section 5.3.

From the description of dislocation movement in Chapter 3, it may be deduced that motion of a dislocation whose Burgers vector is not a lattice vector leaves behind an imperfect crystal containing a stacking fault. Thus, when a stacking fault ends inside a crystal, the boundary in the plane of the fault, separating the faulted region from the perfect region of the crystal, is a partial dislocation. Two of the important partial dislocations, recognised in face-centred cubic metals, are the

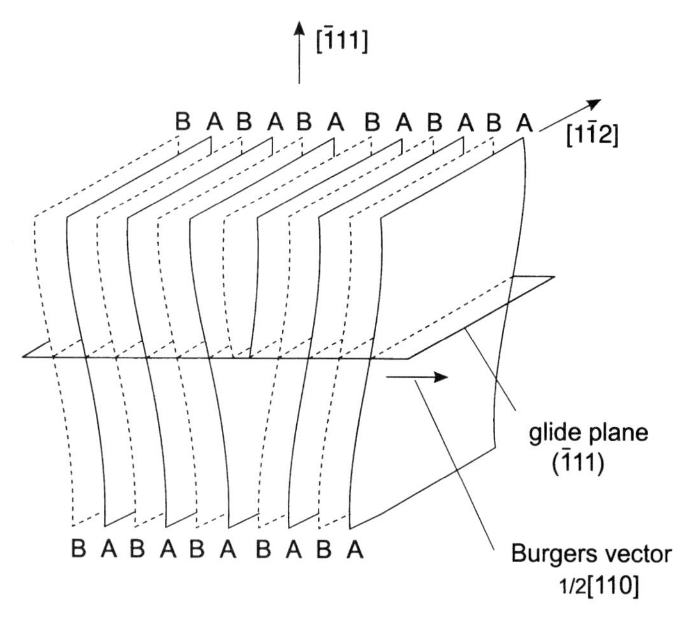

Figure 5.1 Unit edge dislocation $\frac{1}{2}[110]$ in a face-centred cubic crystal. (After Seeger (1957), *Dislocations and Mechanical Properties of Crystals*, p. 243, Wiley.)

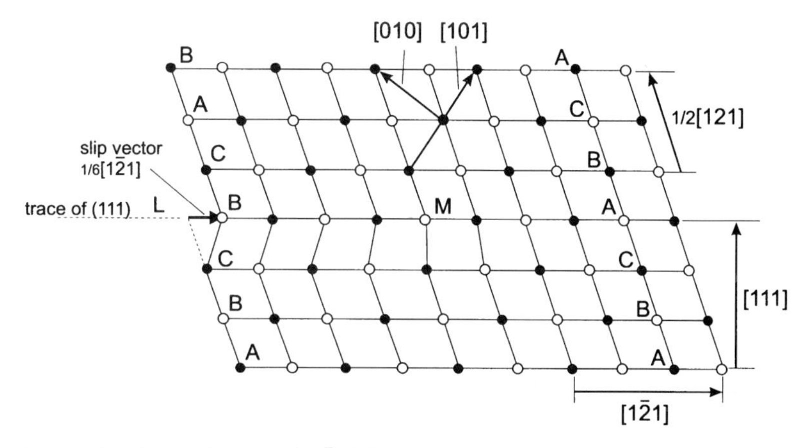

Figure 5.2 Formation of a $\frac{1}{6}[1\bar{2}1]$ Shockley partial dislocation at *M* due to slip along *LM*. The open circles represent the positions of atoms in the $(10\bar{1})$ plane of the diagram and the filled circles the positions of the atoms in the $(10\bar{1})$ planes immediately above and below the plane of the diagram. Some lattice vectors that lie in the $(10\bar{1})$ plane are indicated. (After Read (1953), *Dislocations in Crystals*, McGraw-Hill.)

Shockley partial, which is associated with slip, and the *Frank partial* (see section 5.5). The formation of a Shockley partial edge dislocation is illustrated in Fig. 5.2 and can be compared with the formation of an edge dislocation in an elastic model (Fig. 3.3). The diagram represents a $(10\bar{1})$ section through the lattice. The close-packed (111) planes lie at right angles to the plane of the diagram. At the right of the diagram the (111) layers are stacked in the sequence *ABCABC*... and the lattice is perfect. At the left of the diagram the *A* layer atoms above *LM* have slipped in the $[1\bar{2}1]$ direction to a *B* layer position, the *B* atoms have slipped to *C* and the *C* atoms have slipped to *A*, and have produced a stacking fault and a partial dislocation. The fault vector, which is in the (111) slip plane, is $\mathbf{b} = \frac{1}{6}[1\bar{2}1]$, and the magnitude of the vector is $a/\sqrt{6}$; this compares with $a/\sqrt{2}$ for the $\frac{1}{2}\langle 110\rangle$ perfect dislocation, as shown by the following:

$$\begin{aligned} &\text{Shockley } \left(\mathbf{b} = \frac{1}{6}\langle 112\rangle\right): b^2 = \frac{a^2}{36}(1^2 + 1^2 + 2^2) = \frac{a^2}{6} \\ &\text{Perfect } \left(\mathbf{b} = \frac{1}{2}\langle 110\rangle\right): b^2 = \frac{a^2}{4}(1^2 + 1^2 + 0) \;\; = \frac{a^2}{2} \end{aligned} \tag{5.1}$$

The Burgers vector of a partial dislocation is described in the same way as that of a perfect dislocation, except that the Burgers circuit must start and finish in the surface of the stacking fault; if the circuit started at any other position it would be necessary to cross the fault plane and the one-to-one correspondence of the circuits in the perfect and imperfect lattices would not be maintained. Since the Burgers vector of a partial dislocation is not a unit lattice vector, the 'finish' position of the circuit in the perfect lattice (Fig. 1.19(b)) is not a lattice site.

5.3 Slip

Slip occurs between close-packed {111} atomic planes and the observed slip direction is $\langle 110\rangle$. Only rarely does glide occur on other planes. Since slip involves the sliding of close-packed planes of atoms over each other, a simple experiment can be made to see how this can occur. The close-packed planes can be simulated by a set of hard spheres, as illustrated schematically in Fig. 5.3, and have a three-fold stacking sequence *ABCABC*... (see also Fig. 1.7).

One layer is represented by the full circles, *A*, the second identical layer rests in the sites marked *B* and the third takes the positions *C*. Consider the movement of the layers when they are sheared over each other to produce a displacement in the slip direction. It will be found that the *B* layer of atoms, instead of moving from one *B* site to the next *B* site over the top of the *A* atoms (vector $\mathbf{b}_1$), will move first to the nearby *C* site along the 'valley' between the two *A* atoms (vector $\mathbf{b}_2$) and then to the new *B* site via a second valley (vector $\mathbf{b}_3$). Thus, the *B* plane will slide over the *A* plane in a zig-zag motion.

This simple hard-sphere description has been confirmed by computer simulation (section 2.7). As one part of a crystal slides over another across a {111} plane to form a stacking fault, the energy varies from minima at translations corresponding to perfect stacking to maxima when atoms across the fault are directly over each other. The extra crystal energy per unit area of stacking fault is the *stacking fault energy*

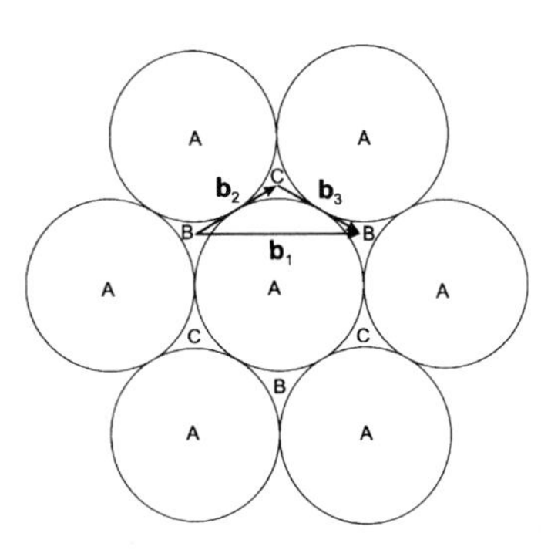

Figure 5.3 Slip of {111} planes in face-centred cubic metals.

γ. A plot of γ versus fault translation vector is a γ *surface* and is shown for a model of copper in Fig. 5.4. The positions of perfect stacking ($\gamma = 0$) are labelled P and are connected by translations of the type $\frac{1}{2}\langle 110\rangle$. The locations F correspond to the *intrinsic stacking fault* (section 1.3) and are reached from P by vectors of the form $\frac{1}{6}\langle 112\rangle$. It can be seen that these translations follow low energy paths. Furthermore, F is a local minimum (with $\gamma \simeq 40\,\text{mJ}\,\text{m}^{-2}$ for copper) so that the fault is stable.

In terms of glide of a perfect dislocation with Burgers vector $\mathbf{b}_1 = \frac{1}{2}\langle 110\rangle$, this demonstration suggests that it will be energetically more favourable for the B atoms to move to B via the C positions. This implies that the dislocation passes as two partial dislocations, one immediately after the other. The first has a Burgers vector $\mathbf{b}_2$ and the second a Burgers vector $\mathbf{b}_3$, each of which has the form $\frac{1}{6}\langle 112\rangle$. The perfect dislocation with Burgers vector $\mathbf{b}_1$ therefore *splits up* or *dissociates* into two dislocations $\mathbf{b}_2$ and $\mathbf{b}_3$ according to the reaction:

$$\mathbf{b}_1 \rightarrow \mathbf{b}_2 + \mathbf{b}_3 \tag{5.2}$$

or

$$\frac{1}{2}\langle 110\rangle \rightarrow \frac{1}{6}\langle 211\rangle + \frac{1}{6}\langle 12\bar{1}\rangle$$

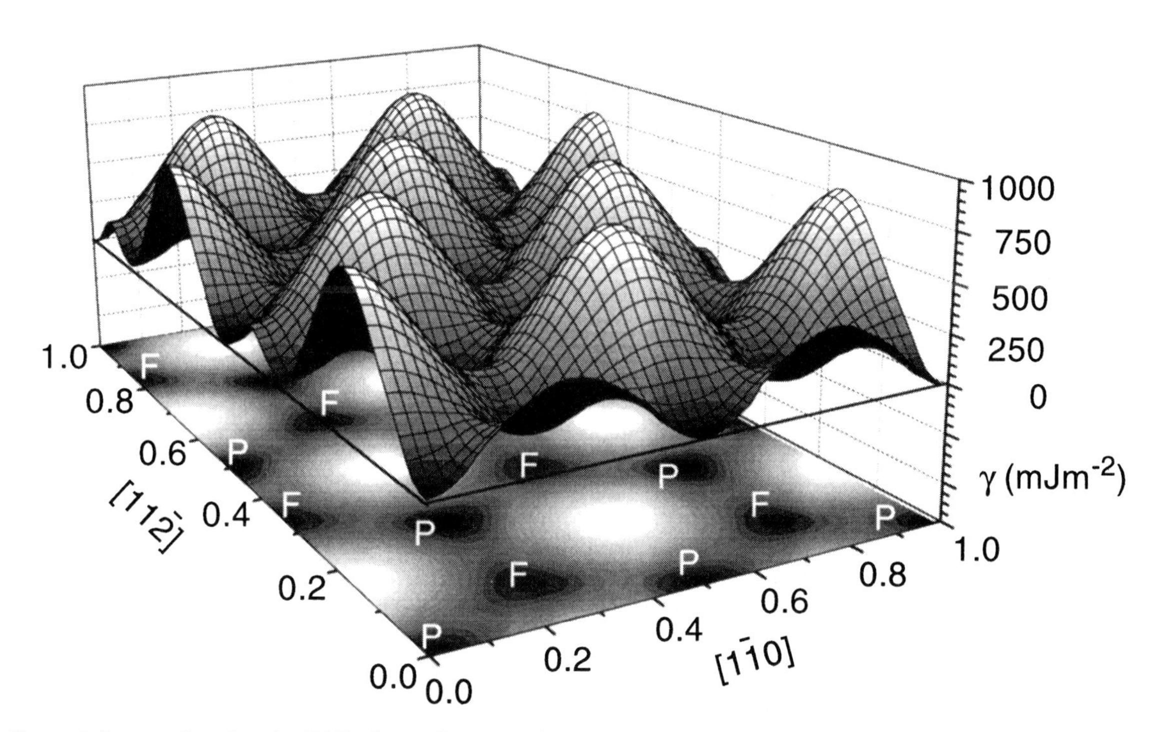

Figure 5.4 γ surface for the (111) plane of copper obtained by computer simulation. γ is zero for perfect stacking (P). Intrinsic fault (F) is metastable (with $\gamma \simeq 40\,\text{mJ}\,\text{m}^{-2}$) and corresponds to fault vectors of the form $\frac{1}{6}\langle 112\rangle$. (Courtesy Yu. N. Osetsky.)

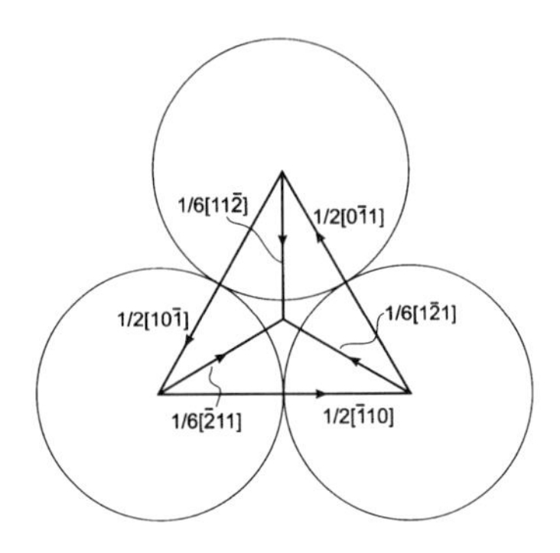

Figure 5.5 Burgers vectors of perfect and Shockley partial dislocations in the (111) plane.

In any of the four $\{111\}$ slip planes there are three $\langle 110\rangle$ directions and three $\langle 112\rangle$ directions, as shown for the (111) plane in Fig. 5.5. Dissociation of a perfect dislocation into two partials on, say, the (111) plane can therefore be one of the Burgers vector reactions given by

$$\begin{aligned}\frac{1}{2}[\bar{1}10] &\rightarrow \frac{1}{6}[\bar{2}11] + \frac{1}{6}[\bar{1}2\bar{1}] \\ \frac{1}{2}[\bar{1}01] &\rightarrow \frac{1}{6}[\bar{2}11] + \frac{1}{6}[\bar{1}\bar{1}2] \\ \frac{1}{2}[0\bar{1}1] &\rightarrow \frac{1}{6}[1\bar{2}1] + \frac{1}{6}[\bar{1}\bar{1}2]\end{aligned} \tag{5.3}$$

and the reactions produced by complete reversal of each vector. It is necessary in dislocation reactions to ensure that the total Burgers vector is unchanged, as explained in section 1.4. The right-hand side of the first of reactions (5.3), for example, is $\frac{1}{6}[\bar{2}+\bar{1}, 1+2, 1+\bar{1}]$, which equals the left-hand side. The same result is demonstrated diagrammatically by the vector triangles in Fig. 5.5.

Frank's rule (section 4.4) shows that the splitting reaction (5.2) is energetically favourable, for from (5.1) $b_1^2 = a^2/2$ which is greater than $b_2^2 + b_3^2 = a^2/3$. Since the two Burgers vectors $\mathbf{b}_2$ and $\mathbf{b}_3$ are at 60° to each other, the two partial dislocations repel each other with a force due to their elastic interaction (see section 4.6). The force may be calculated from the separate forces between their screw components (equation (4.38)) and their edge components (equation (4.36) with $y = 0$). If the spacing of the partials is d, the repulsive force per unit length between the partials of either pure edge or pure screw perfect dislocations is

$$\begin{aligned}F &= \frac{Gb^2(2+\nu)}{8\pi(1-\nu)d} \quad \text{(edge)} \\ F &= \frac{Gb^2(2-3\nu)}{8\pi(1-\nu)d} \quad \text{(screw)}\end{aligned} \tag{5.4}$$

respectively, where b $(=a/\sqrt{6})$ is the magnitude of $\mathbf{b}_2$ and $\mathbf{b}_3$. For the special case $\nu = 0$, F is the same in all line orientations:

$$F = \frac{G\mathbf{b}_2 \cdot \mathbf{b}_3}{2\pi d} = \frac{Gb^2}{4\pi d} \tag{5.5}$$

and this is a reasonable approximation to the more general results. Since $\mathbf{b}_2$ and $\mathbf{b}_3$ are the Burgers vectors of Shockley partial dislocations, it follows that if they separate there will be a *ribbon* of stacking fault between them. The stacking sequence of $\{111\}$ planes outside the dislocation will be *ABCABCABC*... and between the partial dislocations *ABCACABC*... This is the intrinsic fault discussed above and is equivalent to four layers of close-packed hexagonal stacking in a face-centred cubic crystal. The stacking fault energy (J m^{-2} or N m^{-1}) provides a force γ per unit length of line ($\mathrm{N\,m^{-1}}$) tending to pull the dislocations together. An equilibrium separation will be established when the repulsive and attractive forces balance. The approximate equilibrium separation is obtained by equating γ to F in equation (5.5).

$$d = \frac{Gb^2}{4\pi\gamma} \tag{5.6}$$

The unit edge dislocation illustrated in Fig. 5.1 will split up as illustrated in Fig. 5.6. The configuration is called an *extended dislocation*. Note that the width, d, is inversely proportional to the stacking fault energy.

Except for the size of the partial separation d, the dissociation of a perfect dislocation is independent of its character (edge, screw or mixed). Screw dislocations can form a similar configuration to Fig. 5.6. During glide under stress, a dissociated dislocation moves as a pair of partials bounding the fault ribbon, the leading partial creating the fault and the trailing one removing it: the total slip vector is $\mathbf{b}_1 = \frac{1}{2}\langle 110\rangle$. Experimental observations of extended dislocations in thin foils have confirmed that this geometry is correct. Figure 5.7 shows sets of extended dislocations lying in parallel slip planes. The stacking fault ribbon between two partials appears as a parallel fringe pattern. The individual partials are not always visible and their positions are illustrated in Fig. 5.7(b). Numerous estimates of γ have been made from direct observation of

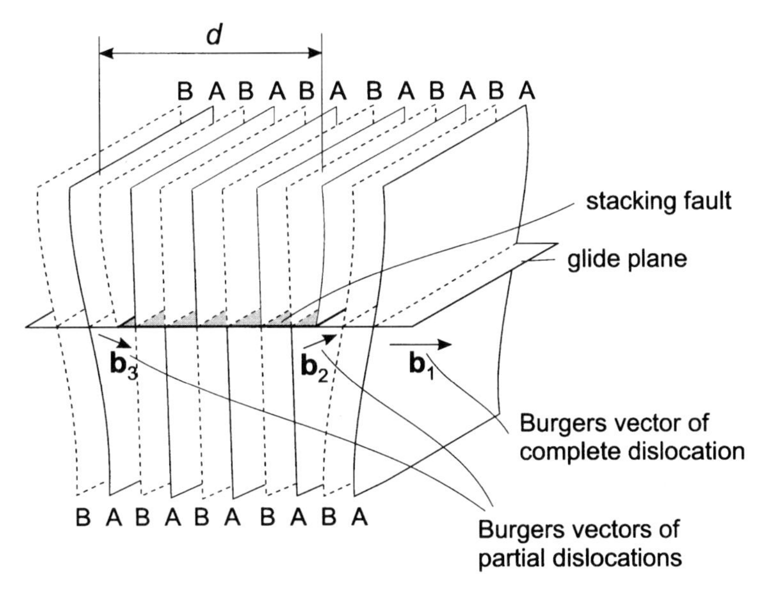

Figure 5.6 Formation of an extended dislocation by dissociation of a unit edge dislocation (Fig. 5.1) into two Shockley partials of Burgers vectors $\mathbf{b}_2$ and $\mathbf{b}_3$ separated by a stacking fault. (After Seeger (1957), *Dislocations and Mechanical Properties of Crystals*, p. 243, Wiley.)

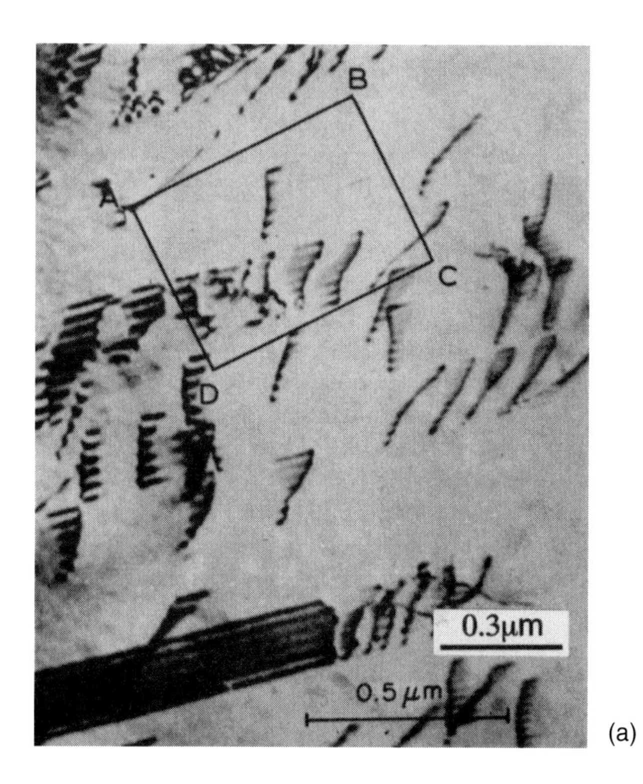

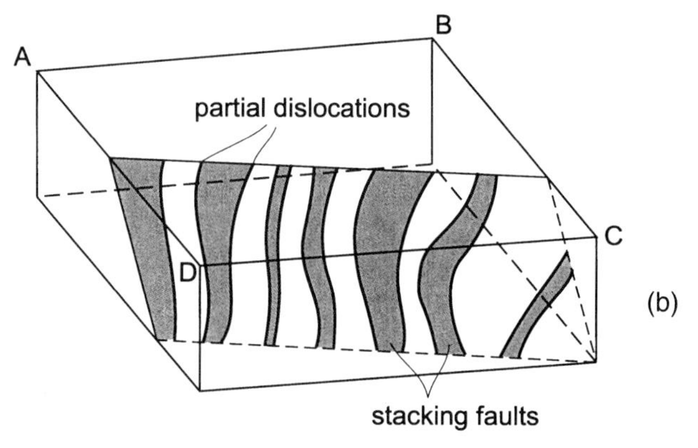

Figure 5.7 (a) Transmission electron micrograph of extended dislocations in a copper–7 per cent aluminium alloy. (From Howie, *Metallurgical Reviews*, **6**, 467, 1961.) (b) Arrangement of dislocations in the inset in (a).

the spacing of partial dislocations in the transmission electron microscope (see sections 2.4 and 7.8), from the shrinkage rate of faulted prismatic loops such as those shown in Fig. 3.19, and indirectly from the temperature dependence of the flow stress of single crystals. It is probable that $\gamma \approx 140\,\mathrm{mJ\,m^{-2}}$ ($140\,\mathrm{ergs\,cm^{-2}}$) for aluminium, $\approx 40\,\mathrm{mJ\,m^{-2}}$

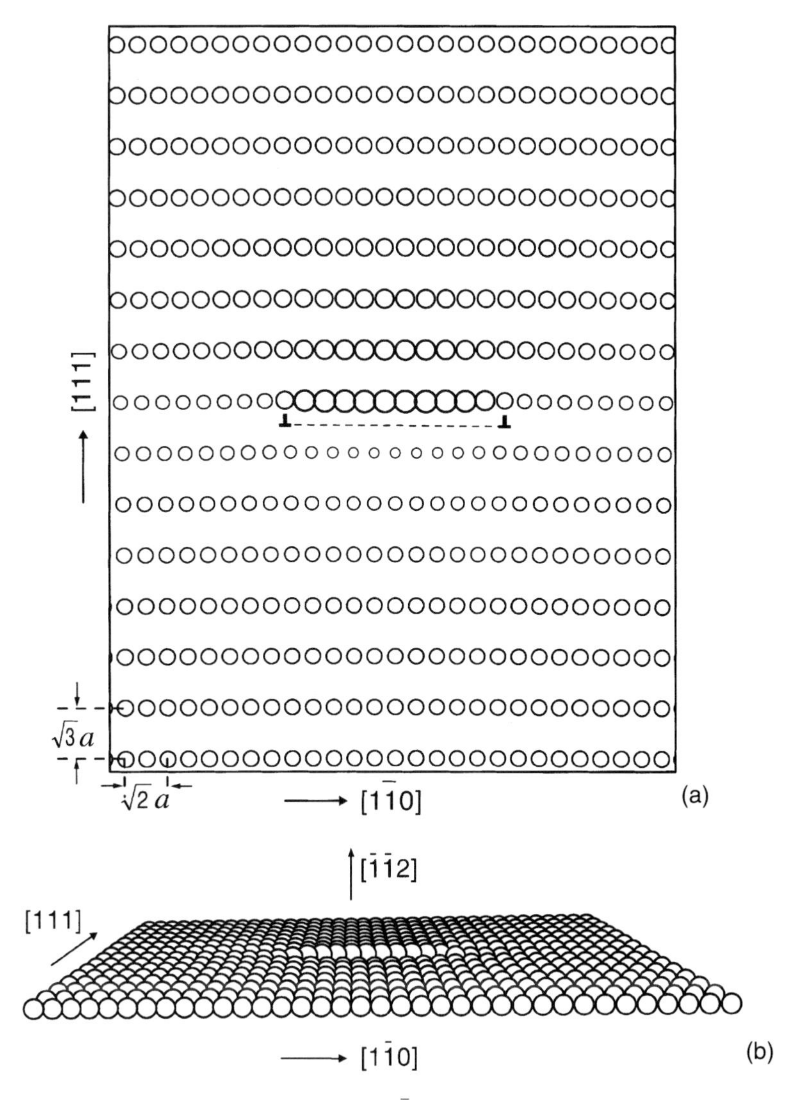

Figure 5.8 (a) Atom positions in a $(11\bar{2})$ plane perpendicular to a pure edge dislocation lying in the $[11\bar{2}]$ direction and having Burgers vector $\frac{1}{2}[1\bar{1}0]$ in a model crystal of copper. The dislocation has dissociated into two Shockley partials at the positions shown. Atom displacements either into or out of the plane of the paper are indicated by smaller or larger circles, respectively. (b) The $(11\bar{2})$ plane viewed at a shallow angle in order to see the edge and screw components of the two partials more clearly. (Courtesy Yu. N. Osetsky.)

for copper and $\approx$20 mJ m^{-2} for silver. The corresponding width of the stacking-fault ribbons given by equation (5.6) are about a, $5a$ and $7a$, respectively. The widths predicted by equations (5.4) are rather greater than these for edge dislocations and less for screws. For comparison

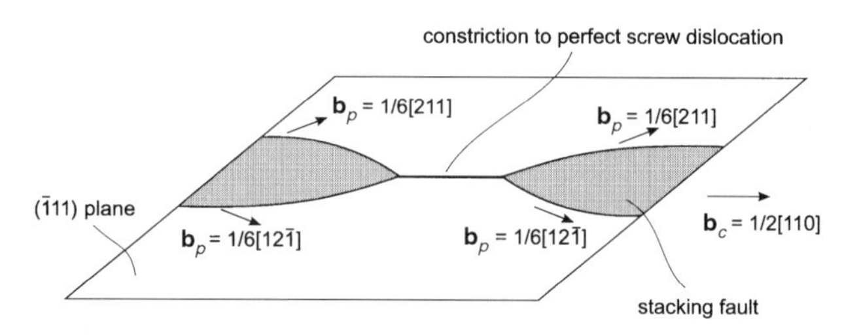

Figure 5.9 Constriction to a perfect screw segment in an extended dislocation in a face-centred cubic lattice. The Burgers vector of a Shockley partial is denoted by $\mathbf{b}_p$, that of the perfect screw by $\mathbf{b}_c$. (After Seeger (1957), *Dislocations and Mechanical Properties of Crystals*, p. 243, Wiley.)

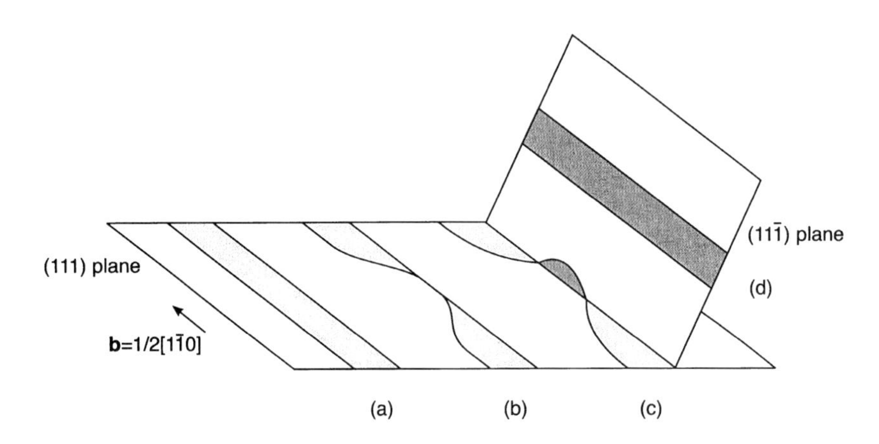

Figure 5.10 Four stages in the cross slip of a dissociated dislocation (a) by the formation of a constricted screw segment (b). The screw has dissociated in the cross-slip plane at (c).

with the sketch in Fig. 5.6, the atomic positions around the dissociated edge dislocation in a model crystal of copper as obtained by computer simulation (see section 2.7) are shown in Fig. 5.8.

Cross slip (see section 3.4) is more difficult to achieve when dissociation occurs, for a $\frac{1}{6}\langle 112 \rangle$ vector lies in only one $\{111\}$ plane and so an individual Shockley partial cannot cross slip. An extended dislocation is therefore constrained to glide in the $\{111\}$ plane of its fault. Although extended screw dislocations cannot cross slip it is possible to form a *constriction* in the screw dislocation and then the perfect dislocation at the constriction will be free to move in other planes as in Fig. 3.9. A constriction is illustrated in Fig. 5.9. Energy is required to form a constriction, since the dislocation is in its lowest energy state when dissociated, and this occurs more readily in metals with a high stacking fault energy such as aluminium. It follows that cross slip will be most difficult in metals with a low stacking fault energy and this produces significant effects on the deformation behaviour. Formation of a constriction can be assisted by thermal activation and hence the ease of cross slip decreases with decreasing temperature.

The sequence of events envisaged during the cross-slip process is illustrated in Fig. 5.10. An extended $\frac{1}{2}\langle 110\rangle$ dislocation, say $\mathbf{b} = \frac{1}{2}[1\bar{1}0]$, lying in the (111) slip plane in (a), has constricted along a short length parallel to the $[1\bar{1}0]$ direction in (b). The constricted dislocation has a pure screw orientation but is unstable with respect to redissociation. A constriction is likely to form at a region in the crystal, such as a barrier provided by non-glissile dislocations, where the applied stress tends to push the partials together. By stage (c) the unit dislocation has dissociated into two different partial dislocations with a stacking fault but on the $(11\bar{1})$ plane rather than (111). This plane intersects the original glide plane along $[1\bar{1}0]$ and is therefore a possible cross-slip plane. The new extended dislocation is free to glide in the cross-slip plane and has transferred totally to this plane by stage (d).

5.4 Thompson's Tetrahedron

Thompson's tetrahedron is a convenient notation for describing all the important dislocations and dislocation reactions in face-centred cubic metals. It arose from the appreciation that the four different sets of $\{111\}$ planes lie parallel to the four faces of a regular tetrahedron and the edges of the tetrahedron are parallel to the $\langle 110\rangle$ slip directions (Fig. 5.1(a)). Note that the ribbons of stacking fault and the bounding Shockley partial dislocations of extended dislocations are confined to the $\{111\}$ planes. The corners of the tetrahedron (Fig. 5.11(b)) are denoted by A, B, C, D, and the mid-points of the opposite faces by α, β, γ, δ. The Burgers vectors of dislocations are specified by their two end points on the tetrahedron. Thus, the Burgers vectors $\frac{1}{2}\langle 110\rangle$ of the perfect dislocations are defined both in magnitude and direction by the edges of the tetrahedron and will be **AB**, **BC**, etc. Similarly, Shockley partial Burgers vectors $\frac{1}{6}\langle 112\rangle$ can be represented by the line from the corner to the centre of a face, such as $\mathbf{A\beta}$,$\mathbf{A\gamma}$, etc. The dissociation of an

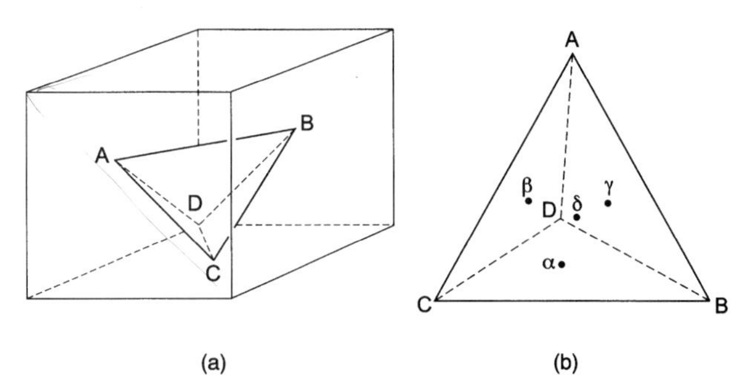

Figure 5.11 (a) Tetrahedron formed by joining four nearest-neighbour sites *ABCD* in a face-centred cubic structure. (b) Thompson's tetrahedron.

$\frac{1}{2}\langle 110\rangle$ dislocation described by relation (5.2) can be expressed alternatively by reactions of the type:

$$\begin{aligned} \mathbf{AB} &= \mathbf{A}\boldsymbol{\delta} + \boldsymbol{\delta}\mathbf{B} \quad \text{(on slip plane } ABC\text{)} \\ \mathbf{AB} &= \mathbf{A}\boldsymbol{\gamma} + \boldsymbol{\gamma}\mathbf{B} \quad \text{(on slip plane } ABD\text{)} \end{aligned} \tag{5.7}$$

Some caution must be exercised in using this notation for analysing dislocation reactions, for it is implicit that the two partials only enclose an intrinsic fault when taken in the correct order. For the Burgers circuit construction used here (section 1.4), the rule is as follows. When a perfect dislocation ($\mathbf{b} = \mathbf{AB}$, say) is viewed along the direction of its positive line sense by an observer outside the tetrahedron, an intrinsic fault is produced by a dissociation in which the partial on the left has a Greek-Roman Burgers vector ($\boldsymbol{\delta}\mathbf{B}$ or $\boldsymbol{\gamma}\mathbf{B}$) and that on the right a Roman-Greek one ($\mathbf{A}\boldsymbol{\delta}$ or $\mathbf{A}\boldsymbol{\gamma}$). For an observer inside the tetrahedron, this order is reversed.

5.5 Frank Partial Dislocation

There is an alternative arrangement by which a stacking fault can end in a crystal. The *Frank partial dislocation* is formed as the boundary line of a fault formed by inserting or removing one close-packed $\{111\}$ layer of atoms. The latter is illustrated in Fig. 5.12. Removal of a layer results in the *intrinsic* fault with stacking sequence *ABCACABC*... whereas insertion produces the *extrinsic* fault with *ABCABACAB*... (See Fig. 1.13.) Geometrically this intrinsic fault is identical to the intrinsic fault produced by dissociation of a perfect dislocation (section 5.3), but the bounding partial is different. The Frank partial has a Burgers vector *normal* to the $\{111\}$ plane of the fault and the magnitude of the vector is equal to the change in spacing produced by one close-packed layer, i.e. $\mathbf{b} = \frac{1}{3}\langle 111\rangle$. In Thompson's notation $\mathbf{b} = \mathbf{A}\boldsymbol{\alpha}$ for a stacking fault in plane *BCD*. The Frank partial is an edge dislocation and since the Burgers

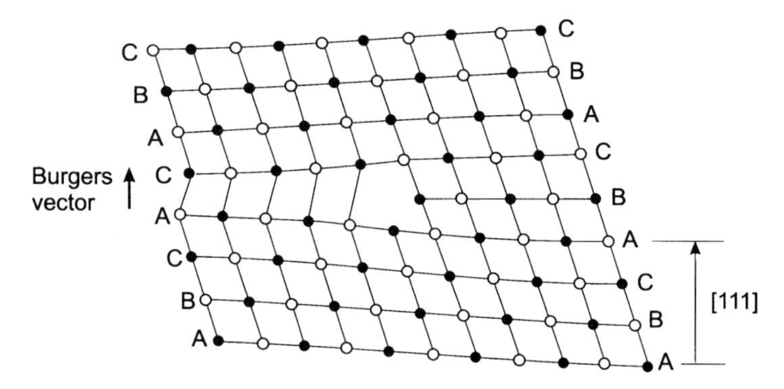

Figure 5.12 Formation of a $\frac{1}{3}[111]$ Frank partial dislocation by removal of part of a close-packed layer of atoms. The projection and directions are the same as Fig. 5.2. (After Read (1953), *Dislocation in Crystals*, McGraw-Hill.)

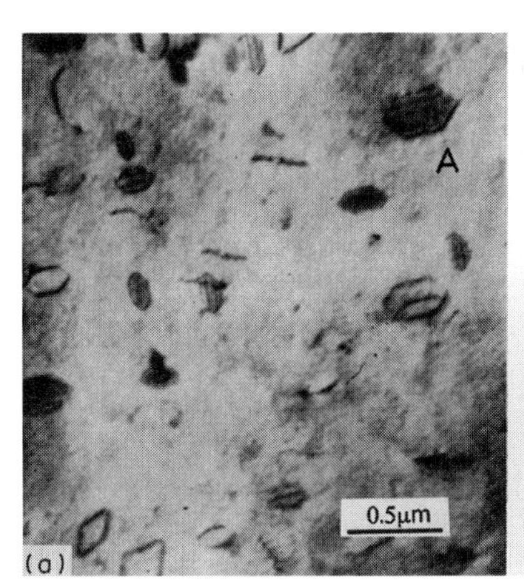

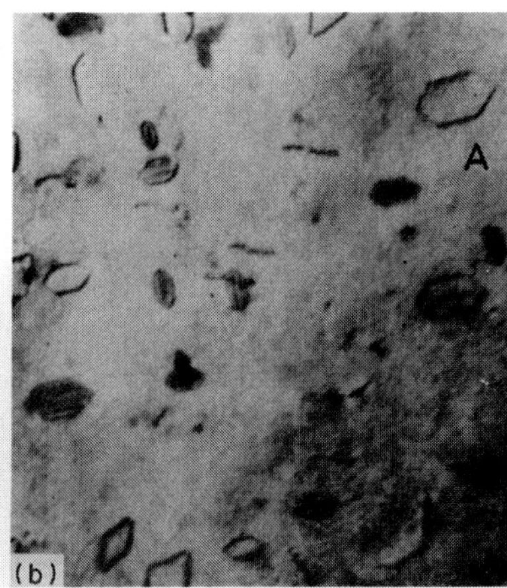

Figure 5.13 Prismatic and sessile dislocation loops in an aluminium 3.5 per cent magnesium alloy quenched from 550°C into silicone oil at −20°C.
(a) Immediately after quenching; some of the loops, e.g. *A*, contain stacking faults and are Frank sessile dislocations. (b) After being heated slightly; stacking fault in one of the loops has disappeared indicating that the loop is now a perfect dislocation. (From Westmacott, Barnes, Hull and Smallman, *Phil. Mag.* **6**, 929, 1961.)

vector is not contained in one of the {111} planes, it cannot glide and move conservatively under the action of an applied stress. Such a dislocation is said to be *sessile*, unlike the *glissile* Shockley partial. However, it can move by *climb*.

A closed dislocation loop of a Frank partial dislocation can be produced by the collapse of a platelet of vacancies as illustrated in Figs 1.13(a) and 3.18: it may arise from the local supersaturation of vacancies produced by rapid quenching (section 1.3) or by the displacement cascades formed by irradiation with energetic atomic particles. By convention this is called a *negative Frank dislocation*. A *positive Frank dislocation* may be formed by the precipitation of a close-packed platelet of interstitial atoms (Fig. 1.13(b)), as produced by irradiation damage. Both positive and negative Frank loops contain stacking faults. Diffraction fringes due to stacking faults (section 2.4) are sometimes observed when dislocation loops are examined in the electron microscope. An example is given in Fig. 5.13(a); this shows hexagonal loops formed in an aluminium alloy containing 3.5 per cent magnesium. The sides of the loop are parallel to ⟨110⟩ close-packed directions in the fault plane. In some cases no stacking fault contrast is observed. All the loops initially grow as negative Frank loops nucleated from collapsed vacancy discs on {111} planes, but the stacking fault can be removed by a dislocation reaction as follows. Considering the negative Frank sessile dislocation in Fig. 5.12, the fault will be removed if the lattice above the fault is sheared so that $C \rightarrow B$, $A \rightarrow C$, $B \rightarrow A$, etc. This $\frac{1}{6}\langle 112\rangle$ displacement

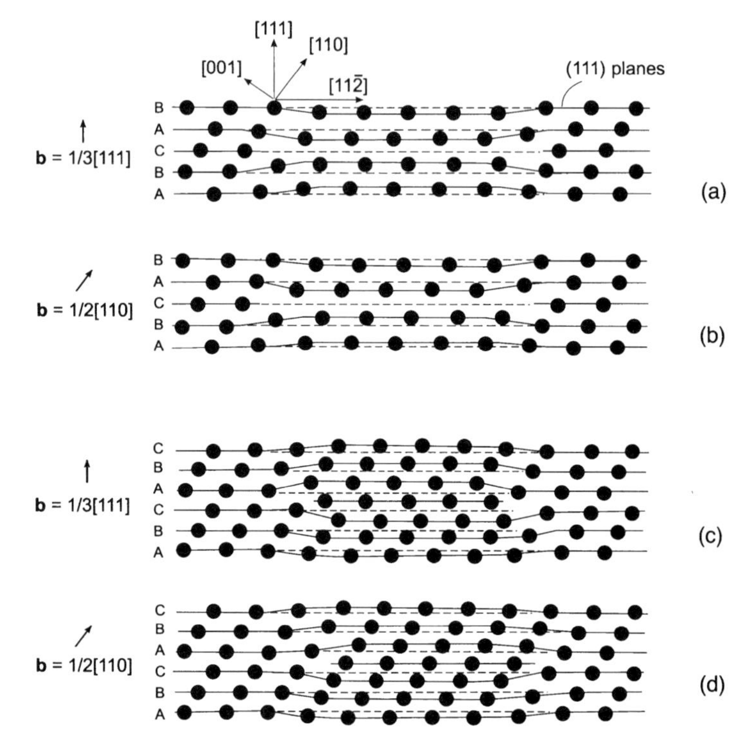

Figure 5.14 (a), (c). Atomic structure through vacancy and interstitial Frank loops on the (111) plane of a face-centred-cubic metal, and (b), (d) the perfect loops formed by the unfaulting reactions (5.8), (5.9). (After Ullmaier and Schilling (1980), in *Physics. of Modern Materials*, p. 301, IAEA, Vienna.)

corresponds to the glide of a Shockley partial dislocation across the fault. The Shockley partial may have one of three $\frac{1}{6}\langle 112\rangle$ type vectors lying in the fault plane. It is envisaged that this partial dislocation forms inside the loop and then spreads across the loop removing the fault; at the outside it will react with the Frank partial dislocation to produce a perfect dislocation. One of the three possible reactions for a loop on the (111) plane (or on the *BCD* plane in the Thompson tetrahedron notation) is

$$\frac{1}{6}[11\bar{2}] + \frac{1}{3}[111] \rightarrow \frac{1}{2}[110] \tag{5.8}$$

$\mathbf{B}\alpha$	$\alpha\mathbf{A}$	$\mathbf{BA}$
Shockley partial	Frank partial	Perfect dislocation

Figure 5.13(b) shows the same field as Fig. 5.13(a) after the foil had been in the microscope for some time; the fringe contrast in loop A has disappeared due to a reaction of the type described above and the Burgers vector of the loop has changed from $\frac{1}{3}\langle 111\rangle$ to $\frac{1}{2}\langle 110\rangle$.

For the interstitial loop, two Shockley partials are required to remove the extrinsic fault. With reference to Fig. 1.13(b), one partial glides below the inserted layer transforming $A \rightarrow C$, $C \rightarrow B$, $A \rightarrow C$, $B \rightarrow A$, etc., leaving an intrinsic fault $CABCBCA\ldots$, and the other sweeps above the layer with the same result as in the vacancy case. One possible reaction in, say, the (111) plane is

$$\underset{\alpha\mathbf{C}}{\frac{1}{6}[\bar{1}2\bar{1}]} + \underset{\alpha\mathbf{D}}{\frac{1}{6}[2\bar{1}\bar{1}]} + \underset{\alpha\mathbf{A}}{\frac{1}{3}[111]} \rightarrow \underset{\mathbf{BA}}{\frac{1}{2}[110]} \tag{5.9}$$

The atom positions before and after reactions of types (5.8) and (5.9) are shown schematically in Fig. 5.14. The prismatic loops thus formed are rings of perfect dislocation and can slip on their cylindrical glide surfaces (Fig. 3.17) to adopt new positions and orientations.

Dislocation reactions (5.8) and (5.9) will occur only when the stacking fault energy is sufficiently high. The essential problem is whether or not the prevailing conditions in the Frank loop result in the nucleation of a Shockley partial dislocation and its spread across the stacking fault. A necessary condition is that the energy of the Frank loop with its associated stacking fault is greater than the energy of the perfect dislocation loop, i.e. there is a reduction in energy when the stacking fault is removed from the loop. To a good approximation, the elastic energy of a circular edge loop of radius r in an isotropic solid with Burgers vector $\mathbf{b}_e$ perpendicular to the loop plane is

$$E = \frac{Gb_e^2 r}{2(1-\nu)} \ln\left(\frac{2r}{r_0}\right) \tag{5.10}$$

and for a circular shear loop with Burgers vector $\mathbf{b}_s$ lying in the loop plane

$$E = \frac{Gb_s^2 r}{2(1-\nu)} \left(1 - \frac{\nu}{2}\right) \ln\left(\frac{2r}{r_0}\right) \tag{5.11}$$

These relations are readily obtained from equations (4.22) by noting that the line length is $2\pi r$, the shear loop is a mixture of dislocation of edge and screw character, and the stress fields of dislocation segments on opposite sides of a loop will tend to cancel at distances $\approx 2r$ from the loop, so that the outer cut-off parameter R in (4.22) is $\approx 2r$. For the Frank loop with $\mathbf{b} = \frac{1}{3}\langle 111\rangle$, $b_e^2 = a^2/3$ and $b_s^2 = 0$, and for the perfect loop with $\mathbf{b} = \frac{1}{2}\langle 110\rangle$, $b_e^2 = a^2/3$ and $b_s^2 = a^2/6$. Thus, the difference in energy between the Frank loop containing the stacking fault and the perfect, unfaulted loop is

$$\Delta E = \pi r^2 \gamma - \frac{rGa^2}{24}\left(\frac{2-\nu}{1-\nu}\right) \ln\left(\frac{2r}{r_0}\right) \tag{5.12}$$

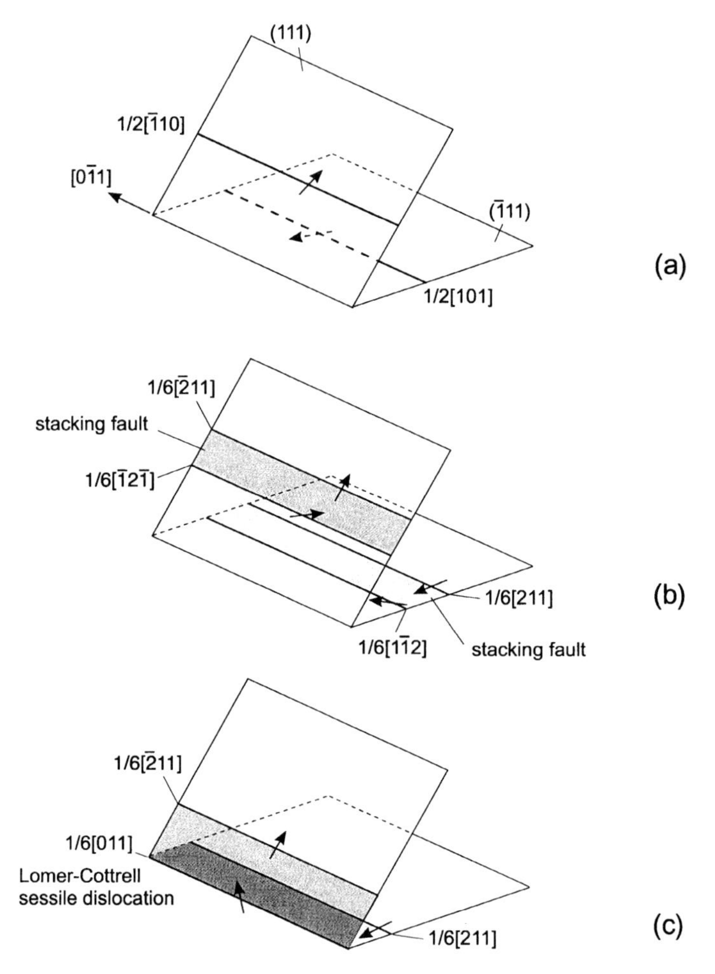

Figure 5.15 Formation of a Lomer–Cottrell sessile dislocation. The perfect dislocations in (a) have dissociated in (b) and reacted favourably to form a Lomer–Cottrell dislocation in (c).

Therefore, the unfaulting reaction will be energetically favourable if

$$\gamma > \frac{Ga^2}{24\pi r}\left(\frac{2-\nu}{1-\nu}\right)\ln\left(\frac{2r}{r_0}\right) \tag{5.13}$$

Thus the lower limit to the value of γ for removal of a fault depends on the size of the loop. Taking $a = 0.35\,\text{nm}$, $r = 10\,\text{nm}$, $G = 40\,\text{GN}\,\text{m}^{-2}$, $r_0 = 0.5\,\text{nm}$, and $\nu = 0.33$, the critical stacking fault energy is about $60\,\text{mJ}\,\text{m}^{-2}$. Since $r = 10\,\text{nm}$ is close to the minimum size for resolving loops in the electron microscope it is not surprising that stacking fault loops are rarely observed in metals with γ values larger than this.

The necessary condition based on initial and final energy values may not be a sufficient one for unfaulting, for it neglects the fact that the Shockley partial must be nucleated somewhere within the Frank loop, and there is almost certainly an energy barrier for this process. If a loop grows by the absorption of point defects, it will become increasingly less stable, but if the Shockley nucleation energy is independent of r, the equilibrium unfaulted state may not be achieved. The probability of nucleation is increased by increasing temperature and the presence of external and internal sources of stress. The removal of the fault in Fig. 5.13 may have occurred due to local shear stresses in the foil. Thus, there is no hard-and-fast rule governing loop unfaulting.

Finally, it is noted that if a second platelet of vacancies nucleates within a negative Frank loop adjacent to the intrinsic fault with stacking sequence *ABCACABC*..., a disc of extrinsic fault *ABCACBC*... is formed. If a third platelet nucleates against the second, the stacking sequence through all three is perfect *ABCABC*... Concentric loops of Frank partial dislocation containing alternating rings of intrinsic, extrinsic and perfect stacking have been observed under certain conditions.

5.6 Lomer–Cottrell Sessile Dislocation

Strain hardening in metals can be attributed to the progressive introduction during straining of barriers to the free movement of dislocations. Several barriers have been proposed; one specific barrier proposed and observed in face-centred cubic metals is the *Lomer–Cottrell lock*. It can be formed in the following way. Consider two perfect dislocations gliding in different {111} planes (Fig. 5.15(a)). In most metals, each dislocation will be dissociated in its glide plane into two Shockley partial dislocations bounding a stacking-fault ribbon (Fig. 5.15(b)). If the dislocations meet at the line of intersection of the two planes, the leading partials repel or attract each other according to the particular directions of their Burgers vectors. There are three possible $\frac{1}{6}\langle 112\rangle$ vectors plus their reverses on each plane (Fig. (5.5)), and there are therefore 36 combinations to consider. It is easy to see from Thompson's tetrahedron of Fig. 5.11(b) that some reactions are favourable and others unfavourable according to Frank's rule (section 4.4). The most favourable gives a product dislocation with Burgers vector of the form $\frac{1}{6}\langle 110\rangle$, i.e. $\boldsymbol{\alpha\beta}$, $\boldsymbol{\alpha\gamma}$, etc., on the tetrahedron. This reaction for the (111) and $(\bar{1}11)$ planes is shown in Fig. 5.15(c) and is

$$\frac{1}{6}[\bar{1}2\bar{1}] + \frac{1}{6}[1\bar{1}2] \rightarrow \frac{1}{6}[011] \tag{5.14}$$

Using the b^2 criterion for dislocation energy per unit length:

$$\frac{a^2}{6} + \frac{a^2}{6} \rightarrow \frac{a^2}{18} \tag{5.15}$$

which represents a considerable reduction. In Thompson's notation for Burgers vectors, two perfect dislocations dissociate on different planes:

$$\begin{aligned} &\mathbf{DA} \rightarrow \mathbf{D}\boldsymbol{\beta} + \boldsymbol{\beta}\mathbf{A} \quad (\text{on } ACD) \\ &\mathbf{BD} \rightarrow \mathbf{B}\boldsymbol{\alpha} + \boldsymbol{\alpha}\mathbf{D} \quad (\text{on } BCD) \end{aligned} \tag{5.16}$$

and one Shockley from each combine:

$$\boldsymbol{\alpha}\mathbf{D} + \mathbf{D}\boldsymbol{\beta} \rightarrow \boldsymbol{\alpha}\boldsymbol{\beta} \tag{5.17}$$

The product partial dislocation forms along one of the six $\langle 110 \rangle$ directions at the intersection of the stacking faults on two $\{111\}$ planes. By analogy with carpet on a stair, it is called a *stair-rod dislocation*. For some reactions with different resultant vectors, the angle between the stacking faults is obtuse.

The Burgers vector $\boldsymbol{\alpha\beta}$ of the stair-rod partial is perpendicular to the dislocation line and does not lie in either of the two $\{111\}$ planes of the adjacent faults. Thus, it cannot glide on these planes, and the $\{100\}$ plane which contains the lines and its Burgers vector is not a slip plane. The dislocation is *sessile*. It exerts a repulsive force on the two remaining Shockley partials according to Frank's rule, and these three partial dislocations (Fig. 5.15(c)) form a stable, sessile arrangement. It acts as a strong barrier to the glide of further dislocations on the two $\{111\}$ planes, and is known as a *Lomer–Cottrell lock*.

5.7 Stacking Fault Tetrahedra

Another dislocation arrangement has been observed in metals and alloys of low stacking-fault energy following treatment that produces a supersaturation of vacancies. It consists of a tetrahedron of intrinsic stacking faults on $\{111\}$ planes with $\frac{1}{6}\langle 110 \rangle$ type stair-rod dislocations along the edges of the tetrahedron. Such defects are believed to form by the *Silcox–Hirsch mechanism*. According to the discussion in section 5.5, when a platelet of vacancies (produced by quenching from a high temperature) collapses to form a loop of Frank partial dislocation, the stacking fault will be stable if the fault energy is sufficiently low. The Frank partial may dissociate into a low-energy stair-rod dislocation and a Shockley partial on an intersecting slip plane according to a reaction of the type

$$\begin{aligned} &\frac{1}{3}[111] \rightarrow \frac{1}{6}[101] + \frac{1}{6}[121] \\ &b^2\!: \; \frac{a^2}{3} \rightarrow \frac{a^2}{18} + \frac{a^2}{6} \end{aligned} \tag{5.18}$$

Discounting the energy of the stacking fault there is a reduction in the dislocation energy, and the reaction is energetically favourable. With reference to the Thompson tetrahedron notation (Fig. 5.11(b)), suppose that the vacancies condense on the $\{111\}$ plane BCD in the form of an equilateral triangle with edges parallel to $\langle 110 \rangle$ directions BC, CD, DB. The triangular Frank partial with Burgers vector $\boldsymbol{\alpha}\mathbf{A}$ can dissociate to produce a stair-rod along each edge and a Shockley partial on each of the three inclined $\{111\}$ planes as shown in Fig. 5.16(a) by Burgers vector reactions of the type (5.18), namely

$$\begin{aligned} \boldsymbol{\alpha}\mathbf{A} &\rightarrow \boldsymbol{\alpha\beta} + \boldsymbol{\beta}\mathbf{A} \quad (\text{on } ACD) \\ \boldsymbol{\alpha}\mathbf{A} &\rightarrow \boldsymbol{\alpha\gamma} + \boldsymbol{\gamma}\mathbf{A} \quad (\text{on } ABD) \\ \boldsymbol{\alpha}\mathbf{A} &\rightarrow \boldsymbol{\alpha\delta} + \boldsymbol{\delta}\mathbf{A} \quad (\text{on } ABC) \end{aligned} \tag{5.19}$$

The partial dislocations $\boldsymbol{\beta}\mathbf{A}$, $\boldsymbol{\alpha}\mathbf{A}$ and $\boldsymbol{\gamma}\mathbf{A}$ will be repelled by the stair-rod dislocations $\boldsymbol{\alpha\beta}$, $\boldsymbol{\alpha\gamma}$ and $\boldsymbol{\alpha\delta}$ respectively and will bow out in their slip planes. Taking account of dislocation line sense, it is found that the partials attract each other in pairs to form another set of stair-rods along DA, BA and CA (Fig. 5.16(b)) according to the reactions

$$\begin{aligned} \boldsymbol{\beta}\mathbf{A} + \mathbf{A}\boldsymbol{\gamma} &\rightarrow \boldsymbol{\beta\gamma} \quad (\text{along } DA) \\ \boldsymbol{\gamma}\mathbf{A} + \mathbf{A}\boldsymbol{\delta} &\rightarrow \boldsymbol{\gamma\delta} \quad (\text{along } BA) \\ \boldsymbol{\delta}\mathbf{A} + \mathbf{A}\boldsymbol{\beta} &\rightarrow \boldsymbol{\delta\beta} \quad (\text{along } CA) \end{aligned} \tag{5.20}$$

In Miller index notation the reactions are of the type (5.14).

The shape of these tetrahedra observed in thin foils by transmission microscopy depends on the orientation of the tetrahedra with respect to the plane of the foil. An example is shown in Fig. 5.17. The complex

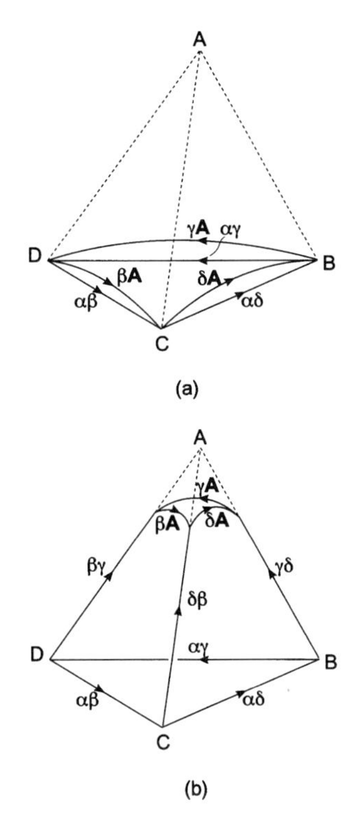

Figure 5.16 Formation of a stacking-fault tetrahedron by the Silcox–Hirsch mechanism. The arrows show the positive line sense used to define the Burgers vectors, which are denoted by the directions on the Thompson tetrahedron.

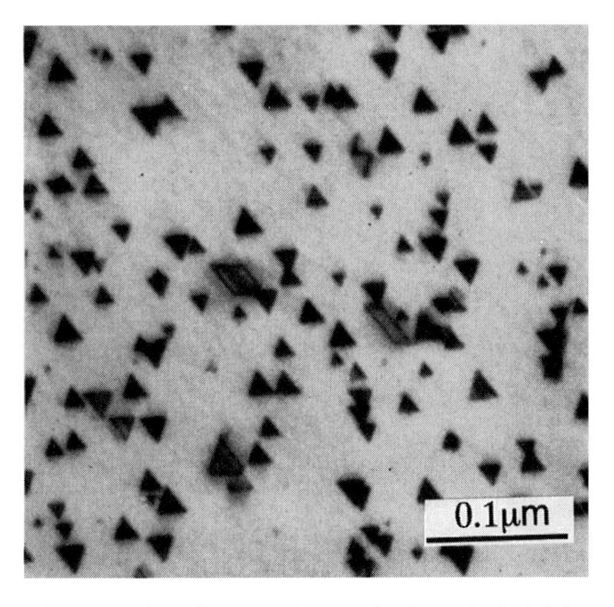

Figure 5.17 Transmission electron micrograph of tetrahedral defects in quenched gold. The shape of the tetrahedra viewed in transmission depends on their orientation with respect to the plane of the foil, (110) foil orientation. (From Cottrell, *Phil. Mag*. **6**, 1351, 1961.)

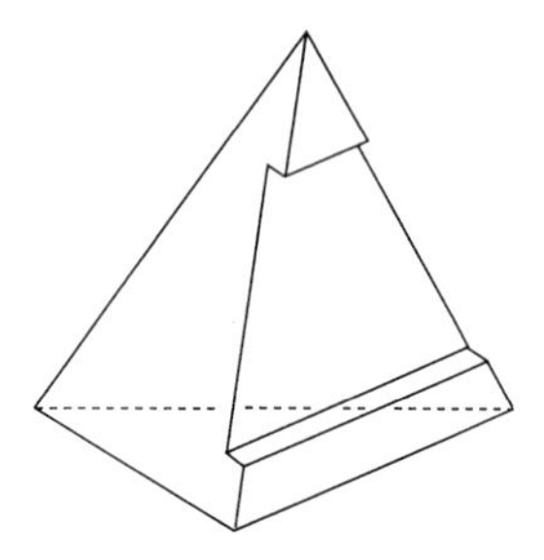

Figure 5.18 Jog lines on a {111} face of a stacking-fault tetrahedron.

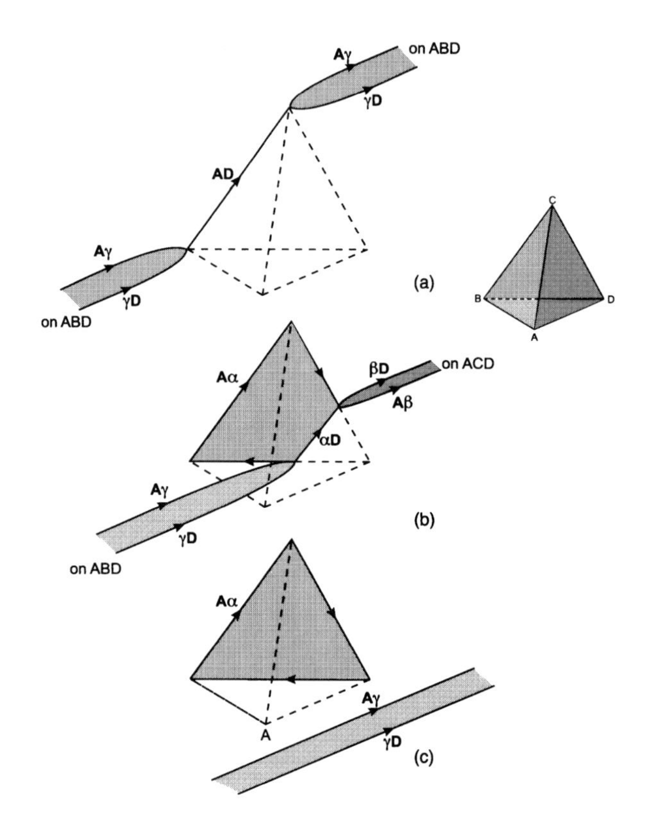

Figure 5.19 Three stages in the formation of a triangular Frank loop by the cross slip of a jogged screw dislocation. Pairs of letters denote Burgers vectors in the notation of Fig. 5.11.

contrast patterns inside the faults arise from overlapping stacking faults in different faces of the tetrahedron. The increase in energy due to the formation of stacking faults places a limit on the size of the fault that can be formed. If the fault energy is relatively high, the Frank loop is stable, or it may only partly dissociate, thereby forming a truncated tetrahedron as in Fig. 5.16(b).

Stacking-fault tetrahedra have been observed in metals following quenching (Fig. 5.17) or radiation damage (Fig. 2.16). Once nucleated, they can grow in a supersaturation of vacancies by the climb of ledges ('*jog lines*') on the {111} faces due to vacancy absorption (Fig. 5.18). They can also result from plastic deformation due to the cross slip of a segment of jogged screw dislocation, as illustrated in Fig. 5.19. In (a), the screw dislocation with Burgers vector **AD** (Thompson tetrahedron notation) is dissociated in the plane *ABD* ($\mathbf{AD} = \mathbf{A}\boldsymbol{\gamma} + \boldsymbol{\gamma}\mathbf{D}$) and contains a jog. The right-hand segment of the screw has cross slipped in (b) onto plane *ACD* ($\mathbf{AD} = \mathbf{A}\boldsymbol{\beta} + \boldsymbol{\beta}\mathbf{D}$) and the jog has dissociated on plane *BCD* into a Frank partial ($\mathbf{A}\boldsymbol{\alpha}$) and a Shockley partial ($\boldsymbol{\alpha}\mathbf{D}$). The jog has been removed in (c): the screw dislocation glides away leaving a triangular Frank loop (Burgers vector **A**$\boldsymbol{\alpha}$) on *BCD* which can form a stacking-fault tetrahedron by the Silcox–Hirsch mechanism of Fig. 5.16.

Further Reading

Amelinckx, S. (1979) 'Dislocations in particular structures', *Dislocations in Solids*, vol. 2, p. 66 (ed. F. R. N. Nabarro), North-Holland.

Cottrell, A. H. (1953) *Dislocations and Plastic Flow in Crystals*, Oxford University Press.

Eyre, B. L., Loretto, M. H. and Smallman, R. E. (1977) 'Electron microscopy studies of point defect clusters in metals', *Vacancies* '76, p. 63 (ed. R. E. Smallman and J. E. Harris), The Metals Society, London.

Hirth, J. P. and Lothe, J. (1982) *Theory of Dislocations*, Wiley.

Kelly, A., Groves, G. W. and Kidd, P. (2000) *Crystallography and Crystal Defects*, John Wiley.

Read, W. T. (1953) *Dislocations in Crystals*, McGraw-Hill, New York.

Thompson, N. (1953) 'Dislocation nodes in face-centred cubic lattices', *Proc. Phys. Soc. B*, **66**, 481.

Veysièrre, P. (1999) 'Dislocations and the plasticity of crystals', *Mechanics and Materials: Fundamental Linkages*, p. 271 (eds M. A. Meyers, R. W. Armstrong and H. Kirchner), John Wiley.

Whelan, M. J. (1959) 'Dislocation interactions in face-centred cubic metals', *Proc. Roy. Soc. A*, **249**, 114.

6 Dislocations in Other Crystal Structures

6.1 Introduction

The face-centred cubic metals have been treated in a separate chapter because although many of the dislocation reactions and properties presented have counterparts in other structures, they are more readily described in the face-centred cubic system. Also, these metals have been more extensively studied than other solids. In general, reducing the crystal symmetry, changing the nature of the interatomic bonding and increasing the number of atom species in the lattice make dislocation behaviour more complex. Nevertheless, many of the features of the preceding chapter carry over to other structures, as will be seen in the following. The two other major metallic structures are discussed first, and then some important compounds and non-metallic cases are considered.

6.2 Dislocations in Hexagonal Close-packed Metals

Burgers Vectors and Stacking Faults

Some of the important metals of this structure are given in Table 6.1. As explained in section 1.2, the (0001) basal plane is close-packed and the close-packed directions are $\langle 11\bar{2}0\rangle$. The shortest lattice vectors are $\frac{1}{3}\langle 11\bar{2}0\rangle$, the unit cell generation vectors **a** in the basal plane. It may be anticipated, therefore, that dislocation glide will occur in the basal plane with Burgers vector $\mathbf{b} = \frac{1}{3}\langle 11\bar{2}0\rangle$: this slip system is frequently observed. None of the metals is ideally close-packed, which would require the lattice parameter ratio $c/a = (8/3)^{1/2} = 1.633$, although magnesium and cobalt have a c/a ratio close to ideal. This suggests that directionality occurs in interatom bonding. In support of this, it is found that some metals slip most easily with $\mathbf{b} = \frac{1}{3}\langle 11\bar{2}0\rangle$ on the first-order prism planes $\{10\bar{1}0\}$ (see Table 6.1).

Burgers vectors for the structure may be described in a similar fashion to the Thompson tetrahedron for face-centred-cubic metals by using the bi-pyramid shown in Fig. 6.1. The important dislocations and their Burgers vectors are as follows.

Table 6.1 Properties of some hexagonal close-packed metals at 300 K

Metal	*Be*	*Ti*	*Zr*	*Mg*	*Co*	*Zn*	*Cd*
c/a ratio	1.568	1.587	1.593	1.623	1.628	1.856	1.886
Preferred slip	basal	prism	prism	basal	basal	basal	basal
Plane for $\mathbf{b} = \mathbf{a}$	(0001)	$\{10\bar{1}0\}$	$\{10\bar{1}0\}$	(0001)	(0001)	(0001)	(0001)

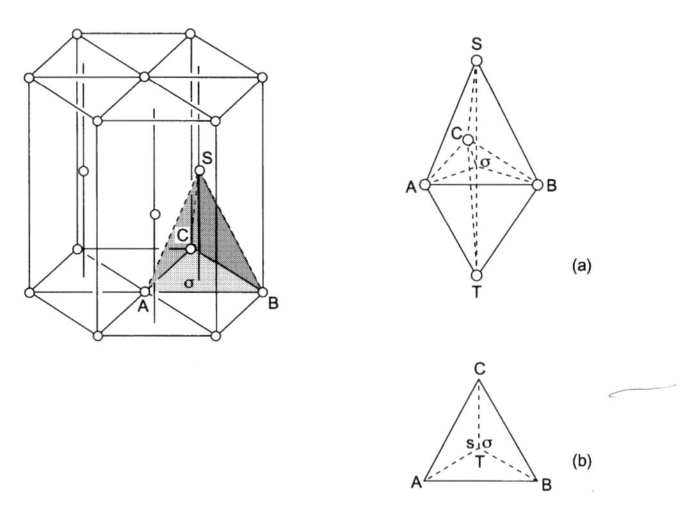

Figure 6.1 Burgers vectors in the hexagonal close-packed lattice. (From Berghezan, Fourdeux and Amelinckx, *Acta Metall.* **9**, 464, 1961.)

(a) Perfect dislocations with one of six Burgers vectors in the basal plane along the sides of the triangular base *ABC* of the pyramid, represented by **AB**, **BC**, **CA**, **BA**, **CB** and **AC**.
(b) Perfect dislocations with one of two Burgers vectors perpendicular to the basal plane, represented by the vectors **ST** and **TS**.
(c) Perfect dislocations with one of twelve Burgers vectors represented by symbols such as **SA/TB**, which means either the sum of the vectors **ST** and **AB** or, geometrically, a vector equal to twice the join of the midpoints of *SA* and *TB*.
(d) Imperfect basal dislocations of the Shockley partial type with Burgers vectors **A**$\boldsymbol{\sigma}$, **B**$\boldsymbol{\sigma}$, **C**$\boldsymbol{\sigma}$, $\boldsymbol{\sigma}$**A**, $\boldsymbol{\sigma}$**B** and $\boldsymbol{\sigma}$**C**.
(e) Imperfect dislocations with Burgers vectors perpendicular to the basal plane, namely, $\boldsymbol{\sigma}$**S**, $\boldsymbol{\sigma}$**T**, **S**$\boldsymbol{\sigma}$ and **T**$\boldsymbol{\sigma}$.
(f) Imperfect dislocations which are a combination of the latter two types given by **AS**, **BS**, etc. Although these vectors represent a displacement from one atomic site to another the associated dislocations are not perfect. This is because the sites do not have identical surroundings and the vectors are not translations of the lattice.

The Miller–Bravais indices and length b of these Burgers vectors **b** are given in Table 6.2. The value of b^2 for ideal close-packing ($c^2 = \frac{8}{3}a^2$) is also given: appropriate adjustments are required when dealing with non-ideal packing.

A number of different stacking faults are associated with the partial dislocations listed in Table 6.2. According to the hard-sphere model of atoms, three basal-plane faults exist which do not affect nearest-neighbour arrangements of the perfect stacking sequence *ABABAB*...

Table 6.2 Dislocations in hexagonal close-packed structures

Type	*AB*	*TS*	*SA/TB*	$A\sigma$	σS	*AS*
b	$\frac{1}{3}\langle 11\bar{2}0\rangle$	[0001]	$\frac{1}{3}\langle 11\bar{2}3\rangle$	$\frac{1}{3}\langle \bar{1}100\rangle$	$\frac{1}{2}[0001]$	$\frac{1}{6}\langle \bar{2}203\rangle$
b	a	c	$(c^2+a^2)^{\frac{1}{2}}$	$a/\sqrt{3}$	$c/2$	$\left(\frac{a^2}{3}+\frac{c^2}{4}\right)^{\frac{1}{2}}$
b^2	a^2	$\frac{8}{3}a^2$	$\frac{11}{3}a^2$	$\frac{1}{3}a^2$	$\frac{2}{3}a^2$	a^2

Two are intrinsic and conventionally called I_1 and I_2. Fault I_1 is formed by removal of a basal layer, which produces a very high energy fault, followed by slip of $\frac{1}{3}\langle 10\bar{1}0\rangle$ of the crystal above this fault to reduce the energy:

$$ABABABABA\ldots \rightarrow ABABBABA\ldots \rightarrow ABABCBCB\ldots (I_1) \tag{6.1}$$

Fault I_2 results from slip of $\frac{1}{3}\langle 10\bar{1}0\rangle$ in a perfect crystal:

$$ABABABAB\ldots \rightarrow ABABCACA\ldots (I_2) \tag{6.2}$$

The extrinsic fault (E) is produced by inserting an extra plane:

$$ABABABAB\ldots \rightarrow ABABCABAB\ldots (E) \tag{6.3}$$

These faults introduce into the crystal a thin layer of face-centred cubic stacking (ABC) and so have a characteristic stacking-fault energy γ. The main contribution to γ arises from changes in the second-neighbour sequences of the planes. There is one change for I_1, two for I_2 and three for E, and so to a first approximation $\gamma_E \approx \frac{3}{2}\gamma_{I_2} \approx 3\gamma_{I_1}$. Experimental estimates of γ for the basal-slip metals are generally higher than those quoted earlier for the face-centred cubic metals. Stable faults for the basal plane of the prism-slip metals such as titanium and zirconium either have very high energy or may not even exist.

There is no direct experimental evidence of stacking faults on other planes, which are not close-packed; but the incidence of non-basal glide and results of simulations using computer models suggest that stable faults may exist on prism and pyramidal planes, although possibly with high energy.

Basal and Non-basal Slip

Slip by the system $\frac{1}{3}\langle 11\bar{2}0\rangle(0001)$ in metals such as beryllium, magnesium, cadmium and zinc is similar to $\frac{1}{2}\langle 1\bar{1}0\rangle\{111\}$ slip in the face-centred cubic metals, in that the critical resolved shear stress is low ($\lesssim 1\mathrm{MN\,m^{-2}}$) and the perfect dislocation dissociates into two Shockley partials bounding a ribbon of stacking fault. The Burgers vector reaction is

$$\mathbf{AB} \rightarrow \mathbf{A}\boldsymbol{\sigma} + \boldsymbol{\sigma}\mathbf{B} \tag{6.4}$$

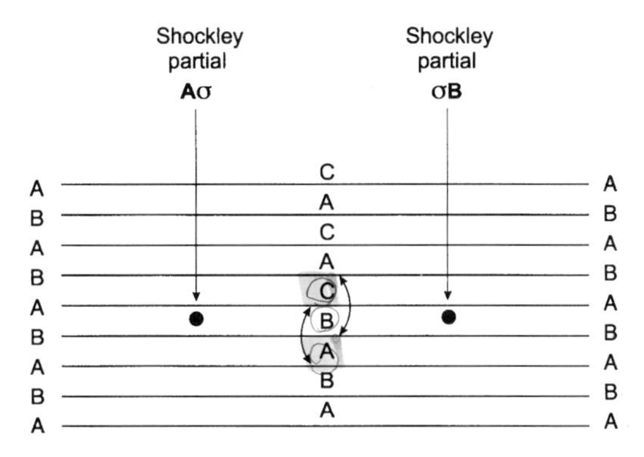

Figure 6.2 Dissociation of a perfect dislocation with Burgers vector **AB** (Fig. 6.1) into two Shockley partial dislocations separated by a stacking fault I_2. Double arrows indicate the two errors in the two-fold stacking sequence of the basal planes.

e.g.

$$\frac{1}{3}[11\bar{2}0] \rightarrow \frac{1}{3}[10\bar{1}0] + \frac{1}{3}[01\bar{1}0]$$

$$b^2: \; a^2 \qquad a^2/3 \qquad a^2/3$$

The geometry is the same as that of the face-centred cubic case, in that the partial vectors lie at $\pm 30°$ to the perfect vector and the fractional reduction in dislocation energy given by b^2 is 1/3. The spacing of the partials may be calculated from equations (5.4) and (5.5) by replacing b^2 by $a^2/3$. The fault involved is the intrinsic fault I_2, as shown schematically in Fig. 6.2.

Basal slip also occurs in the metals for which prism slip is most favoured, but the critical resolved shear stress is high ($\approx$100 MN m^{-2}). It is probable, therefore, that interatomic forces prevent formation of a stable basal fault and that dissociation (6.4) does not take place.

Slip on non-basal systems occurs most commonly with $\mathbf{b} = \frac{1}{3}\langle 11\bar{2}0\rangle$. It is seen from Fig. 6.3 that dislocations with this slip vector can glide on the first-order prism and pyramidal planes in addition to the basal plane. However, in metals for which basal slip is preferred, the resolved shear stress required for non-basal slip is one to two orders of magnitude higher. This is explained by the absence of low-energy stacking faults on other planes and the subsequent difficulty of dislocation glide. Furthermore, a dislocation extended in the basal plane by reaction (6.4) can only cross-slip onto other planes by first constricting, as explained in section 5.3. This will be hindered by the absence of other faults except at high values of γ_{I_2}, stress and temperature.

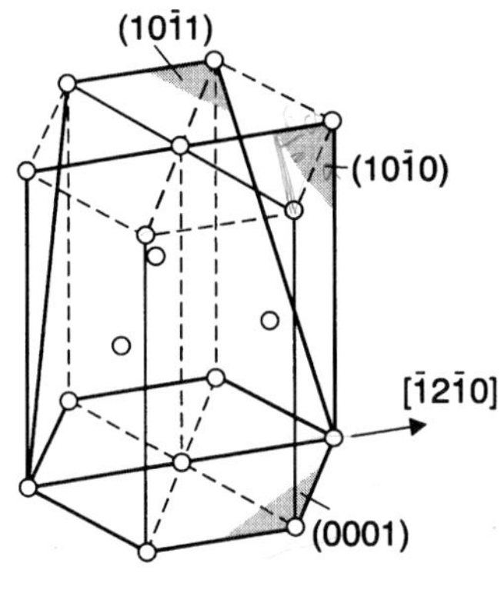

Figure 6.3 Planes in an hexagonal lattice with a common $[\bar{1}2\bar{1}0]$ direction.

Even in metals for which it is the favoured slip system, the critical resolved shear stress for $\frac{1}{3}\langle 11\bar{2}0\rangle\{1\bar{1}00\}$ prism slip is high ($\gtrsim$10 MN m^{-2}). An obvious consideration for basal and prism slip is to compare the inter-planar spacings of the two sets of slip planes, for in simple terms the lattice resistance to glide tends to be smaller for planes with wide spacings (section 10.3). The spacings of the basal and corrugated prism planes are $c/2$ and $\sqrt{3}a/2$ respectively, and thus the prism-plane spacing is greater when $c/a < \sqrt{3}$. This approach fails to explain the preference for basal slip in several metals (Table 6.1) and points to the importance of stacking faults in determination of slip systems. Two dissociations have been proposed to account for prism slip in titanium and zirconium, namely

$$\frac{1}{3}\langle 11\bar{2}0\rangle \rightarrow \frac{1}{18}\langle 42\bar{6}3\rangle + \frac{1}{18}\langle 24\bar{6}3\rangle \tag{6.5}$$

$$\frac{1}{3}\langle 11\bar{2}0\rangle \rightarrow \frac{1}{9}\langle 11\bar{2}0\rangle + \frac{2}{9}\langle 11\bar{2}0\rangle \tag{6.6}$$

The first produces a fault which is stable in a lattice consisting of hard spheres. The second results in prism planes adjacent to the fault ribbon adopting the stacking of {112} planes in body-centred cubic metals, the relevance being that this is the stable crystal structure for titanium and zirconium at high temperature. There is no evidence that these faults exist in real metals, however, and lattice imaging using transmission electron microscopy shows that dislocation cores do not extend widely on the prism plane. This is supported by atomic-scale computer modelling, which also indicates that the core with a narrow width on the prism plane is favoured energetically over the dissociated core on the basal plane (equation 6.4) because of the high energy of the I_2 basal fault in these metals.

In polycrystalline metals, basal and prism slip do not supply sufficient slip modes to satisfy von Mises' criterion that every grain should be able to plastically deform generally to meet the shape changes imposed by its neighbours (section 10.9). This requires five independent slip systems, whereas basal and prism slip provide only two each. Consequently, twinning and occasionally other slip systems play an important role in the plasticity of these metals.

A variety of twin modes with different habit plane and shear direction occur, depending on the metal, temperature and nature of loading. A deformation twin grows or shrinks under a resolved shear stress by the motion of steps along its boundary. Such steps have dislocation character and are known as *twinning dislocations*. They are described in more detail in section 9.7.

Slip with Burgers vector $\frac{1}{3}\langle 11\bar{2}3\rangle$ has been widely reported, although only under conditions of high stress and orientations in which more favoured slip vectors cannot operate. The glide planes are $\{10\bar{1}1\}$ and $\{11\bar{2}2\}$. The magnitude of the Burgers vector is large and the planes are atomically rough, which explains the high glide stress. It has been proposed on the basis of the hard-sphere lattice model that the dislocation on $\{11\bar{2}2\}$ may dissociate into four partials with approximately equal

Burgers vectors. One has a core spread over three successive $\{11\bar{2}2\}$ planes, within which the atomic displacements are different. These three adjacent faults are bounded by the three remaining partials. Significantly, the first dislocation is a *zonal* twinning dislocation, a multi-layer step in a coherent twin boundary which generates by its movement a twin. Thus, there may be a close relationship between $\langle 11\bar{2}3\rangle\{11\bar{2}2\}$ slip and twinning, although not necessarily of the hard-sphere lattice form.

Vacancy and Interstitial Loops

As in the face-centred cubic metals, vacancies and interstitials in excess of the equilibrium concentration can precipitate as platelets to form dislocation loops. The situation is more complicated, however, because although the basal plane is close-packed, the relative density of atoms in the different crystallographic planes varies with c/a ratio. Also, stacking faults occur in some metals and not others, as discussed in the preceding section. The simplest geometries are considered here, starting with basal-plane loops.

Condensation of vacancies in a single basal plane (Fig. 6.4(a)) results in two similar atomic layers coming into contact (Fig. 6.4(b)). This unstable situation of high energy is avoided in one of two ways. In one, the stacking of one layer adjacent to the fault is changed; for example *B* to *C* as in Fig. 6.4(c). This is equivalent to the glide below the layer of one Shockley partial with Burgers vector $\frac{1}{3}\langle 10\bar{1}0\rangle$ followed by glide above of a second Shockley of opposite sign. The Burgers vector reaction is

$$\boldsymbol{\sigma}\mathbf{S} + \boldsymbol{\sigma}\mathbf{A} + \mathbf{A}\boldsymbol{\sigma} \rightarrow \boldsymbol{\sigma}\mathbf{S}$$

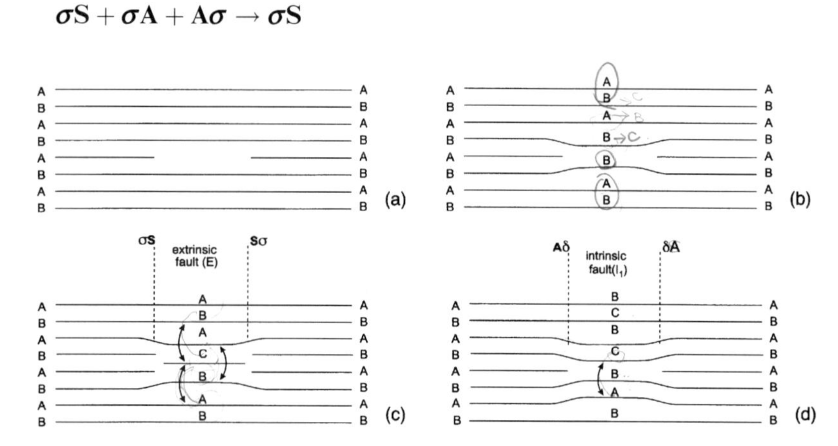

Figure 6.4 Formation of prismatic dislocation loops as a result of the precipitation of one layer of vacancies. (a) Disc-shaped cavity. (b) Collapse of the disc bringing two *B* layers together. (c) Formation of a high-energy stacking fault *E*. (d) Formation of a low-energy stacking fault I_1. The actual sequence of planes formed will depend on the plane in which the vacancies form, i.e. *A* or *B*. It is possible to have loops of the same form with different Burgers vectors lying on adjacent planes. (After Berghezan, Fourdeux and Amelinckx, *Acta Metall.* **9**, 464, 1961.)

e.g.

$$\frac{1}{2}[0001] + \frac{1}{3}[1\bar{1}00] + \frac{1}{3}[\bar{1}100] \rightarrow \frac{1}{2}[0001] \qquad (6.7)$$

The resultant sessile Frank partial with $b^2 = c^2/4 \simeq 2a^2/3$ surrounds the extrinsic (E) stacking fault described earlier in this section. In the alternative mechanism, a single Shockley partial sweeps over the vacancy platelet, displacing the atoms above by $\frac{1}{3}\langle 10\bar{1}0\rangle$ relative to those below. The Burgers vector reaction is

$$\mathbf{A}\boldsymbol{\sigma} + \boldsymbol{\sigma}\mathbf{S} \rightarrow \mathbf{AS}$$

e.g.

$$\frac{1}{3}[\bar{1}100] + \frac{1}{2}[0001] \rightarrow \frac{1}{6}[\bar{2}203] \qquad (6.8)$$

This sessile Frank partial with $b^2 \simeq a^2$ surrounds the type I_1 intrinsic fault (Fig. 6.4(d)). The E-type loop of reaction (6.7) can transform to the I_1 form by reaction (6.8), and since γ_E is expected to be approximately three times γ_{I_1}, the $\frac{1}{6}\langle 20\bar{2}3\rangle$ loops may be expected to dominate. The dislocation energy is proportional to b^2, however, and the total energy change accompanying reaction (6.8) is dependent on the γ values and loop size, in a similar manner to the unfaulting of loops in face-centred cubic metals (section 5.5). There is therefore a critical loop size for the

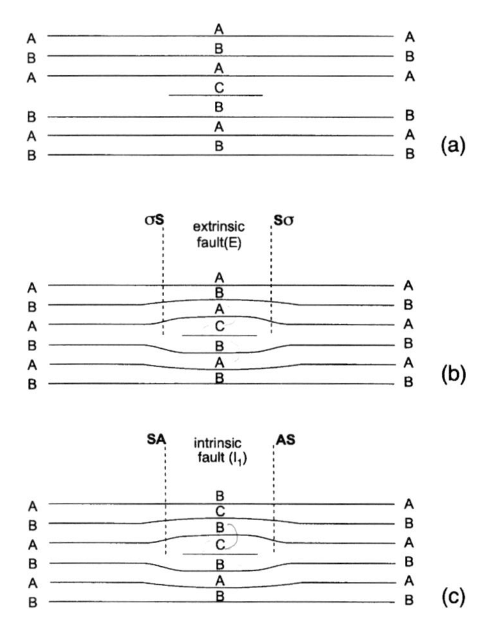

Figure 6.5 (a) Precipitation of a layer of interstitials. (b) Prismatic loop resulting from the layer of interstitials: the loop contains a high-energy stacking fault E. (c) Prismatic loop containing low-energy stacking fault I_1. (After Berghezan, Fourdeux and Amelinckx, *Acta Metall.* **9**, 464, 1961.)

reaction, which may be influenced by factors such as stress, temperature and impurity content. Experimentally, both forms are observed in quenched or irradiated metals.

Precipitation of a basal layer of interstitials as in Figs 6.5(a) and (b) produces an E-type fault surrounded by a Frank partial loop with Burgers vector $\frac{1}{2}[0001]$. Again, provided the loop is large enough, this can transform to an I_1 loop by the nucleation and sweep of a Shockley partial according to reaction (6.8). The stacking sequence is shown in Fig. 6.5(c). Interstitial loops with Burgers vectors $\frac{1}{2}[0001]$ and $\frac{1}{6}\langle 20\bar{2}3\rangle$ have been seen in irradiated magnesium, cadmium and zinc. In the latter two metals, perfect loops with Burgers vector [0001] are also observed. They result from a double layer of interstitials, and as they grow during irradiation, some loops transform into two concentric $\frac{1}{2}[0001]$ loops, as shown in Fig. 6.6.

The atomic density of the basal planes is only greater than that of the corrugated $\{10\bar{1}0\}$ prism planes when $c/a > \sqrt{3}$, suggesting that the existence of basal stacking faults aids the stability of the basal vacancy and interstitial loops in magnesium. These faults are not believed to occur in titanium and zirconium, however, and large basal-plane loops are not expected. This is confirmed experimentally by transmission-electron microscopy of irradiated specimens, which reveals that basal-

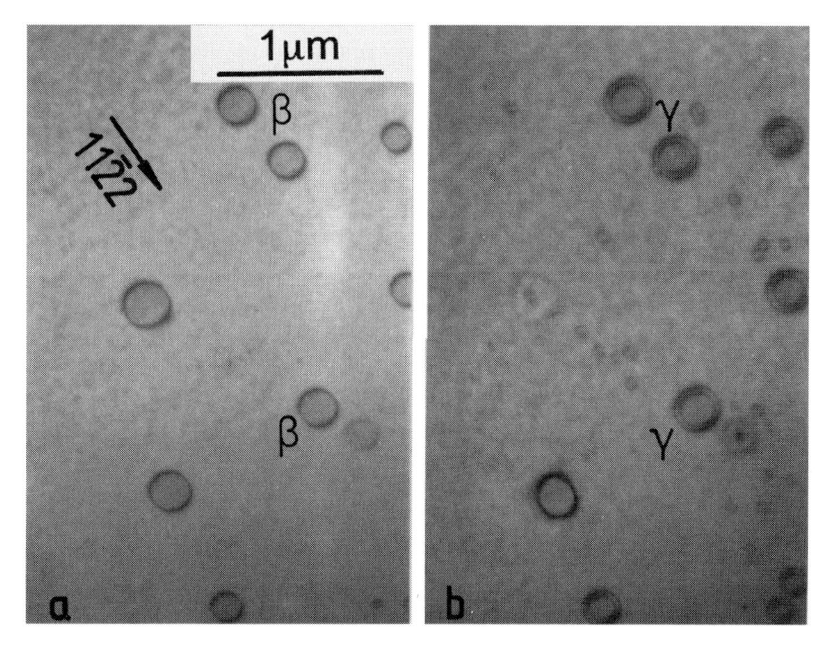

Figure 6.6 Transmission electron micrographs of interstitial loops with $\mathbf{b} = [0001]$ in zinc. (a) Below the threshold voltage at which electrons displace zinc atoms and (b) above. Loops β split into concentric loops γ as they grow. (From Eyre, Loretto and Smallman (1976), *Proc. Vacancies '76*, p. 63, The Metals Society, London.)

plane loops with a **c** component in their Burgers vector occur only rarely, for example at high levels of radiation damage. Generally, both vacancy and interstitial loops with the perfect Burgers vector $\frac{1}{3}\langle 11\bar{2}0\rangle$ are seen. They lie on planes of the [0001] zone at angles up to 30° from the pure-edge $\{11\bar{2}0\}$ orientation. The $\{11\bar{2}0\}$ planes are neither widely-spaced nor densely-packed, and it is possible that the point defects precipitate initially as single-layer loops on the $\{10\bar{1}0\}$ prism planes. The resulting stacking fault probably has a high energy and is removed by shear when the loops are small by the Burgers vector reaction

$$\frac{1}{2}\langle 10\bar{1}0\rangle + \frac{1}{6}\langle \bar{1}2\bar{1}0\rangle \rightarrow \frac{1}{3}\langle 11\bar{2}0\rangle \tag{6.9}$$

The glissile loops thus produced can adopt the variety of orientations observed in practice. The small loops shown in Fig. 2.15 are examples of vacancy loops with $\mathbf{b} = \frac{1}{2}\langle 10\bar{1}0\rangle$ or $\frac{1}{3}\langle 11\bar{2}0\rangle$ in ruthenium irradiated with heavy ions.

6.3 Dislocations in Body-centred Cubic Metals

In body-centred cubic metals (e.g. iron, molybdenum, tantalum, vanadium, chromium, tungsten, niobium, sodium and potassium) *slip* occurs in close-packed $\langle 111\rangle$ directions. The shortest lattice vector, i.e. the Burgers vector of the perfect slip dislocation, is of the type $\frac{1}{2}\langle 111\rangle$. The crystallographic slip planes are $\{110\}$, $\{112\}$ and $\{123\}$. Each of these planes contains $\langle 111\rangle$ slip directions and it is particularly significant that three $\{110\}$, three $\{112\}$ and six $\{123\}$ planes intersect along the same $\langle 111\rangle$ direction. Thus, if cross slip is easy it is possible for screw dislocations to move in a haphazard way on different $\{110\}$ planes or combinations of $\{110\}$ and $\{112\}$ planes, etc., favoured by the applied stress. For this reason slip lines are often wavy and ill-defined. It has been found that the apparent slip plane varies with composition, crystal orientation, temperature and strain rate. Thus, when pure iron is deformed at room temperature the slip plane appears to be close to the maximum resolved shear stress plane irrespective of the orientation, whereas when it is deformed at low temperatures, or alloyed with silicon, slip tends to be restricted to a specific $\{110\}$ plane.

An interesting feature of yielding is the *asymmetry of slip*. It is found, for example, that the slip plane of a single crystal deformed in uniaxial compression may be different from the slip plane which operates in tension for the same crystal orientation. In other words, the shear stress to move a dislocation lying in a slip plane in one direction is not the same as the shear stress required to move it in the opposite direction in the same plane. Slip is easier when the applied stress is such that a dislocation would move in the twinning sense (see below) on $\{112\}$ planes rather than the anti-twinning sense, even when the actual slip plane is not $\{112\}$. Electron microscopy of metals deformed at low temperature reveals long screw dislocations, implying that non-screw dislocations are more mobile and that screw dislocations dictate the slip characteristics.

It is a feature of the body-centred cubic metals that stacking faults have not been observed experimentally, and the ease of cross slip suggests that faults are at best of very high energy. Consequently, simple elastic calculations of dislocation dissociation are inadequate and much of our knowledge about core structure and behaviour has been obtained from computer models of crystals, in which atoms are allowed to interact by suitable interatomic potentials (see section 2.7). By computing the energy of such crystals when stacking faults are deliberately created on the low-index planes, it has been confirmed that stable faults are unlikely to exist in this structure.

Computer simulations of the screw dislocation with Burgers vector $\frac{1}{2}\langle 111 \rangle$ show that the core has a non-planar character. An example of two equivalent atomic configurations for a $\frac{1}{2}[111]$ screw are shown in Figs 6.7(a) and (b). The small circles represent atom positions projected on the (111) plane: the dislocation is perpendicular to the plane at the centre of each diagram. The orientation of the traces of the $\{110\}$ and $\{112\}$ planes are shown in Fig. 6.7(c). The only atomic displacements which are not negligible are parallel to the dislocation line [111], and the atom projections are the same as for a perfect crystal. In order to represent the [111] *disregistry* of atoms in the core, arrows are drawn between pairs of neighbouring atoms on the projections. The length of an arrow is proportional to the difference of the [111] displacements of the two atoms, and scaled such that for a displacement difference of $b/3$, the arrow just runs from one atom to the other. When the difference falls between $b/2$ and b, it is reduced by b. For the isotropic elastic solution (equation (4.11)), the length of the arrows would decrease in inverse proportion to the distance from the core centre and would exhibit complete radial symmetry. In the atomic model, the displacements are concentrated on the three intersecting $\{110\}$ planes, each of which contains an unstable fault produced by a $\frac{1}{6}[111]$ displacement. Although the $\frac{1}{2}[111]$ dislocation spreads into three $\frac{1}{6}[111]$ cores, these *fractional* dislocations do not bound stable stacking faults, unlike Shockley partial dislocations. Close examination of the displacements reveals that the fractional cores also spread asymmetrically, but on three $\{112\}$ planes in the twinning sense. This accounts for the slip asymmetry referred to above.

By applying stress to the model crystals, it has been found that the core structure changes before slip occurs. For example, under an increasing shear stress on $(\bar{1}01)$ tending to move the screw dislocation in Fig. 6.7(a) to the left, the fractional dislocation on $(\bar{1}01)$ extends the core to the left and the two others constrict towards the core centre, as shown in Fig. 6.8(a). As the stress is increased (Fig. 6.8(b)), the fractional dislocation on $(0\bar{1}1)$ disappears to be replaced by another on $(\bar{1}10)$ before glide of the whole core occurs on $(\bar{1}01)$. Movement of the core to the right under stress takes a different form. Although the detailed core changes are dependent on the interatomic potentials, it is found that under pure shear stress slip occurs on the $\{110\}$ planes with the

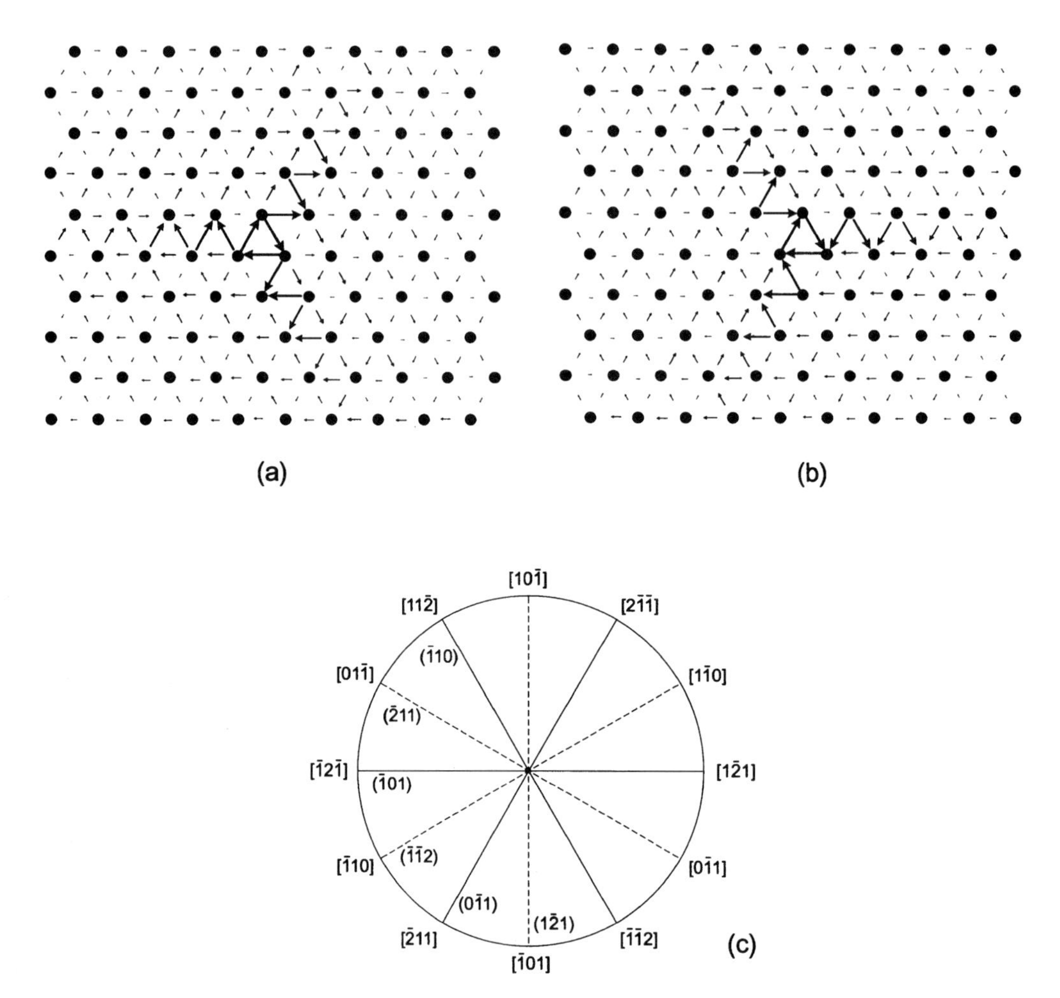

Figure 6.7 (a) Atomic positions and displacement differences (shown by arrows) for a screw dislocation with Burgers vector $\frac{1}{2}[111]$. (b) Alternative to (a). (c) Orientation of the $\{110\}$ and $\{112\}$ planes of the [111] zone. (After Vitek, *Proc. Roy. Soc.* **A352**, 109, 1976.)

asymmetry described above. Furthermore, computer simulation has shown that the screw core responds differently to stresses with different non-shear components, in good agreement with the effects of compression and tension found in experiment.

Similar studies of non-screw dislocations, on the other hand, show they have cores which are planar in form on either $\{110\}$ or $\{112\}$ but do not contain stable stacking faults. Like their face-centred cubic counterparts, they are not sensitive to the application of non-shear stresses, and they glide at much lower shear stresses than the screw dislocation.

Another set of perfect dislocations in the body-centred cubic metals are those with Burgers vector $\langle 001\rangle$. They are occasionally observed in

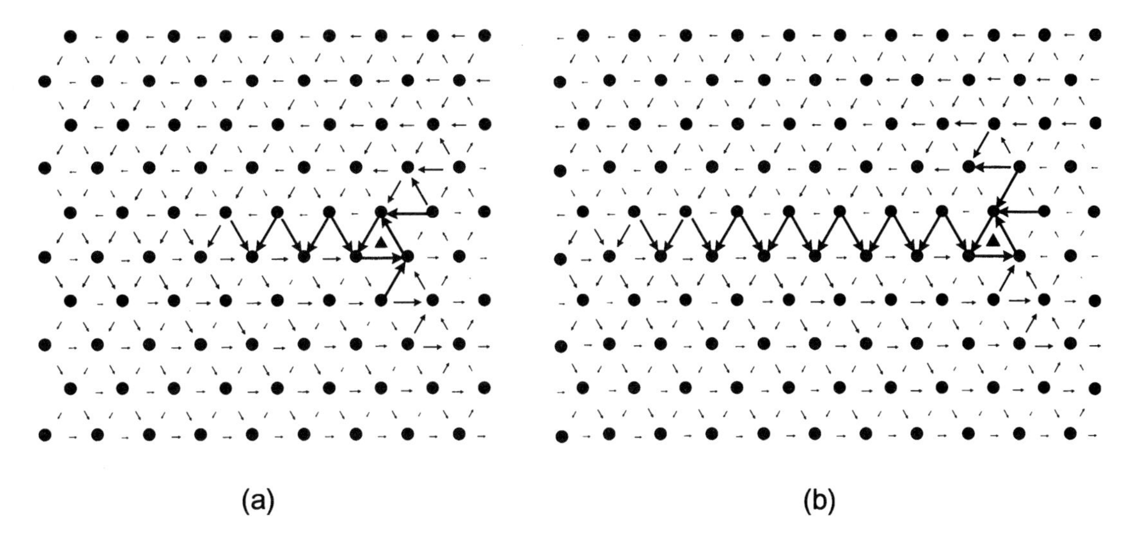

Figure 6.8 Core structures of the dislocation of Fig. 6.7(a) under applied shear stress of (a) 0.0115G and (b) 0.0265G, where G is the shear modulus. The whole dislocation moves at a stress of 0.0275 G. (After Duesbery, Vitek and Bowen, *Proc. Roy. Soc.* **A332**, 85, 1973.)

dislocation networks and are believed to occur from the reaction of two perfect $\frac{1}{2}\langle 111\rangle$ dislocations:

$$\text{e.g.} \quad \frac{1}{2}[111] + \frac{1}{2}[11\bar{1}] \rightarrow [100] \tag{6.10}$$

Computer modelling of the edge dislocation has shown that large tensile forces in the core region just below the extra half-plane lead to severing of atomic bonds and the formation of a microcrack there. This dislocation is therefore unlikely to take part in plastic deformation, but since the main cleavage planes are $\{100\}$, it may play a role in crack nucleation.

Deformation twinning is observed in all the body-centred cubic transition metals when they are deformed at low temperature and/or high strain rate. As noted with respect to the hexagonal close-packed metals in section 6.2, and described in more detail in section 9.7, the atomic displacements that allow growth of a twin occur by the glide of steps (twinning dislocations) over the twin boundary plane. For the body-centred cubic metals twinning occurs on $\{112\}\langle 111\rangle$ systems. In section 1.2 it was shown that the stacking sequence of $\{112\}$ planes in the body-centred cubic structure is *ABCDEFAB*... The homogeneous shear required to produce a twin is $1/\sqrt{2}$ in a $\langle 111\rangle$ direction on a $\{112\}$ plane. This shear can be produced by a displacement of $\frac{1}{6}\langle 111\rangle$ on every successive $\{112\}$ plane i.e. the Burgers vector of a twinning dislocation has the form $\frac{1}{6}\langle 111\rangle$ and *the step height* is $a/\sqrt{6}$. It is seen from Fig. 1.6(b) that if all the atoms in, say, an E layer and above are translated by $\frac{1}{6}[111]$, then E shifts to C, F shifts to D, etc., and the new sequence is *ABCDCDE*... A second translation on the adjacent plane displaces D to B, E to C, etc., resulting in *ABCDCBC*... Repetition of this translation on successive planes gives *ABCDCBA*..., which is the stacking of a twinned crystal. (This process is described in more detail in section 9.7.)

From the atom positions in Fig. 1.6(b), it is clear that the single translation $\frac{1}{6}[\bar{1}11]$ which displaces *E* to *C*, *F* to *D*, etc., produces a different result from the translation $\frac{1}{6}[1\bar{1}\bar{1}]$ of opposite sense. The latter moves *E* to *D* and produces an untwinned structure of high energy. There is thus an asymmetry with respect to $\frac{1}{6}\langle 111\rangle$ translations in the *twinning* and *anti-twinning sense* on $\{112\}$ planes.

Dislocation loops formed by interstitials and vacancies are an important product of radiation damage. There are no close-packed planes in the body-centred cubic structure, and it has been suggested that the loops nucleate with Burgers vector $\frac{1}{2}\langle 110\rangle$ on the $\{110\}$ planes, which are the most densely packed. In the absence of stable stacking faults on these planes, the partial dislocation loops would shear at an early stage of growth to become perfect by one of two reactions:

$$\frac{1}{2}\langle 110\rangle + \frac{1}{2}\langle 001\rangle \rightarrow \frac{1}{2}\langle 111\rangle \tag{6.11}$$

$$\frac{1}{2}\langle 110\rangle + \frac{1}{2}\langle 1\bar{1}0\rangle \rightarrow \langle 100\rangle \tag{6.12}$$

The resultant dislocation in reaction (6.11) has the lower energy (i.e. b^2) and loops with this Burgers vector have been observed in many metals. Computer simulation (section 2.7) of damage production in displacement cascades shows that small clusters of self-interstitial atoms nucleate directly as loops with Burgers vector $\frac{1}{2}\langle 111\rangle$. A cluster consists of $\langle 111\rangle$ *crowdion* interstitials, i.e. each interstitial is an extra atom inserted in a close-packed $\langle 111\rangle$ atomic row, as illustrated by the computer-generated image of a cluster of 19 interstitials in Fig. 6.9. The distortion is focused along the crowdion axis and simulation predicts that these small loops move easily along their glide prism. The one in Fig. 6.9 has moved by about 7a.

In α-iron and its alloys, however, a high proportion of loops are of $\langle 100\rangle$ type: this somewhat surprising effect is as yet unexplained.

6.4 Dislocations in Ionic Crystals

An important feature of dislocations in ionic solids is that electrical charge effects can be associated with them. For example, compressive deformation increases electrical conductivity and electric current is produced during plastic deformation by the motion of charged dislocations. Ionic crystals contain atoms of elements from different sides of the periodic table which transfer electrons from one species to the other, producing sets of *cations* and *anions*. One of the simplest forms is the *rocksalt* (*NaCl*) structure shown in Fig. 1.12, in which each anion is surrounded by six cations and vice versa. It has a face-centred cubic Bravais lattice with an anion–cation pair for each lattice point, one ion at 0, 0, 0 and the other at $\frac{1}{2}$, 0, 0. MgO, LiF and AgCl also have this

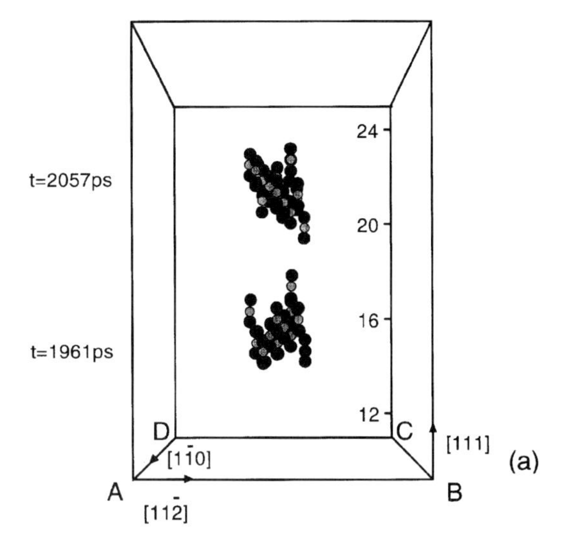

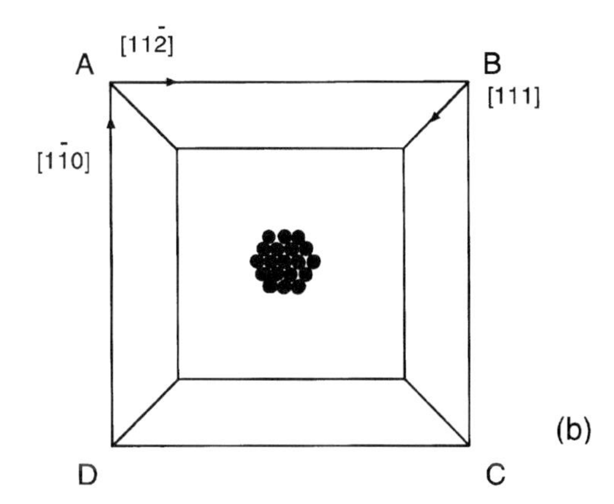

Figure 6.9 (a) Computer simulation of a cluster of 19 self-interstitial atoms in iron at 260 K showing the position of the cluster at two different times (1 ps = 10^{-12} seconds). The light and dark spheres are vacant sites and interstitial atoms respectively. The cluster consists of [111] crowdions and moves along its [111] axis by forward and backward jumps of the individual crowdions, which remain bound together during this one-dimensional migration. (b) Cluster in projection along the migration axis. (Courtesy Yu. N. Osetsky.)

structure. It has been widely studied and its description provides a basis for more complicated systems.

The shortest lattice vector is $\frac{1}{2}\langle 110\rangle$, and this is the Burgers vector of the dislocations responsible for slip. The principal slip planes are $\{110\}$. Slip steps are also observed on $\{100\}$ and (occasionally) $\{111\}$ and $\{112\}$ planes after high stresses, particularly at high temperatures and in crystals of high polarizability, where the ionic nature of bonding decreases. Cross slip of screw dislocations can occur only by glide on

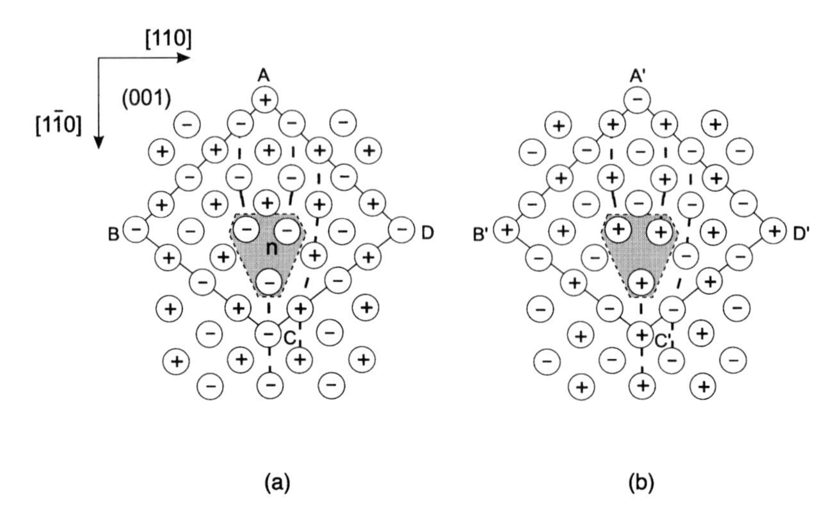

Figure 6.10 Edge dislocation in the sodium chloride structure, with the Na^+ cations represented by + and the Cl^- anions by −. (a) Initial configuration of surface ions. (b) Configuration after removal of one surface layer. (From Amelinckx (1979), *Dislocations in Solids*, vol. II, p. 66, North-Holland.)

planes other than {110}, for only one ⟨110⟩ direction lies in a given {110} plane.

Figure 6.10 shows a pure *edge* dislocation with a $\frac{1}{2}[110]$ Burgers vector and $(1\bar{1}0)$ slip plane emerging on the (001) surface. The extra half-plane actually consists of two supplementary half-planes, as shown. The ions in the planes below the surface alternate between those shown in Figs 6.10(a) and (b). The figure serves to illustrate that there is an *effective charge* associated with the point of emergence of the dislocation on the (001) surface. Intuitively, if it is $-q$ in (a) it must be $+q$ in (b). It is readily shown that $q = e/4$, where e is the electronic charge, as follows. In any cube of NaCl in which the corner ions are of the same type, as in Fig. 1.12, the ions of that sign exceed in number those of opposite sign by one. This excess charge of $\pm e$ may be considered as an effective charge $\pm e/8$ associated with each of the eight corners. In a cube with equal numbers of anions and cations, the positive and negative corners neutralise each other. Consider the dislocation of Fig. 6.10(a) to be in a block of crystal $ABCD$ bounded by {100} faces. The effective charge of the four corners A, B, C, D is $(+e/8 - 3e/8) = -e/4$, and so the net effective charge at the (001) surface is $(-e/4 - q)$. Remove a single layer of ions to expose the new face $A'B'C'D'$ (Fig. 6.10(b)). The net effective charge is now $(-e/8 + 3e/8 + q)$, which has been achieved by removing sixteen anions and fifteen cations, i.e. $-e$. Since the initial charge must equal the final charge plus the charge removed,

$$-\frac{e}{4} - q = \frac{e}{4} + q - e \tag{6.13}$$

and $q = e/4$. The same result holds for emergence on {110} planes.

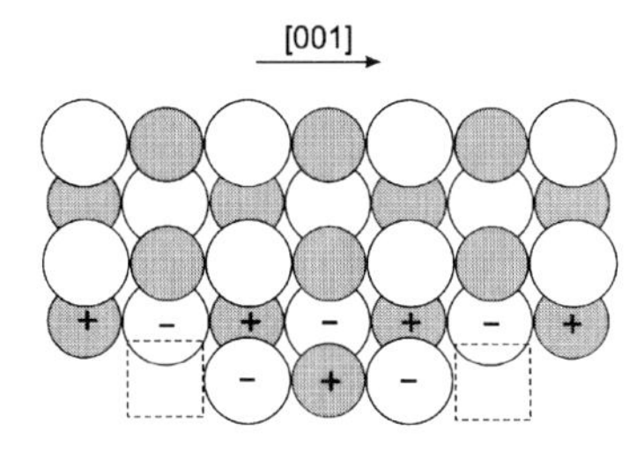

Figure 6.11 Extra half-planes of the edge dislocation in Fig. 6.10 with a jog at each end denoted by the squares. (From Amelinckx, Supplement Vol. 7, Series X, *Nuovo Cimento*, p. 569, 1958.)

When the edge dislocation glides on its $(1\bar{1}0)$ slip plane, there is no displacement (and hence transport of charge) along the line, and the effective charge of the emergent point does not change sign. For the same reason, *kinks* in edge dislocations bear no effective charge. If the dislocation climbs by, say, removal of the anion labelled n at the bottom of the extra half-plane in Fig. 6.10(a), the configuration changes to the mirror image of that in Fig. 6.10(b), and so the effective charge at the emergent point changes sign. It follows that *jogs* carry effective charge, as demonstrated by the illustration in Fig. 6.11 of two *elementary* (or *unit*) jogs of one atom height. The bottom row of ions has an excess charge $-e$, which is effectively carried by the two jogs. Thus, depending on the sign of the end ion of the incomplete row, an elementary jog has a charge $\pm e/2$. It cannot be neutralized by point defects of integer charge. Charged jogs attract or repel each other electrostatically, and only jogs of an even number times the elementary height are neutral. (Note that in divalent crystals such as MgO, the effective charges are twice those discussed here.)

The formation energies of anion and cation vacancies are in general different, and this results in a higher probability of a jog being adjacent to a vacant site of the lower energy. Dislocations thus have an effective charge per unit length in thermal equilibrium, although this is neutralised in the crystal overall by an excess concentration of vacancies of opposite sign.

The situation with the $\frac{1}{2}\langle 110\rangle$ *screw* dislocation is more complicated. The ions in any particular $\langle 110\rangle$ row are of the same sign (see Fig. 1.12), and since displacements are parallel to the Burgers vector, motion of the screw results in displacement of charge parallel to the line. Consequently, both kinks and jogs on screw dislocations are charged, the effective charge being $\pm e/4$ in each case. The effective charge of the point of emergence on $\{110\}$ and $\{100\}$ surfaces is $\pm e/8$.

The reason underlying the choice of $\{110\}$ as the principal slip plane is unclear. It has long been considered that the glide system is determined by the strengths of the electrostatic interactions within the dislocation core. This is partly supported by recent calculations. Although stable stacking faults do not exist, the core may spread on the $\{110\}$ planes to a width of about $6b$, and thus consist of two fractional dislocations of Burgers vector $\frac{1}{4}\langle 110\rangle$ bounding an unstable fault. Dissociation on the $\{100\}$ planes does not occur and is small on $\{111\}$. It has also been suggested, however, that since the ions of the row at the bottom of the extra half-plane of the $\frac{1}{2}\langle 110\rangle$ edge dislocation all have the same sign for $\{100\}$ and $\{111\}$ slip, but alternate in sign for $\{110\}$ slip, interaction between edge dislocations and charged impurities may be an important factor.

6.5 Dislocations in Superlattices

In many substitutional solid solutions of one element, A, in another, B, the different species of atoms are arranged at random on the atomic positions of the lattice. At a composition, A_xB_{1-x}, for example, any

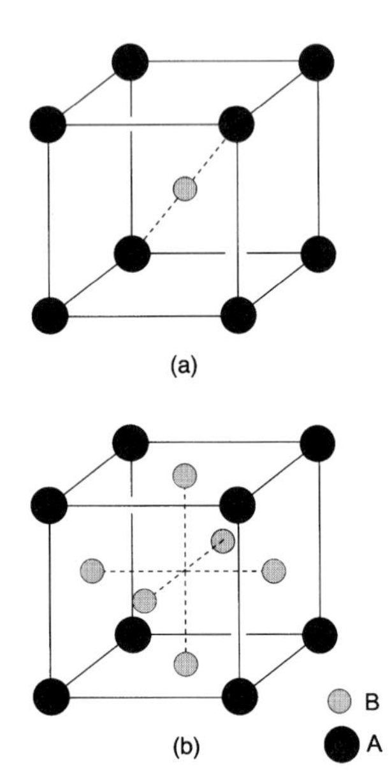

Figure 6.12 Unit cells of the *B2* and $L1_2$ superlattices.

given lattice point is occupied indifferently by either *A* or *B* atoms. There are some solid solutions, however, particularly near stoichiometric compositions such as *AB*, AB_2,AB_3, etc., in which a specific distribution of the atom species can be induced. Below a critical temperature, atoms of one kind segregate more or less completely on one set of lattice positions, leaving atoms of the other kind to the remaining positions. The resulting arrangement can be described as a lattice of *A* atoms interpenetrating a lattice of *B* atoms. A random solid solution is changed to an *ordered alloy* with a *superlattice*. Many ordered structures exist. Two possible cubic superlattices produced by alloys of composition *AB* (e.g., CuZn, NiAl) and AB_3 (e.g., Cu_3Au, Ni_3Al) are shown in Figs 6.12(a) and (b). These structures are given the crystallographic identification *B2* and $L1_2$, respectively. In the disordered state, CuZn (β-brass) is body-centred cubic and Cu_3Au is face-centred cubic. In the ordered form, both are based on a simple cubic Bravais lattice, one with two atoms per unit cell, the other with four. During the nucleation and growth of ordered domains in a disordered crystal, the lattice parameter change is usually sufficiently small for the atomic planes to remain continuous. Thus, when domains meet, the *A* and *B* sublattices are either in phase, i.e. in 'step', with each other or out of phase. The latter condition results in an *antiphase boundary* (*APB*). It has a characteristic energy because the nearest-neighbour coordination of the superlattice is destroyed: typical energies are similar to those of stacking faults, i.e. $\sim$10–100 mJ m^{-2}.

Antiphase boundaries also arise in the core of dislocations in ordered alloys and this is of considerable technological significance because of their influence on the high temperature mechanical properties of these materials. At low temperature, for example, $L1_2$ alloys behave like the face-centred cubic metals in that the critical resolved shear stress (CRSS) is almost independent of temperature, but as the temperature is increased the CRSS actually increases, an effect known as the *yield stress anomaly*. If the order–disorder transition temperature for the alloy is high enough, the yield stress reaches a peak. This occurs between 800 and 1000 K in the case of Ni_3Al and allows Ni-based superalloys that contain Ni_3Al precipitates (γ' phase) in a disordered matrix (γ phase) to be used in high temperature applications, e.g., turbine blades. The key to this property is the core structure of the dislocations responsible for slip.

In disordered $L1_2$ alloys, dislocation behaviour is similar to that described for face-centered cubic metals in Chapter 5, e.g., the unit slip vector is $\frac{1}{2}\langle 110\rangle$. In the ordered state, $\frac{1}{2}\langle 110\rangle$ vectors are not lattice translation vectors and so gliding dislocations leave behind a surface of disorder (*APB*). This is illustrated schematically for the $L1_2$ superlattice in Fig. 6.13(a), which shows the atomic arrangement in two adjacent $\{111\}$ planes. Displacement of one layer by $\frac{1}{2}\langle 110\rangle$ with respect to the other shifts *X* to *Y* and creates an *APB* in which *A* atoms occupy nearest-neighbour sites to each other. Order may be restored, however, by a second displacement $\frac{1}{2}\langle 110\rangle$ which takes the atom originally *X* to *Z*.

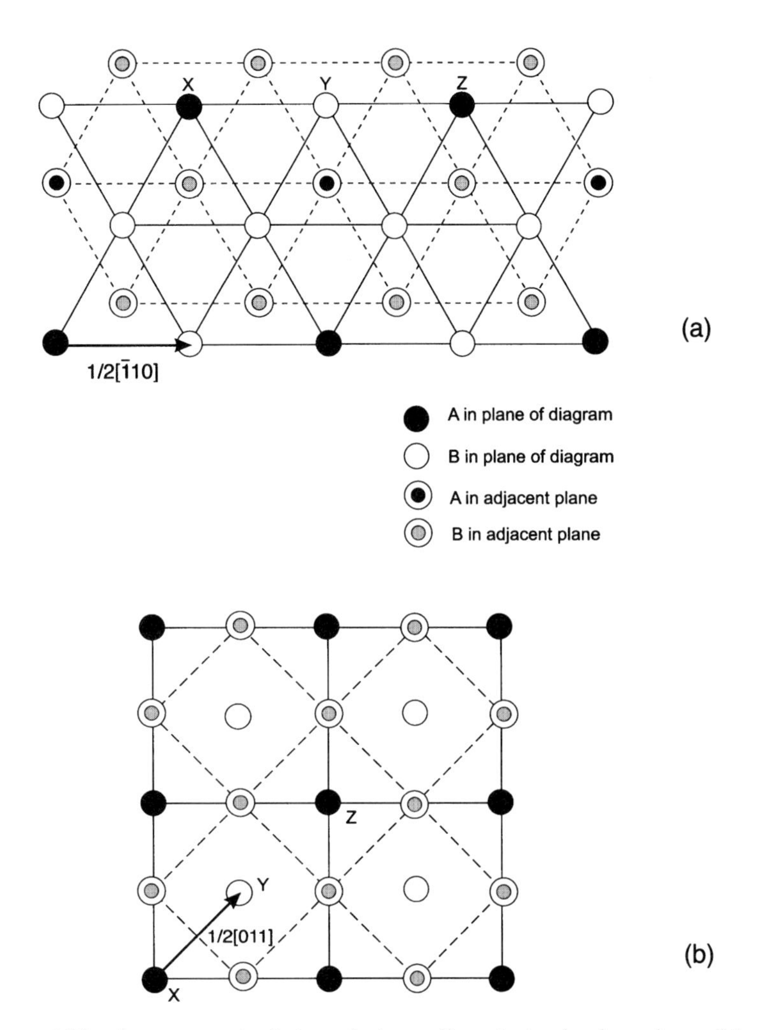

Figure 6.13 Arrangement of atoms in two adjacent atomic planes in an $L1_2$ superlattice: (a) (111) planes and (b) (100) planes.

Unlike the face-centred cubic metals, the $L1_2$ alloys are also commonly observed to slip on the $\{100\}$ planes. In this case, see Fig. 6.13(b), the shift X to Y leaves nearest-neighbour bonds across the APB unchanged and the second-neighbour changes are the major contribution to the APB energy, implying that the APB energy is lower on $\{100\}$ than $\{111\}$. Again, the second shift Y to Z restores order.

Thus, the perfect dislocation moving on either the $\{111\}$ or $\{100\}$ planes of the $L1_2$ superlattice consists of two $\frac{1}{2}\langle 110\rangle$ *superpartial* dislocations joined by an APB. This *superdislocation* has a Burgers vector $\langle 110\rangle$, which is a lattice translation vector. The superdislocation is similar to two Shockley partials joined by a stacking fault (section 5.3), for the spacing in equilibrium is given by a balance between the elastic repulsive force between the two superpartials and the opposing force due to the *APB* energy. Furthermore, if the two dislocations are each dissociated into $\frac{1}{6}\langle 112\rangle$ Shockley partials in the disordered phase, as is possible in the face-centred cubic structure, they may retain this form in the ordered $L1_2$ lattice. The superdislocation would then consist of two extended dislocations connected by an *APB*. The slip system of the ordered structure is apparently stabilised by the dislocation behaviour in the disordered state, for the Burgers vector of the superdislocation is not the shortest lattice vector, which is $\langle 100\rangle$ in both structures in Fig. 6.12. In the alloy of Fig. 6.12(a), which is body-centred cubic in the disordered state, the Burgers vector is $\langle 111\rangle$ given by two $\frac{1}{2}\langle 111\rangle$ components. The requirement that dislocations in superlattices travel in pairs separated by an *APB* provides the strengthening mechanism referred to above. It arises because of a 'locking' mechanism that affects the superpartial pair in the screw orientation.

It was seen in section 5.3 that when the perfect $\frac{1}{2}\langle 110\rangle$ dislocation in a face-centred cubic metal dissociates into two Shockley partials, it is restricted to glide in one $\{111\}$ plane because the Burgers vector $\frac{1}{6}\langle 112\rangle$ of a partial lies in only one $\{111\}$ plane. The glide plane can only change when constriction occurs. The same behaviour will apply to each superpartial in an $L1_2$ alloy if there is a tendency to split into Shockley partials. The two $\frac{1}{2}\langle 110\rangle$ dislocations will be separated by a $\{111\}$ *APB*, as shown for a screw superdislocation in Fig. 6.14(a). If one of the superpartials is not dissociated, it can glide on a $\{100\}$ plane, as illustrated in Fig. 6.14(b). This cross slip from the $\{111\}$ ('octahedral') plane to the $\{100\}$ ('cube') plane may be energetically favourable if the *APB* energy is lower on $\{100\}$ than $\{111\}$, which is the implication of the simple geometrical picture of nearest-neighbour coordination described above. Furthermore, depending on the elastic anisotropy of the crystal, the force (or torque) one partial exerts on the other as a result of its stress field can enhance the stability of this arrangement.

When cross slip of a screw superdislocation onto $\{100\}$ planes occurs, glide of the remainder of the dislocation on the $\{111\}$ system is restricted because the Peierls stress resisting glide on $\{100\}$ is much higher than that on $\{111\}$. Thus, the applied stress has to be increased to maintain plastic flow. The $\{100\}$ cross-slipped segment is known as a *Kear–Wilsdorf lock*. The effect of temperature that results in the yield stress anomaly referred to above probably arises from the ease with which a $\frac{1}{2}\langle 110\rangle$ superpartial can cross slip onto a $\{100\}$ plane. If it is dissociated into two $\frac{1}{6}\langle 112\rangle$ Shockley partials on a $\{111\}$ plane, it will have to constrict, as mentioned above, and the energy barrier for that can be overcome by thermal activation. The result is increased hardening due to cross slip with increasing temperature, as seen experimentally. An example of the dislocation structure in a crystal of $Ni_3(Al, Hf)$ deformed at 400 °C by slip on a single slip system is shown in the transmission electron microscope image taken under dark-field conditions (section 2.4) in Fig. 6.15. Not all $L1_2$ alloys exhibit anomalous hardening, however, and this probably results from a higher ratio of the *APB* energy on $\{100\}$ to that on $\{111\}$ and a lower value

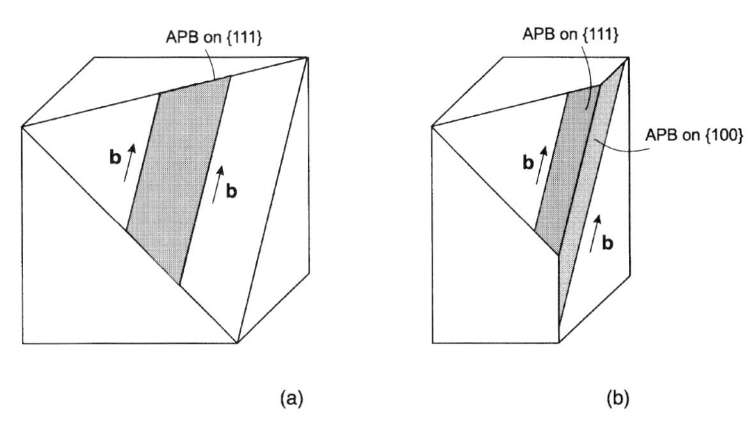

Figure 6.14 (a) Screw superpartial pair with an *APB* on a {111} plane. (b) Cross slip onto a {100} plane, starting the formation of a Kear–Wilsdorf lock.

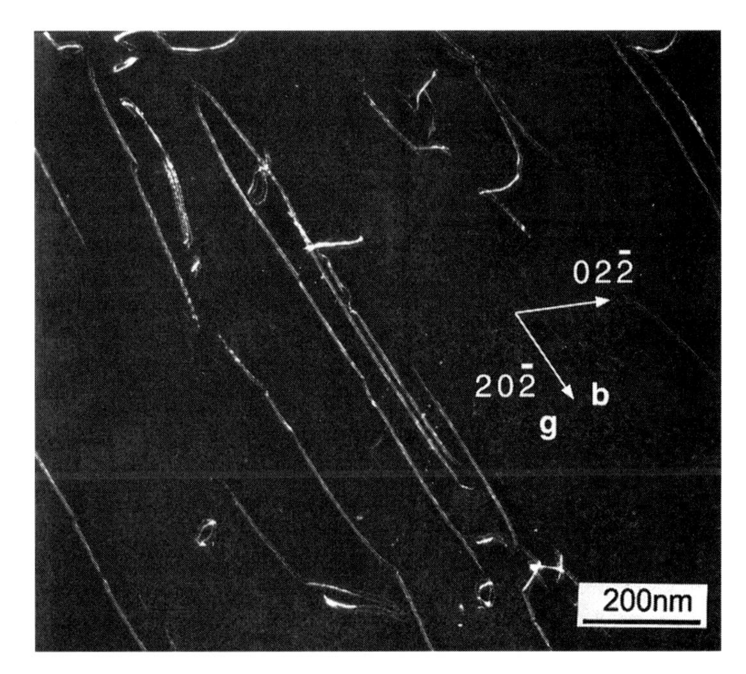

Figure 6.15 Dislocation structure in $Ni_3(Al,Hf)$ deformed at 400 °C. The (111) slip plane is the plane of the paper and the $[10\bar{1}]$ slip direction is indicated by **b**. The superdislocations are elongated in the screw direction and are bowed out in the (010) cube cross-slip plane. The edge segments are shorter. The superdislocations consist of two $\frac{1}{2}[10\bar{1}]$ superpartials with a spacing of a few nm. (Courtesy P. Veyssiere.)

of the elastic torque. Finally, it should be noted that sessile screw configurations equivalent to the Kear–Wilsdorf lock have been found in ordered alloys with other crystal structures.

6.6 Dislocations in Covalent Crystals

The covalent bond formed by two atoms sharing electrons is strongly localized and directional, and this feature is important in determining the characteristics of dislocations. Of the many covalent crystals, the cubic structure of diamond, silicon and germanium is one of the simplest and most widely studied. Compounds such as gallium arsenide (GaAs) have the same atomic arrangement. Dislocations in these semiconductors affect both mechanical and electrical properties.

The space lattice is face-centred cubic with two atoms per lattice site, one at 0, 0, 0 and the other at $\frac{1}{4}, \frac{1}{4}, \frac{1}{4}$ (Fig. 6.16). In GaAs one atom would be gallium and the other arsenic. Each atom is tetrahedrally bonded to four nearest-neighbours, and the shortest lattice vector $\frac{1}{2}\langle 110\rangle$ links a second-neighbour pair. The close-packed $\{111\}$ planes have a six-fold stacking sequence *AaBbCcAaBb*... as shown in Fig. 6.17. Atoms of adjacent layers of the same letter such as *Aa* lie directly over each other, and planar stacking faults arising from insertion or removal of such pairs do not change the tetrahedral bonding. By reference to the face-centred cubic metals, the intrinsic fault has stacking sequence *AaBbAaBbCc*... and the extrinsic fault has *AaBbAaCcAaBb*... Faults formed between adjacent layers of the same letter do not restore tetrahedral bonding and have high energy.

Perfect dislocations have Burgers vector $\frac{1}{2}\langle 110\rangle$ and slip on $\{111\}$ planes. They usually lie along $\langle 110\rangle$ directions at 0° or 60° to the Burgers vector as a result of low core energy in those orientations (see Fig. 8.7). From consideration of dislocations formed by the cutting operations of section 3.2, two dislocation types may be distinguished. The cut may be made between layers of either different letters, e.g., *aB*, or the same letter, e.g., *bB*. Following Hirth and Lothe the dislocations produced belong to either the *glide set* or the *shuffle set*, as denoted in Fig. 6.17. Diagrammatic illustrations of the two sets are shown in Fig. 6.18. The *dangling bonds* formed by the free bond per atom along the core are apparent.

Dislocations of both sets are glissile and can dissociate, but the mechanism of dissociation is different for the two cases. In the glide

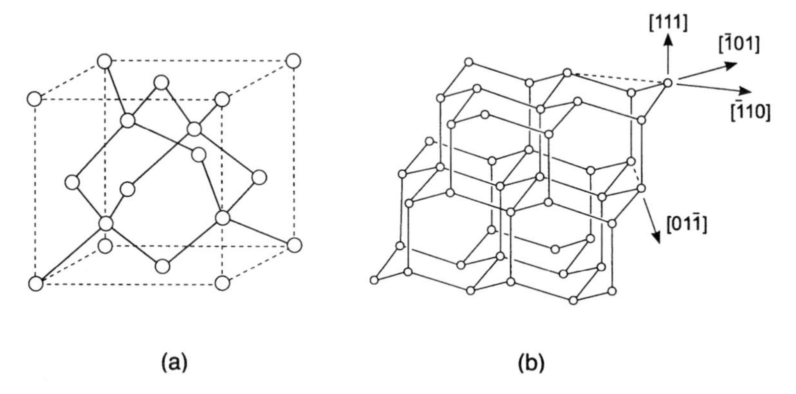

Figure 6.16 (a) Diamond-cubic unit cell. (b) Illustration of the structure showing the tetrahedral bonds and important crystallographic directions.

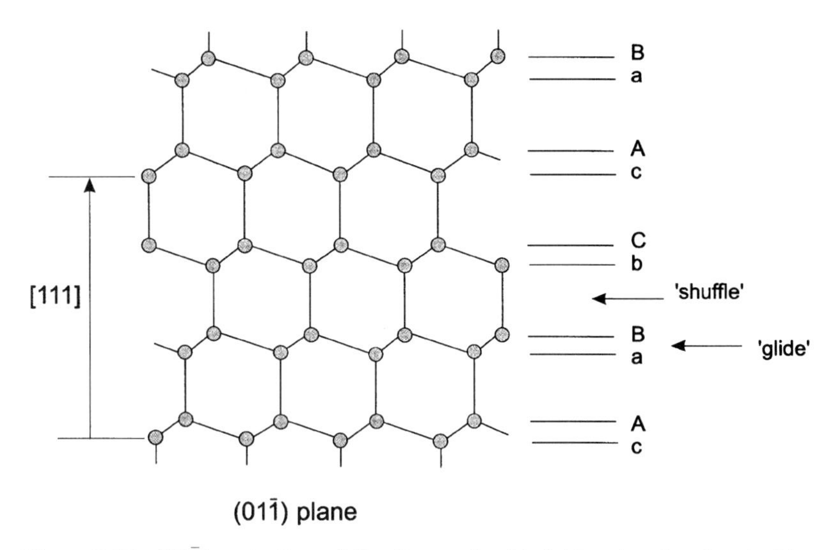

Figure 6.17 (01$\bar{1}$) projection of the diamond-cubic lattice showing the stacking sequence of the (111) planes and the shuffle and glide planes defined in the text.

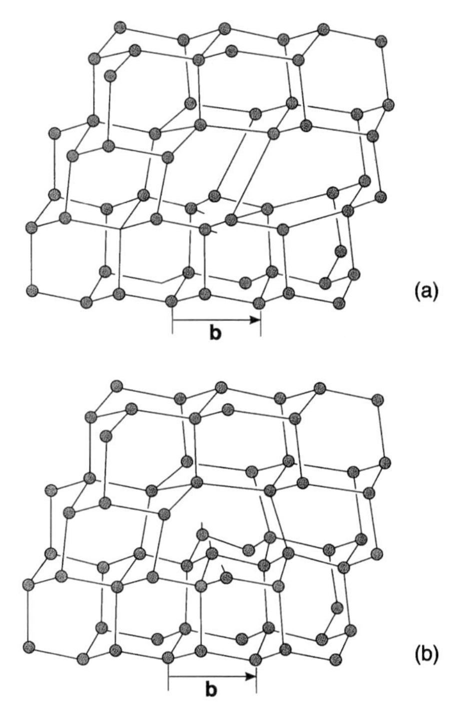

Figure 6.18 Perfect 60° dislocations of (a) the glide set and (b) the shuffle set. The extra half-planes are in the lower part of each drawing. (After Hirth and Lothe (1968), *Theory of Dislocations*, McGraw-Hill.)

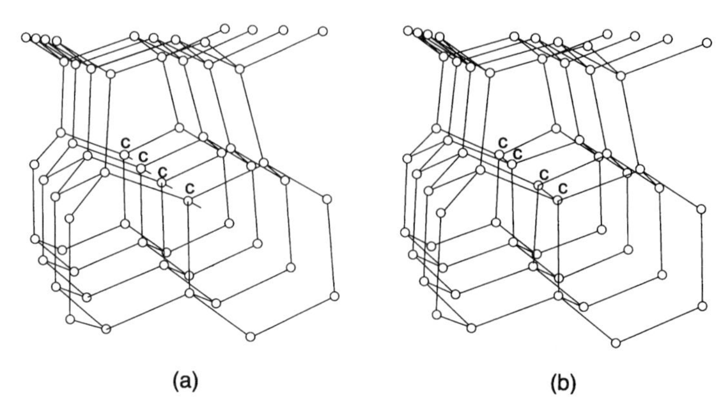

Figure 6.19 Two possible core configurations for the 30° partial of the glide set in silicon. (a) Dangling bonds occur at atoms *CCCC*. (b) Reconstructed core. (From Marklund, *Phys. Stat. Sol.* (b), **92**, 83, 1979.)

set, the perfect dislocation dissociates into two $\frac{1}{6}\langle 112\rangle$ Shockley partials separated by the intrinsic stacking fault, as in the face-centred cubic metals. Dissociation of the shuffle dislocation is not so simple because of the absence of low-energy shuffle faults. It occurs by the nucleation and glide of a Shockley partial of the glide type between an adjacent pair of $\{111\}$ layers. This results in a fault of the glide set bounded on one side by a Shockley partial and on the other by a Shockley partial and, depending on whether the glide fault is above or below the shuffle plane, a row of interstitials or vacancies. It is probable that this dislocation is less mobile than the glide-set dislocation because movement of the row of point defects within the core can only occur during slip by shuffling: hence the nomenclature. Climb, which involves point defect absorption or emission, transforms shuffle-set dislocations to glide-set dislocations, and vice versa. The stacking-fault energy in silicon and germanium is sufficiently low ($\simeq 50\,\mathrm{mJ\,m^{-2}}$) for dissociation to be resolved directly in the transmission electron microscope (Fig. 2.10). Computer calculations of the core energy of dislocations in model silicon crystals suggest that the energy is reduced by *bond reconstruction*, a process in which dangling bonds reform with others so that all atoms retain approximately tetrahedral coordination. This is shown schematically in Fig. 6.19 for the glide-set partial with $\mathbf{b} = \frac{1}{6}\langle 11\bar{2}\rangle$ at 30° to the line direction. Reconstruction occurs by the dangling bonds at *CCCC* rebonding in pairs along the core. The structure within the core affects the electron energy levels and, therefore, the electrical properties of crystals containing dislocations.

6.7 Dislocations in Layer Structures

There is a large group of materials with a pronounced layer-type structure which can arise in two ways. Firstly, when the binding forces between atoms in the layers are much stronger than the binding

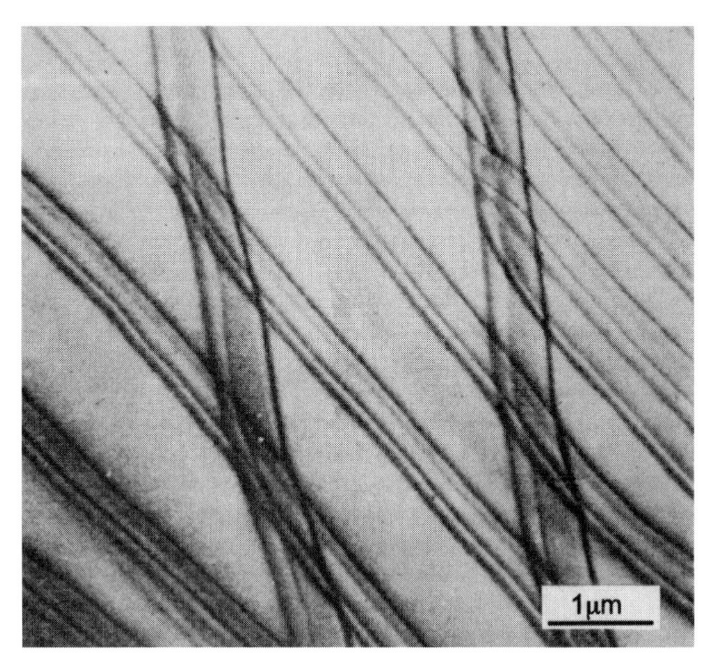

Figure 6.20 Electron transmission micrograph of dislocation ribbons in talc. (From Amelinckx and Delavignette (1962), *Direct Observation of Imperfections in Crystals*, p. 295, Interscience.)

forces between atoms in adjacent layers, as, for example, in graphite. Secondly, when the arrangement of the atoms in complex molecular structures results in the formation of two-dimensional sheets of molecules as, for example, in talc and mica. There are a number of important consequences of the *layer structure*. Slip occurs readily in planes parallel to the layers and is almost impossible in non-layer planes and, therefore, the dislocation arrangements and Burgers vectors are confined mainly to the layer planes. The weak binding between layers can result in stacking faults of low energy and hence unit dislocations are usually widely dissociated into partial dislocations. Figure 6.20 shows an example of dislocations in talc. In this material the unit dislocations dissociate into four component partial dislocations. The dislocations appear as ribbons lying in the layer planes, and in some circumstances the electron diffraction conditions allow all four partials to be observed.

Many studies have been made of crystals with layer structures. They are particularly convenient to study experimentally because uniformly thick specimens for transmission electron microscopy can be obtained simply by cleavage along the layer planes. The geometry of the Burgers vectors and dislocation reactions are described using the methods developed in preceding sections.

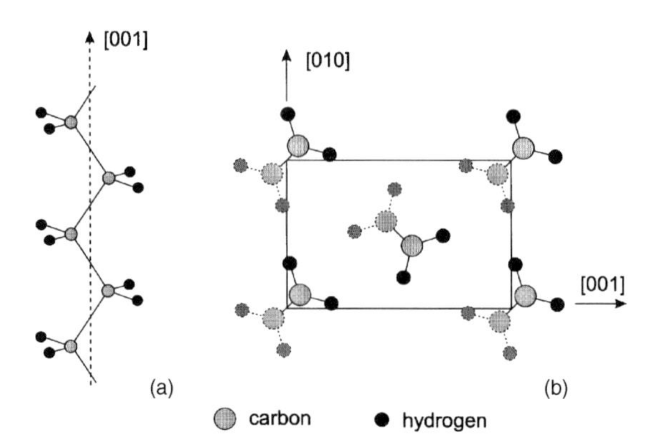

Figure 6.21 (a) The CH_2 chain structure of the polyethylene molecule. (b) (001) projection of the unit cell of orthorhombic polyethylene.

6.8 Dislocations in Polymer Crystals

The plastic deformation of crystalline and semi-crystalline polymers involves the mechanisms well established for other crystalline solids, namely dislocation glide, deformation twinning and, in some cases, martensitic transformations. As in preceding sections, dislocation behaviour can be described after making allowances for the structure. Polymer crystals are distinguished by having strong covalent bonding in the direction of the molecular chain axis, and weak van der Waals bonding in the transverse directions. This is illustrated for the orthorhombic phase of polyethylene in Fig. 6.21, which has been the subject of much investigation. As a result of the relative stiffness of the molecules, the important dislocations lie along the [001] direction. The shortest lattice vector is [001], and the screw dislocation produces chain-axis slip by gliding as a perfect dislocation on the (100), (010) and {110} planes. Transverse slip results from the glide of edge dislocations, and although the shortest lattice vectors in the transverse direction are [010] and [100], the slip vector is $\langle 110 \rangle$. This is because the perfect $\langle 110 \rangle$ dislocation can dissociate into two Shockley partials of Burgers vector approximately $\frac{1}{2}\langle 110 \rangle$ bounding a stacking fault on a {110} plane. The fault is believed to have low energy ($\lesssim 10\,\mathrm{mJ\,m^{-2}}$), but the [001] screw is prevented from dissociating by Frank's rule.

In bulk polyethylene, the (001) surfaces of a crystal consist of the folds formed by the molecules leaving and re-entering the crystal. The folds are aligned along certain crystallographic directions, and this influences the dislocation behaviour within the crystals by favouring slip planes containing the folds.

Further Reading

Body-centred Cubic and Hexagonal Metals

Bacon, D. J. (Ed.) (1991) Adriatico Research Conference on 'Defects in Hexagonal Close-packed Metals', *Phil. Mag. A*, **63**, 819–1113.

Christian, J. W. (1983) 'Some surprising features of the plastic deformation of b.c.c. metals and alloys', *Metall. Trans. A* **14**, 1233.
deCrecy, A., Bourret, A., Naka, S. and Lasalmonie, A. (1983) 'High resolution determination of the core structure of $\frac{1}{3}\langle 11\bar{2}0\rangle\{10\bar{1}0\}$ edge dislocation in titanium', *Phil. Mag. A*, **47**, 245.
Duesbery, M. S. and Vitek, V. (1998) 'Plastic anisotropy in b.c.c. transition metals', *Acta Mater.* **46**, 1481.
Eyre, B. L., Loretto, M. H. and Smallman, R. E. (1977) 'Electron microscopy studies of point defect clusters in metals', *Vacancies* '76, p. 63 (ed. R. E. Smallman and J. E. Harris), The Metals Society, London.
Griffiths, M. (1993) 'Evolution of microstructure in h.c.p. metals during irradiation', *J. Nuclear Mater.* **205**, 225.
Mahajan, S. and Williams, D. F. (1973) 'Deformation twinning in metals and alloys', *Inter. Metall. Rev.* **18**, 43.
Partridge, P. G. (1967) 'The crystallography and deformation modes of hexagonal close-packed metals', *Metall. Rev.* No. 118, 169.
Sigle, W. (1999) 'High-resolution electron microscopy and molecular dynamics study of the $\frac{1}{2}\langle 111\rangle$ screw dislocation in molybdenum', *Phil. Mag. A*, **79**, 1009.
Vitek, V. (1974) 'Theory of the core structures of dislocations in bcc metals', *Crystal Lattice Defects* **5**, 1.
Yoo, M. H. and Wuttig (Eds.) (1994) *Twinning in Advanced Materials*, TMS.

Other Crystal Structures

Alexander, H. (1986) *Dislocation in Solids*, vol. 7, p. 113 (ed. F. R. N. Nabarro), North-Holland.
Amelinckx, S. (1979) 'Dislocations in particular structures', *Dislocations in Solids*, vol. 2, p. 66 (ed. F. R. N. Nabarro), North-Holland, Amsterdam.
Delavignette, P. (1974) 'Dissociation and plasticity of layer crystals', *J. de Physique*, Colloq. C7, **35**, 181.
Haasen, P. (1974) 'Dissociation of dislocations and plasticity of ionic crystals', *J. de Physique*, Colloq. C7, **35**, 167.
Hirth, J. P. and Lothe, J. (1982) *Theory of Dislocations*, Wiley.
Labusch, R. and Schroter, W. (1979) 'Electrical properties of dislocations in semiconductors', *Dislocations in Solids*, vol. 5, p. 127 (ed. F. R. N. Nabarro), North-Holland, Amsterdam.
Puls, M. P. and So, C. B. (1980) 'The core structure of an edge dislocation in NaCl', *Phys. Stat. Sol. (b)* **98**, 87.
Skrotzki, W. and Haasen, P. (1981) 'Hardening mechanisms of ionic crystals on {110} and {100} slip planes', *J. de Physique*, Colloq. C3, **42**, 119.
Sprackling, M. T. (1976) *The Plastic Deformation of Simple Ionic Crystals*, Academic Press.
Thrower, P. A. (1969) 'The study of defects in graphite by transmission electron microscopy', *Chem. Phys. Carbon* **5**, 217.
Young, R. J. (1979) 'Deformation and fracture of polymer crystals', *Developments in Polymer Fracture*, p. 223, Applied Science Publishers, London.
Papers in Section 1 'Structure and properties of dislocations', *Microscopy of Semiconducting Materials*, Conf. Ser. No. 60, Inst. of Phys., London, 1981.
Chapters in '$L1_2$ ordered alloys', *Dislocation in Solids*, vol. 10 (eds F. R. N. Nabarro and M. S. Duesbery), North-Holland, 1996.

7 Jogs and the Intersection of Dislocations

7.1 Introduction

It has been shown that dislocations glide freely in certain planes under the action of an applied shear stress. Since even well-annealed crystals usually contain a network of dislocations, it follows that every slip plane will be threaded by dislocations and a dislocation moving in the slip plane will have to intersect the dislocations crossing the slip plane. The latter are called '*forest dislocations*'. As plastic deformation proceeds, slip occurs on other slip systems and the slip plane of one system intersects slip planes of the other systems, thus increasing the number of forest dislocations. The ease with which slip occurs depends, to a large degree, on the way the gliding dislocations overcome the barriers provided by the forest dislocations. Since the dislocation density in a crystal increases with increasing strain, the intersection processes affect the rate at which the crystal hardens as it is strained. The elementary

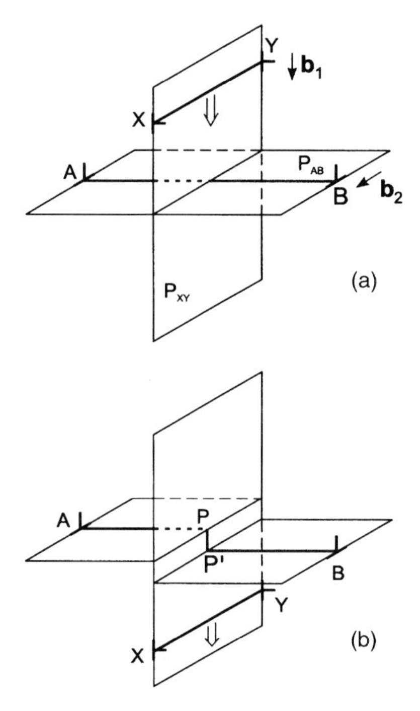

Figure 7.1 Intersection of edge dislocations with Burgers vectors at right angles to each other. (a) A dislocation XY moving on its slip plane P_{XY} is about to cut the dislocation AB lying in plane P_{AB}. (b) XY has cut through AB and produced a jog PP' in AB.

features of the intersection process are best understood by considering the geometry of the intersection of straight dislocations moving on orthogonal slip planes.

7.2 Intersection of Dislocations

The intersection of two edge dislocations with Burgers vectors at right angles to each other is illustrated in Fig. 7.1. An edge dislocation XY with Burgers vector $\mathbf{b}_1$ is gliding in plane P_{XY}. It cuts through dislocation AB with Burgers vector $\mathbf{b}_2$ lying in plane P_{AB}. Since the atoms on one side of P_{XY} are displaced by $\mathbf{b}_1$ relative to those on the other side when XY passes, the intersection results in a jog PP' (see section 3.6) parallel to $\mathbf{b}_1$ in dislocation AB. The jog is part of the dislocation AB and has a Burgers vector $\mathbf{b}_2$, but the length of the jog is equal to the length of $\mathbf{b}_1$. The Burgers vector of dislocation AB is parallel to XY and no jog is formed in the dislocation XY. The overall length of the dislocation AB is increased by b_1. Since the energy per unit length of a dislocation is αGb^2 where $\alpha \approx 1$ (equation (4.24)) the energy of the jog is αGb^3, neglecting the effect of elastic interaction with adjacent dislocation segments. However, a jog in an undissociated dislocation is a short length of dislocation with practically no long-range elastic energy and $\alpha \ll 1$. The energy is largely determined, then, by the core energy of the dislocation (section 4.4). A value of $\alpha = 0.2$ will be used.

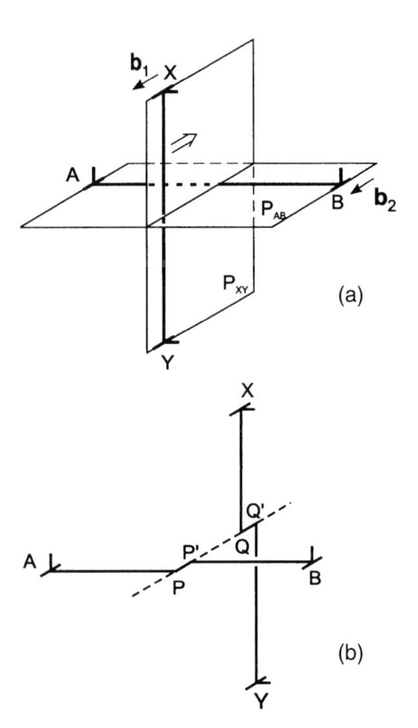

Figure 7.2 Intersection of edge dislocations with parallel Burgers vectors. (a) Before intersection. (b) After intersection.

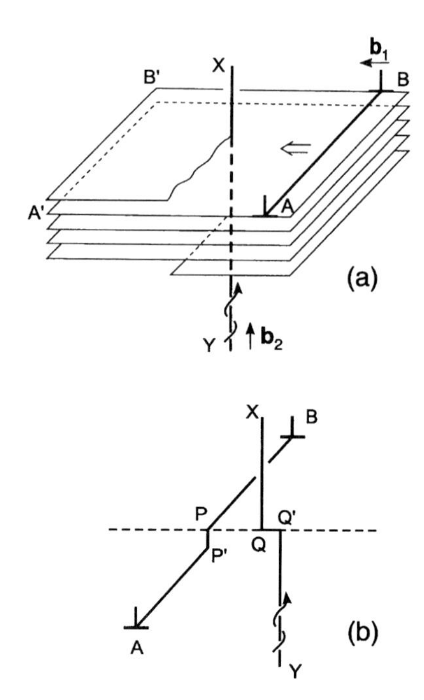

Figure 7.3 Intersection of an edge dislocation AB with a right-handed screw dislocation XY. (a) AB moving in its slip plane is about to cut XY. Planes threaded by XY form a single spiral surface. AB glides over this surface and after crossing to $A'B'$ the ends do not lie on the same plane. Thus the dislocation must contain a jog PP' as shown in (b).

The rule that *when two dislocations intersect each acquires a jog equal in direction and length to the Burgers vector of the other* may be used to analyse other cases. The intersection of two orthogonal edge dislocations with parallel Burgers vectors is illustrated in Fig. 7.2. Jogs are formed on both dislocations. The length of the jog QQ' is equal to b_2 and the length of the jog PP' is equal to b_1. The increase in energy as a result of the interaction is twice that for the example above. The intersections of a screw dislocation with an edge dislocation and a screw dislocation are illustrated in Figs 7.3 and 7.4 respectively. The sign of the screw dislocations is represented by the arrows. For the examples given all the screw dislocations are right-handed, according to the definition given in section 1.4. Jogs are produced on all the dislocations after intersection.

The length, or 'height', of all the jogs described is equal to the shortest lattice translation vector. These are referred to as *elementary* or *unit jogs.*

7.3 Movement of Dislocations Containing Elementary Jogs

Consider the jog segment formed on the edge dislocation AB in Fig. 7.1; the Burgers vector $\mathbf{b}_2$ is normal to PP' and it is therefore an *edge dislocation.* The slip plane defined by dislocation AB has a step in it,

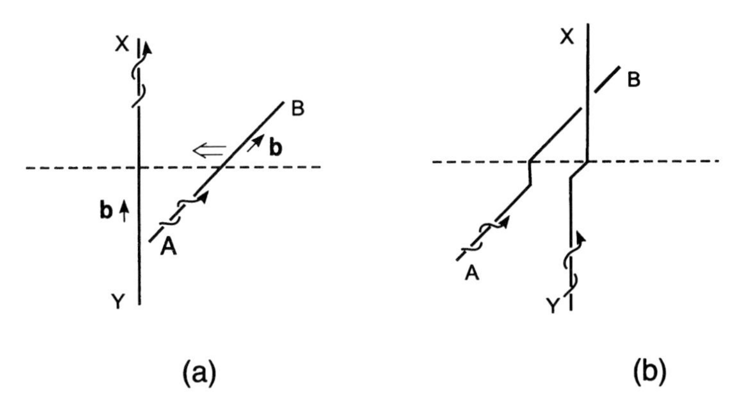

Figure 7.4 Intersection of screw dislocations. (a) Before intersection. (b) After intersection.

but the Burgers vector is at all times in the slip plane. Thus, the jog will glide along with the dislocation. The jogs formed on the edge dislocations XY and AB in Fig. 7.2 are parallel to the Burgers vectors of the dislocation and therefore have a *screw orientation*. They are *kinks* (see Fig. 3.16) lying in the glide planes P_{XY} and P_{AB} and do not impede the motion of the edge dislocations. Thus, an important conclusion is that *jogs in pure edge dislocations do not affect the subsequent glide of the dislocation*.

Consider the jogs in the screw dislocations in Figs 7.3 and 7.4. All the jogs have an *edge character*. Since an edge dislocation can glide freely only in the plane containing its line and Burgers vector, the only way the jog can move by slip, i.e. *conservatively*, is along the axis of the screw dislocation as illustrated in Fig. 7.5. It offers no resistance to motion of the screw provided the screw glides on the same plane, i.e. the jog is a kink (Fig. 3.16(b)). If the glide plane of the dislocation is different from that of the jog (Fig. 3.16(d)), however, the screw dislocation can move forward and take the jog with it only by a *non-conservative process*. As

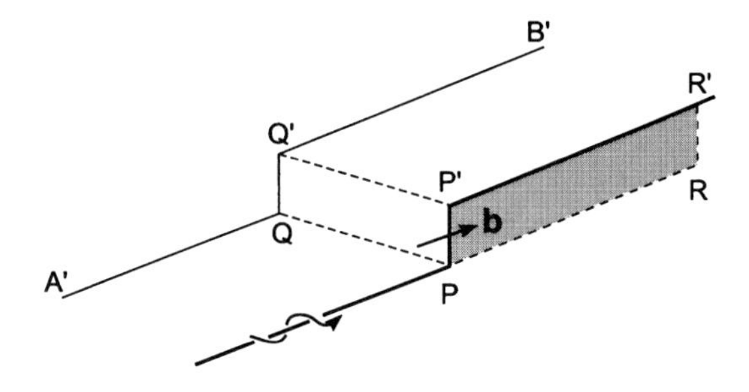

Figure 7.5 Movement of a jog on a screw dislocation. Jog PP' has a Burgers vector normal to PP' and is, therefore, a short length of edge dislocation. The plane defined by PP' and its Burgers vector is $PP'\ RR'$ and is the plane in which PP' can glide. Movement of the screw dislocation to $A'QQ'B'$ would require climb of the jog from PP' to QQ'.

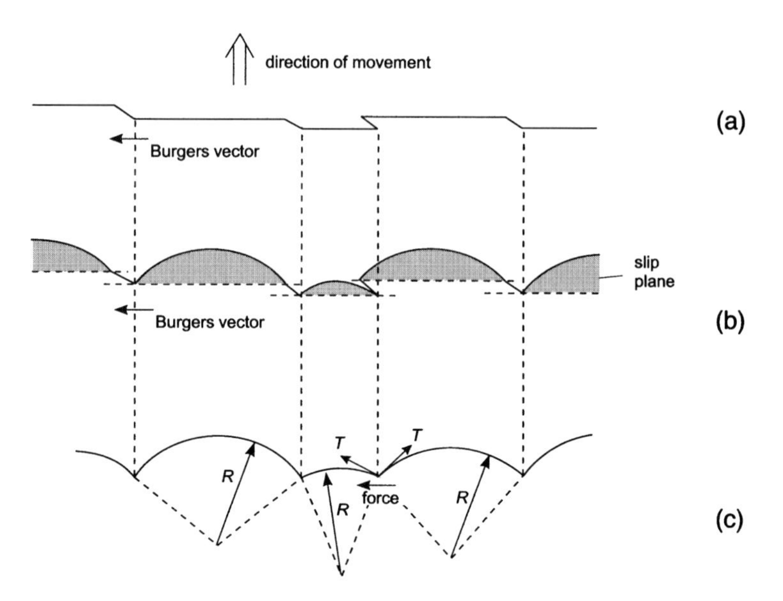

Figure 7.6 Movement of a jogged screw dislocation. (a) Dislocation with no applied stress. (b) Dislocation segments pinned by jogs bowing under stress to radius of curvature R. (c) Projection in slip plane showing force on jog due to unbalanced sideways components of line tension T.

described in section 3.6, this process requires stress and thermal activation, and consequently the movement of the screw dislocation will be temperature dependent. At a sufficiently high stress, movement of the jog will leave behind a trail of vacancies or interstitial atoms depending on the sign of the dislocation and the direction the dislocation is moving. A jog which moves in such a direction that it produces vacancies is called a *vacancy jog*, and if it moves in the opposite direction it is called an *interstitial jog*.

A screw dislocation can acquire both vacancy- and interstitial-producing jogs during plastic deformation. A critical applied stress is required for dislocation movement. Consider a screw dislocation with an array of jogs along its length (Fig. 7.6(a)). Under an applied shear stress τ acting in the slip plane in the direction of the Burgers vector $\mathbf{b}$, the dislocation

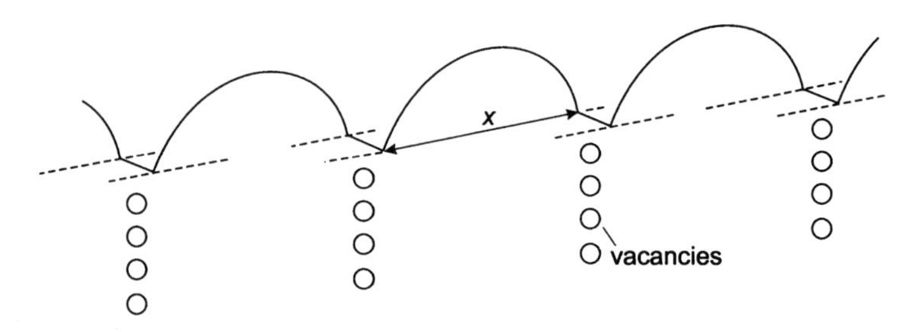

Figure 7.7 Glide of a jogged screw dislocation producing trails of point defects.

bows out between the jogs to a radius of curvature R given by equation (4.30) (Figs 7.6(b) and (c)). Two relatively closely-spaced jogs experience a net sideways force from the line tension of the dislocation segments meeting at the jogs (Fig. 7.6(c)), making them glide together and resulting in annihilation (as in Fig. 7.6) or formation of a jog of twice the unit length. The remaining jogs will be of approximately uniform spacing, x. The forward force on a given jog is that due to the applied stress on the two dislocation segments (of average length $x/2$) on each side, and is τbx from equation (4.27). Thus, when a point defect is created and the jog moves forward one atomic spacing b, the work done by the applied load is $\tau b^2 x$. If the formation energy of a point defect at a jog is E_f, the line moves forward generating point defects (Fig. 7.7) at a critical stress

$$\tau_0 = E_f / b^2 x \tag{7.1}$$

At temperatures greater than 0 K, thermal activation assists in the formation of point defects and reduces the critical stress of equation (7.1).

Since E_f for interstitials is approximately two to four times that for vacancies in metals (section 1.3), it is unlikely that interstitials are formed by jog dragging. More probably, interstitial jogs glide along the line and combine. There is a considerable body of evidence from measurement of physical properties, such as electrical resistivity, that vacancies are generated by plastic deformation.

7.4 Superjogs

A jog more than one atomic slip plane spacing high is referred to as a *super*- (or *long*) *jog*. The movement of dislocations with such jogs can be divided into three categories, depending on the jog height. They are illustrated in Fig. 7.8. For a very small jog (Fig. 7.8(a)) with height of only a few, say n, atom spacings, it may be possible for the screw dislocation to drag the jog along, creating several vacancies at each atomic plane as described in the preceding section: the critical stress will be approximately n times the value given by equation (7.1). For longer jogs, however, the maximum stress the dislocation line can experience may not attain the critical value. The maximum stress is reached when the segments bowing between jogs of spacing x have their minimum radius of curvature, see equation (4.30), which is $x/2$. If the maximum stress $2\alpha Gb/x$ (given by equation (4.30)) is less than the critical stress $nE_f^v/b^2 x$ given by equation (7.1), glide proceeds without vacancy creation. Taking $\alpha = 0.5$ and noting that $Gb^3 \simeq 4$ eV for many face-centred cubic metals and $\simeq 5-10$ eV for many body-centred cubic metals, it is apparent that vacancy generation ceases for n greater than two or three. Under these conditions, the behaviour of the dislocation depends on whether or not the superjogs are of intermediate or large height.

At the maximum stress, the bowing segments have radius of curvature $R = x/2$ and are approximately semicircular in shape. The two dislocation arms meeting at a jog are therefore parallel and both perpendicular to

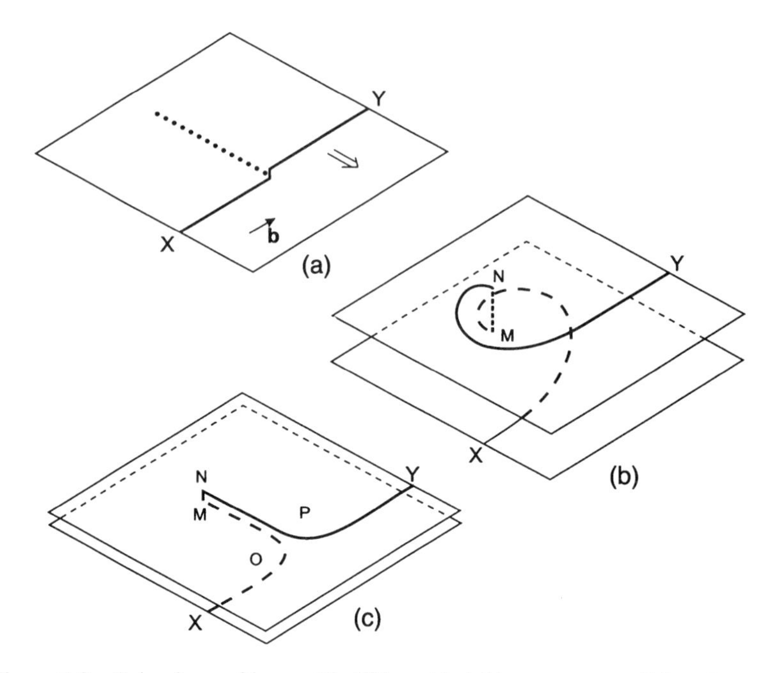

Figure 7.8 Behaviour of jogs with different heights on a screw dislocation moving in the direction shown by the double arrow. (a) Small jog is dragged along, creating point defects as it moves. (b) Very large jog – the dislocations NY and XM move independently. (c) Intermediate jog – the dislocations NP and MO interact and cannot pass by one another except at a high stress. (After Gilman and Johnston, *Solid State Physics*, **13**, 147, 1962.)

the Burgers vector. These edge elements of opposite sign adopt a configuration similar to MO and NP in Fig. 7.8(c), and can only pass each other if the force per unit length τb on each due to the applied stress τ exceeds the maximum value of their mutual repulsion force (Fig. 4.13); that is

$$\tau b > \frac{0.25\,Gb^2}{2\pi(1-\nu)y} \tag{7.2}$$

where y is the jog length MN. Jogs of relatively small length will be unable to meet this criterion, and the gliding screw will draw out a *dislocation dipole* consisting of two edge lines of the same Burgers vector and opposite sign, as shown in Fig. 7.8(c). Long jogs, on the other hand, may have sufficiently large values of y to satisfy condition (7.2). Their two edge arms can effectively behave independently of each other as single-ended sources as illustrated in Fig. 7.8(b) (see Chapter 8 on dislocation multiplication). For $\tau = 10^{-3}G$, the transition between these two forms of behaviour is found from condition (7.2) to occur at $y \simeq 60b$.

Examples of these dislocation arrangements are shown in Fig. 7.9, which is a transmission electron micrograph of a foil from a silicon-iron single crystal deformed about 1% ($\tau = 45\,\mathrm{MN\,m^{-2}}, G = 58\,\mathrm{GN\,m^{-2}}$).

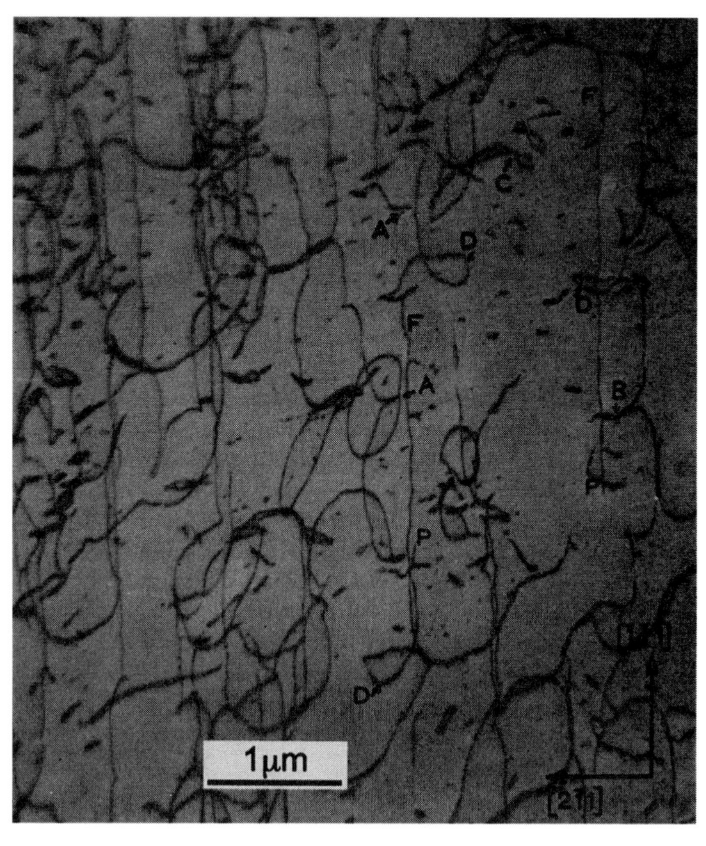

Figure 7.9 Transmission micrograph of thin foil of iron 3 per cent silicon alloy parallel to (011) slip plane. *A*, dipole trails. *B* and *C*, pinching off of dipole trails. *D*, single-ended sources at large jogs. *FP*, jogged screw dislocation. (From Low and Turkalo, *Acta Metall.* **10**, 215, 1962.)

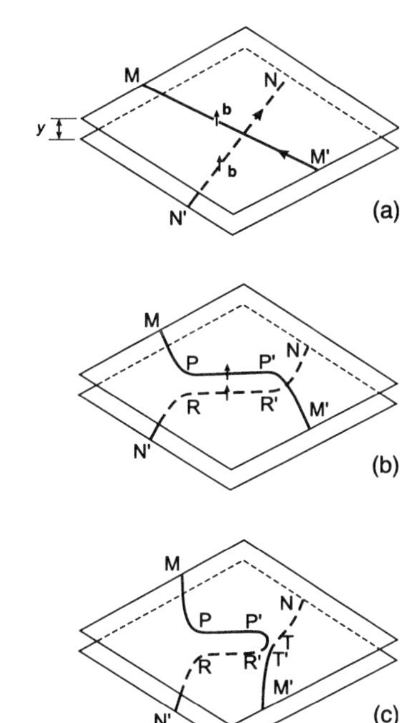

Figure 7.10 A mechanism for dislocation dipole deformation. (After Tetelman, *Acta Metall.* **10**, 813, 1962.)

Slip occurred on only one system, namely $\frac{1}{2}[\bar{1}\bar{1}1](011)$, and the foil is parallel to the (011) slip plane. The $[\bar{1}\bar{1}1]$ and $[2\bar{1}1]$ directions marked on the micrograph are therefore parallel to the screw and edge orientations respectively. The screw dislocations are relatively long and have numerous jogs along their length, e.g. the dislocations *FP*. Examples of dislocation dipoles occur at points *A* and they are aligned approximately in the $[2\bar{1}1]$ edge orientation as expected. Single ended sources have been produced at site *D*.

Double cross slip, which is illustrated in Fig. 3.9, is a ready source of long jogs. The segments of the dislocation which do not lie in the principal slip plane have a predominantly edge character. More generally, any movement of the dislocation out of the slip plane will result in the formation of jogs. Another sequence of events which can lead to superjog and dipole formation during plastic deformation is illustrated in Fig. 7.10. Two dislocations MM' and NN' with the same Burgers vector **b** but almost opposite line sense are gliding on parallel slip planes of spacing y (Fig. 7.10(a)). The force on each due to the stress field of the other acts to reorientate part of their lengths in the glide plane to give parallel edge segments PP' and RR' of opposite sign (Fig. 7.10(b)). If y is sufficiently small, a large shear stress τ is required to separate them (relation (7.2)). Furthermore, if an adjacent part of one line, such as $P'M'$, is close to the screw orientation, it can cross slip to join the other, such as $R'N$. The two segments which join pinch off and leave an edge dipole $PP'R'R$ on dislocation MN' and superjog TT' on dislocation NM' (Fig. 7.10(c)). (This reaction is best understood by placing an arrow on each line to denote positive line sense, as in Fig. 7.10(a), and remembering from the Burgers circuit construction (section 1.4) that dislocations of opposite sign have the same Burgers vector but opposite line sense.) Dislocation dipoles are a feature of the early stages of plastic deformation, when slip is confined to one set of planes.

7.5 Jogs and Prismatic Loops

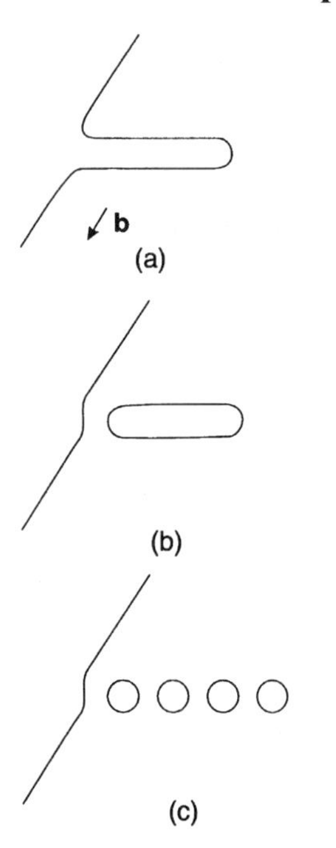

Figure 7.11 Formation of dislocation loops from a dislocation dipole. (a) Dislocation dipole. (b) Elongated loop and jogged dislocation. (c) Row of small loops.

Trails of defects and prismatic loops are often produced during plastic deformation. In Fig. 7.9 numerous small loops, many of them elongated, can be seen. This so-called *debris* is left behind by moving dislocations and is a direct result of edge jogs on screw dislocations.

Two mechanisms of forming loops are possible. Firstly, by the diffusion and coalescence of vacancies formed at a moving jog, as depicted in Fig. 7.7. If the temperature is sufficiently high to allow diffusion of the defects they can collect together to form a dislocation loop. At low temperatures interstitials diffuse more readily than vacancies and providing interstitials can be formed at an elementary jog, it will be possible to form interstitial dislocation loops. Vacancy loops, however, form only at higher temperatures because of the restricted rate of diffusion. The second mechanism represents a further stage in the development of a dislocation dipole formed either from an intermediate sized jog (Fig. 7.8(c)) or by the interaction of dislocations on parallel slip planes (Fig. 7.10). Two stages in this process are illustrated in Fig. 7.11. The dipole may pinch off from the line by a cross slip mechanism similar to that of dislocations $P'M'$ and $R'N$ in Fig. 7.10(b). The pinching off of dipoles can be seen at points C in Fig. 7.9. Furthermore, the dipole may break up because of the mutual attraction of the positive and negative edge dislocations of its two elongated sides. This results in a row of prismatic loops (Fig. 7.11(c)). It requires climb, but can occur by pipe diffusion at temperatures well below that for volume self-diffusion (section 3.8). This enables material to be re-distributed around the dislocation core. The dislocation loops formed by the breaking up of dipoles can be either vacancy or interstitial loops depending on the sign of the initial jog or dipole.

Jogs and prismatic loops can also be formed by the movement of dislocations past impenetrable obstacles in dispersion-strengthened alloys. In the *Orowan mechanism* (discussed further in section 10.7), the gliding dislocation wraps around the particle (Fig. 7.12(a)) and by

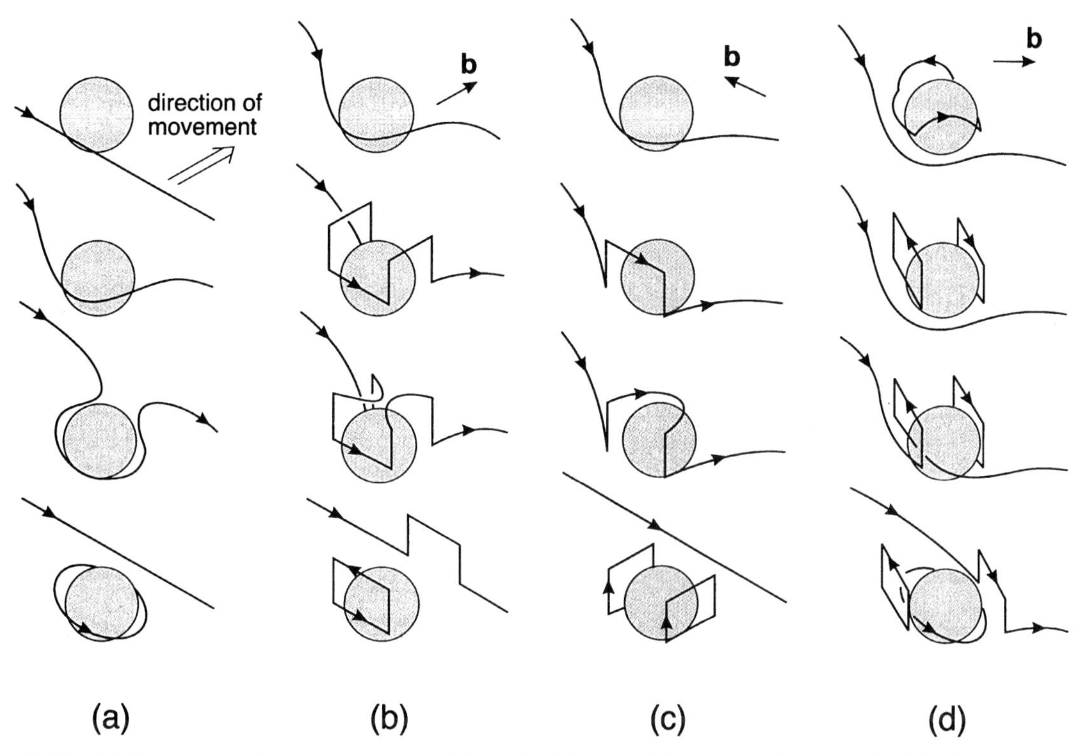

Figure 7.12 Dislocation motion past a particle either (a) without or (b), (c), (d) with cross slip. The Burgers vector in (b), (c), (d) is denoted by **b**. (After Hirsch and Humphreys (1969), *Physics of Strength and Plasticity*, p. 189, M. I. T. Press.)

mutual annihilation of segments on the far side leaves behind a *shear loop*, that is, one with its Burgers vector lying in the loop plane. The *Hirsch mechanism* involves cross slip of screw segments in order for the dislocation to bypass the obstacle. Cross slip occurs two or three times, depending on whether or not the line is predominantly edge or screw in character (Figs 7.12(b) and (c)), and one or two prismatic loops are left near the particle. Prismatic loops are observed in transmission electron microscopy of heavily deformed alloys. An Orowan loop produced as in Fig. 7.12(a) can also be induced to cross slip (Fig. 7.12(d)) and then react with the dislocation to form a prismatic loop and another shear loop. Superjogs are produced in these cross slip processes, but are only permanent features of the line in cases (b) and (d). (Again, a clear understanding of the reactions involved is best obtained by defining the positive line sense, as in the first sketch of each sequence.)

7.6 Intersections of Extended Dislocations and Extended Jogs

The simple geometrical models of intersections illustrated in Figs 7.2–7.5 have to be modified somewhat when the intersecting dislocations are extended into partials, as is expected to be the case in face-centred cubic metals with low stacking fault energy. It is difficult to generalise the

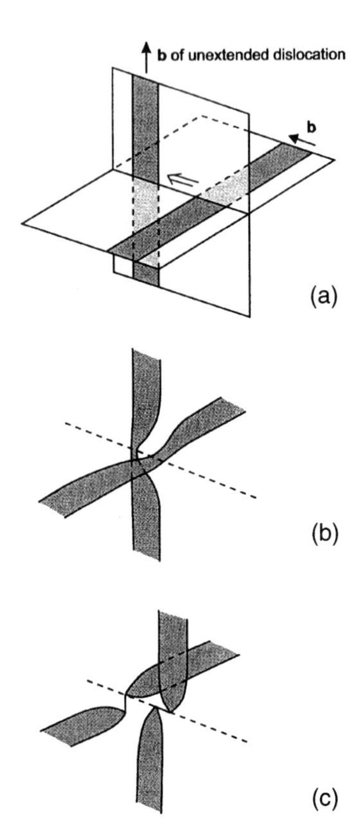

Figure 7.13 Intersection of extended dislocations. (a) Extended dislocations moving towards each other. (b) Constriction of dislocations due to the interaction of the elastic strain fields of the leading partial dislocations. (c) Formation of jogs on extended dislocations.

problem because of the number of variants of intersections that are possible. One schematic example will be considered to illustrate some of the principles involved. In Fig. 7.13 two extended dislocations intersect at right angles, a situation equivalent to Fig. 7.3. The dislocations can only cut through each other by constricting to form perfect unextended dislocations in the region close to the intersection (Fig. 7.13(b)). The dislocations can then separate and elementary jogs are produced on the extended dislocations (Fig. 7.13(c)). It is clear that to form jogs on extended dislocations work will have to be done to constrict the dislocations and so the energy of the jog will depend on the stacking fault energy or, alternatively, on the spacing of the partial dislocations. This problem has been reviewed by Friedel (1964) and Nabarro (1967).

The jogs on the extended dislocations of Fig. 7.13 are shown as single segments for simplicity, but they also can dissociate in crystals where stacking faults are possible. This can have an important effect on dislocation mobility. The configurations are easier to visualise clearly for long jogs, which are able to obey the usual rules for dislocation reactions, and the same forms are often assumed to be valid even for short jogs. The number of possibilities is frequently large for a given crystal structure, and two examples for the face-centred cubic case will suffice to indicate the principles involved. The Burgers vectors and planes will be defined with reference to the Thompson tetrahedron described in section 5.4.

Suppose an edge dislocation of Burgers vector ('bv') **BD** extended on plane *BCD* is intersected by a dislocation with bv **BA** to form two acute bends, as illustrated in Fig. 7.14(a). It acquires a jog in the direction *BA* with bv **BD**, which can dissociate on plane *ABD* into two Shockley partials with bv $\mathbf{B}\gamma$ and $\gamma\mathbf{D}$, as shown in Fig. 7.14(b). Since Burgers vector is conserved at dislocation junctions (equation (1.9)), this dissociation results in the formation of positive and negative stair-rod partials with bv $\gamma\alpha$ along the $\langle 110 \rangle$ lines of intersection of the intrinsic stacking faults. The jog will be glissile on the slip plane *ABD*, however. Suppose now a screw dislocation also with bv **BD** and dissociated on *BCD* is intersected by a dislocation with bv **CA** to form a jog and two acute bends, as illustrated in Fig. 7.15(a). The jog with bv **BD** is parallel to *CA* and can dissociate in plane *ACD* into a sessile Frank partial with bv $\mathbf{B}\beta$ and a glissile Shockley partial with bv $\beta\mathbf{D}$, as shown in Fig. 7.15(b). Furthermore, since dislocation energy is proportional to b^2 (equation 4.24), additional reduction in energy may occur by the Frank partial dissociating into a sessile stair-rod partial with bv $\delta\beta$ and a glissile Shockley partial on plane *ABC* with bv $\mathbf{B}\delta$, as depicted in Fig. 7.15(c). This jog is sessile.

Jogs may be formed with either acute or obtuse bends, and it is also possible for extrinsic faults to occur. The dissociations can be analysed using the methods established here, and full details may be found in the original paper of Hirsch (1962) and the reviews of Hirth and Lothe (1982) and Amelinckx (1979).

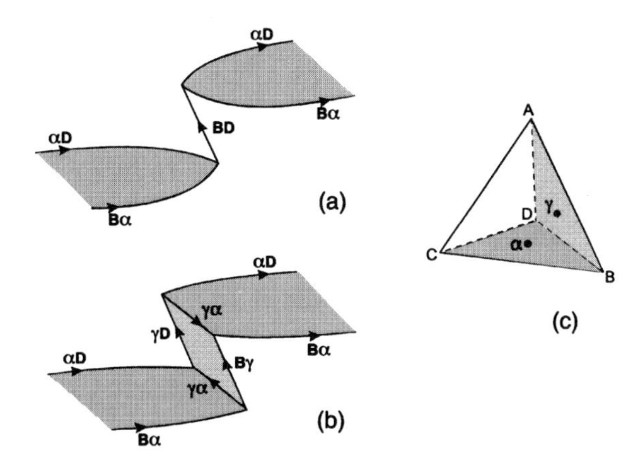

Figure 7.14 Formation of a glissile extended jog on an edge dislocation. The jog is constricted in (a) and extended in (b). Positive line senses are denoted by arrows and the Burgers vectors by the letters next to the lines. Orientation of the planes and directions is shown by the Thompson tetrahedron in (c).

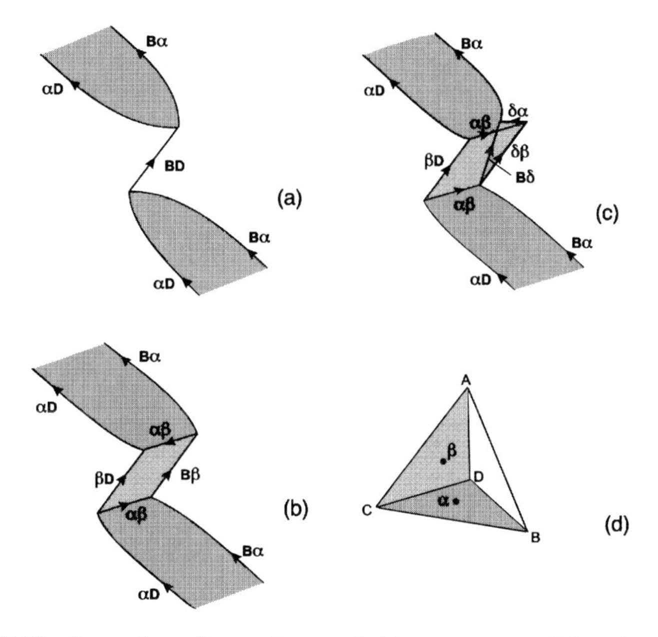

Figure 7.15 Formation of a sessile extended jog on a screw dislocation. Jog is constricted in (a), extended in (b), and fully extended in (c). Orientation of the Thompson tetrahedron is shown in (d). Positive line sense and Burgers vector are given for all segments of line.

7.7 Attractive and Repulsive Junctions

The intersection of gliding dislocations with the forest dislocations that intersect their slip planes was treated in section 7.2 as a short-range effect leading to jog and kink formation in the dislocation core. Dislocations can interact at long range, however, by virtue of the stress fields they produce. The treatment of section 4.6 considered this effect

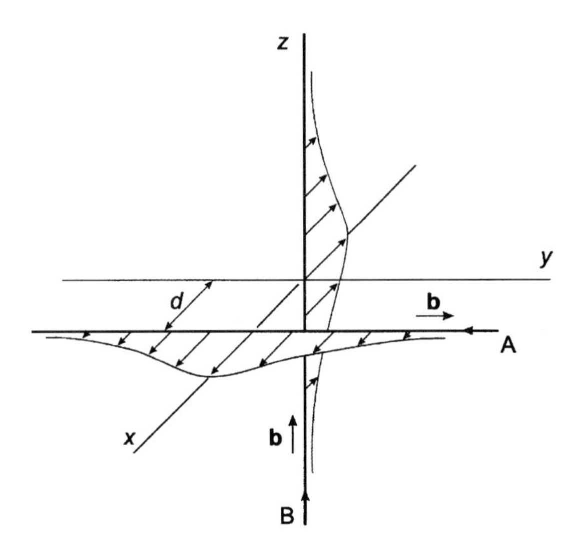

Figure 7.16 Interaction forces on two orthogonal screw dislocations *A* and *B*: *A* is left-handed and *B* right-handed. (After Hartley and Hirth (1965), *Acta Met.* **13**, 79.)

for straight, parallel dislocations only, but the same method may be applied to more general orientations. Although conditions for attractive, repulsive and neutral interactions may be determined, the analyses are more complicated because the forces vary with position along each line. Consider the simplest case of two perpendicular screw dislocations *A* and *B* parallel to the *y*- and *z*-axes, respectively, and having a perpendicular spacing *d* (Fig. 7.16). The glide force per unit length on *A* is given by *b* times the stress σ_{yz} due to *B*, and the force on *B* is *b* times the stress σ_{zy} due to *A*. These forces are readily obtained from equations (4.13) by suitable choice of axes and origin, and are $Gb^2d/2\pi(d^2+y^2)$ and $Gb^2d/2\pi(d^2+z^2)$ respectively. The force distribution on the dislocations when they have opposite sign is shown schematically in Fig. 7.16: the maximum force on each is $Gb^2/2\pi d$. As a crude approximation, the dislocation shape may be assumed to adopt the profile of the force distribution, but external and internal stresses may modify this substantially. In particular, the line tension which tends to keep a dislocation straight and the change in stress due to change of dislocation shape can be very important for some configurations. However, the analysis based on straight-line interactions elucidates two general points.

First, the forces shown in Fig. 7.16 are repulsive, and work would have to be done, in addition to that required to form a jog, to bring about dislocation intersection. It could be estimated by integrating the forces involved. If the sign of one dislocation is reversed, the forces become attractive and extra work would be required to separate the dislocations after intersection. Thus, dislocation intersections can result in *attractive* and *repulsive* junctions. The extra work is the same for both forms for straight, rigid dislocations. However, real dislocations change shape under their mutual interaction, and whereas the strength of the interaction is thereby reduced for a repulsive junction, it is increased for an attractive one. Attractive junctions are therefore stronger. Second, the two dislocations can react at their junction to form a segment with Burgers vector equal to the sum of the two individual vectors. This is energetically favourable if the geometry of the vectors is such that b^2 is reduced. The process is illustrated for one example in the face-centred cubic structure in Fig. 7.17, where the $\frac{1}{2}\langle 110\rangle$ Burgers vectors and $\{111\}$ slip planes are defined by reference to the Thompson tetrahedron (Fig. 7.17(c)). A dislocation of Burgers vector **DC** gliding on plane *BCD* reacts with a forest dislocation with Burgers vector **CB** and lying on plane *ABC*. From the line senses given, it is seen that a segment with Burgers vector **DB** is formed along the line of intersection of the two planes. The segment is glissile on *BCD*, but its ends are restrained by the forest dislocation. The gliding dislocation is therefore held unless the

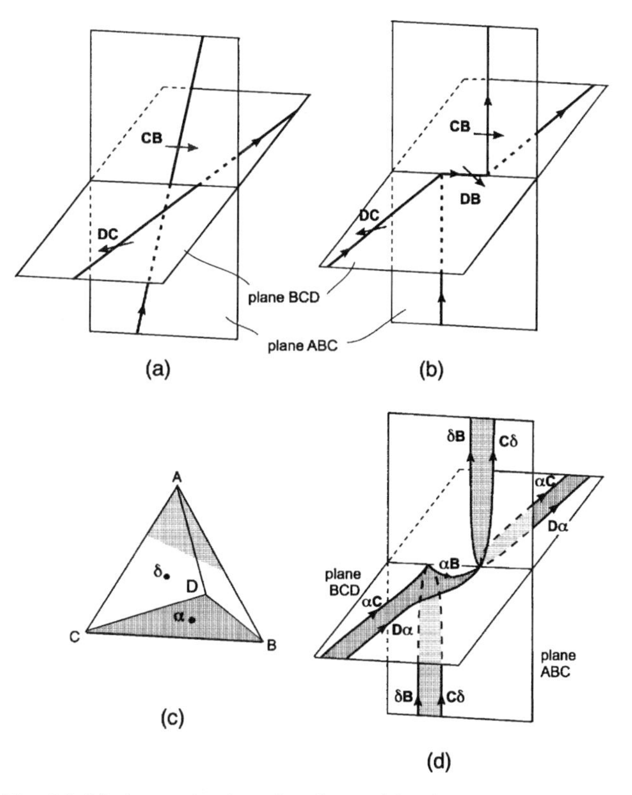

Figure 7.17 (a) (b) Attractive junction formed by the reaction of a dislocation gliding on plane *BCD* with a forest dislocation on plane ABC. Orientation of the Thompson tetrahedron is shown in (c). (d) Same reaction when the dislocations are extended, showing the formation of an extended node (section 7.8). Burgers vectors in (d) are indicated by the letters against each line.

applied stress is sufficient to overcome the reduction in dislocation energy and shrink the segment to zero.

The stress τ to move a dislocation through forest dislocations of spacing x has been estimated for a variety of configurations. The contribution due to jog creation alone arises from the energy to form a single jog E_j. In the absence of thermal activation, this is equal to the work done by the external load when a dislocation segment of length x under force τb per unit length creates a jog and moves forward by b; i.e. $E_j = \tau b^2 x$. Treating the jog as a dislocation element of length b, E_j is approximately $0.2Gb^3$ (section 7.2). The stress is therefore given by

$$\tau \simeq Gb/5x \tag{7.3}$$

The stress required to break attractive junctions has been estimated by various workers to be

$$\tau \simeq Gb/3x \tag{7.4}$$

and is believed to be the dominant contribution to strengthening due to forest dislocations.

7.8 Extended Stacking-fault Nodes

A particularly interesting dislocation structure, which provides a method for direct experimental measurement of stacking-fault energy γ in face-centred cubic metals, is illustrated in Fig. 7.18. It arises when the dislocations forming an attractive junction such as that shown in Fig. 7.17(b) are extended. The dislocation on plane BCD dissociates into two Shockley partials with Burgers vectors (bvs) $\mathbf{D}\boldsymbol{\alpha}$ and $\boldsymbol{\alpha}\mathbf{C}$, and the other on plane ABC dissociates into partials with bvs $\mathbf{C}\boldsymbol{\delta}$ and $\boldsymbol{\delta}\mathbf{B}$. The junction segment with bv $\mathbf{DB}$ dissociates into Shockley partials $\mathbf{D}\boldsymbol{\alpha}$ and $\boldsymbol{\alpha}\mathbf{B}$, as shown in Fig. 7.17(d). The same configuration arises if the two dislocations are considered extended before they intersect, for then partial $\boldsymbol{\alpha}\mathbf{C}$ reacts with partial $\mathbf{C}\boldsymbol{\delta}$ to form a stair-rod partial $\boldsymbol{\alpha\delta}$ along the line of intersection of the two planes: this in turn combines with partial $\boldsymbol{\delta}\mathbf{B}$ to form the new Shockley partial with Burgers vector $\boldsymbol{\alpha}\mathbf{B}$, which is glissile on plane BCD and is caused to bow by surface tension of the stacking fault, as illustrated. Repeated intersections lead to a network of extended and contracted nodes, such as that in a deformed copper–aluminium alloy shown in Fig. 7.19.

The partial dislocation at X in Fig. 7.18 has radius of curvature R and experiences a line tension T tending to straighten it. This results in a force T/R per unit length acting towards the centre of curvature (equation (4.29)) and is opposed by force γ per unit length due to the 'surface tension' of the stacking fault on the convex side. In equilibrium, therefore,

$$\gamma = \frac{T}{R} \simeq \frac{\alpha G b^2}{R} \tag{7.5}$$

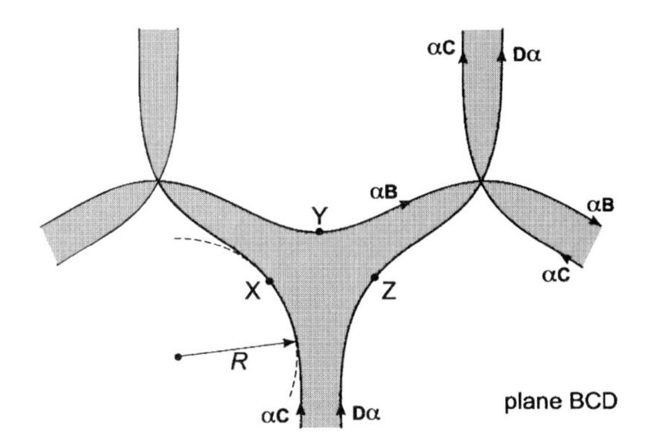

Figure 7.18 Extended node in a face-centred-cubic lattice. Shaded area represents a stacking fault. (After Whelan, *Proc. Roy. Soc.* **A249**, 114, 1959.)

Figure 7.19 Transmission electron micrograph of extended nodes in a copper 8 per cent aluminium alloy deformed 5 per cent at room temperature. (From Swann (1963), *Electron Microscopy and Strength of Crystals*, p. 131, Interscience.)

The equilibrium spacing of Shockley partials well away from the node centre is approximately (equation (5.6))

$$d \simeq \frac{Gb^2}{4\pi\gamma} \tag{7.6}$$

Although relation (7.5) takes no account of the effect on R of interactions with other partials and the variation of T with position along the line, it can be seen that $R \approx 6d$. The usual parameter measured experimentally is not R but the internal dimension of the node, e.g. the radius of a circle passing through X, Y, Z, and this can therefore be an order of magnitude larger than *d*. Hence, it is possible to measure γ from node observations in alloys where direct measurement of partial spacing is normally impossible. The more-accurate equations employed, and the structure of other forms of nodes, have been reviewed by Amelinckx (1979).

Further Reading

Amelinckx, S. (1979) 'Dislocations in particular structures', *Dislocations in Solids*, vol. 2, p. 66 (ed. F. R. N. Nabarro), North-Holland, Amsterdam.

Friedel, J. (1964) *Dislocations*, Pergamon Press, Oxford.

Hirsch, P. B. (1962) 'Extended jogs in face-centred cubic metals', *Phil. Mag.* **7**, 67.

Hirsch, P. B. and Humphreys, F. J. (1969) 'Plastic deformation of two-phase alloys containing small non-deformable particles', *Physics of Strength and Plasticity*, p. 189 (ed. A. S. Argon), M. I. T. Press.

Hirth, J. P. and Lothe, J. (1982) *Theory of Dislocations*, Wiley.

Nabarro, F. R. N. (1967) *Theory of Crystal Dislocations*, Oxford University Press.

Scattergood, R. O. and Bacon, D. J. (1976) 'Symmetric stacking fault nodes in anisotropic crystals', *Acta Met.* **24**, 705.

Seeger, A. (1955) 'Jogs in dislocation lines', *Defects in Crystalline Solids*, p. 391, Physical Soc., London.

Veysièrre, P. (1999) 'Dislocations and the plasticity of crystals', *Mechanics and Materials: Fundamental Linkages* (eds. M. D. Meyers, R. W. Armstrong and H. Kirchner), p. 271, John Wiley.

8 Origin and Multiplication of Dislocations

8.1 Introduction

It has been estimated that an upper limit for the contribution the *entropy* of a dislocation makes to the free energy of a crystal is $\sim -2kT/b$ per unit length (Cottrell, 1953). This compares with the strain energy contribution $\sim Gb^2$ (equation (4.24)). Since Gb^3 is typically ~ 5 eV and kT at 300 K is 1/40 eV, the net free energy change due to dislocations is positive. Thus, unlike the intrinsic point defects discussed in section 1.3, the thermodynamically stable density of dislocations in a stress-free crystal is zero. Nevertheless, apart from crystal *whiskers* and isolated examples in larger, carefully prepared crystals of materials like silicon, dislocations occur in all crystals. The dislocation density in well-annealed crystals (i.e. crystals which have been heated for a long time close to their melting point to reduce the dislocation density to a low value) is usually about $10^4\,\text{mm}^{-2}$ ($10^{10}\,\text{m}^{-2}$) and the dislocations are arranged in networks as previously mentioned in Chapter 1 (Fig. 1.22). A similar density of dislocations is present in crystals immediately after they have been grown from the melt or produced by strain anneal techniques and the origin of these dislocations is described in the next section.

When annealed crystals are deformed there is a rapid multiplication of dislocations and a progressive increase in dislocation density with increasing strain. After large amounts of plastic deformation the dislocation density is typically in the range 10^{14} to $10^{15}\,\text{m}^{-2}$. In the early stages of deformation dislocation movement tends to be confined to a single set of parallel slip planes. Later, slip occurs on other slip systems and dislocations moving on different systems interact. The rapid multiplication leads to work hardening (Chapter 10). In this chapter the main mechanisms of dislocation multiplication will be described.

8.2 Dislocations in Freshly Grown Crystals

It is very difficult to grow crystals with low dislocation density because dislocations are readily introduced during the growing process. There are two main sources of dislocations in freshly grown crystals. Firstly, dislocations or other defects present in the 'seed' crystals or other surfaces used to initiate the growth of the crystal. Any dislocations in a seed crystal which intersect the surface of the seed on which new growth occurs will extend into the growing crystal. Secondly, 'accidental' nucleation during the growth process. The main mechanisms which have been suggested are: (a) heterogeneous nucleation of dislocations due to internal stresses generated by impurity particles, thermal contraction, etc.; (b) impingement of different parts of the growing interface; (c) formation and subsequent movement of dislocation loops formed by the collapse of vacancy platelets.

The basic principle involved in the nucleation of dislocations by local internal stresses during growth and subsequent cooling is embodied in the specific example given in section 8.4. High local internal stresses are produced when neighbouring parts of the crystal are constrained to change their specific volume. This can occur by neighbouring regions expanding or contracting by different amounts due to (a) thermal gradients, (b) change in composition, or (c) change in lattice structure. An additional effect is the adherence of the growing crystal to the sides of the container. When the stress reaches a critical value, about $G/30$, dislocations are nucleated. If this occurs at high temperatures the dislocations created will rearrange themselves by climb. It should also be noted that under normal laboratory conditions there is every possibility of isolated vibrations which will affect the growth process and produce additional random stress effects.

The formation of dislocations by impingement is demonstrated during the coalescence of two adjacent dendrites in the growing interface. Thus, the dendrites may be misaligned or have growth steps on their surfaces so that perfect matching is impossible and dislocations are formed at the interfaces. Dislocations can also form at the interface between crystals of the same orientation but different lattice parameter. The atoms at the interface adjust their positions to give regions of good and bad registry, the latter being *misfit dislocations* (section 9.6). These dislocations are a common feature of solid-state phenomena such as precipitation and epitaxial growth.

The formation of dislocation loops by collapse of vacancy platelets follows directly from previous descriptions of the formation of dislocation loops. In section 1.3 it was shown that, when crystals are rapidly cooled from temperatures close to the melting point, the high-temperature equilibrium concentration of vacancies is retained in a supersaturated state and the vacancies can precipitate to form dislocation loops. A critical supersaturation c_1 is required for the process. This is illustrated by the fact that, in quenched specimens containing a sufficient supersaturation of vacancies to form loops in the centre of grains, no loops are found close to grain boundaries because the degree of supersaturation is reduced by the migration of vacancies to the boundary. The solidification of a melt involves the movement of a liquid–solid interface. There is normally an appreciable temperature gradient in the solid. Immediately below the interface the solid will be very close to the melting point and the equilibrium concentration of vacancies will be high. If this region cools sufficiently rapidly a high density of vacancies will be retained and, providing the supersaturation is greater than c_1, loops will be formed. Speculation about this mechanism revolves around whether or not the supersaturation is sufficient. It is most unlikely to be so when the rate of cooling is slow. However, any loops that are produced in the solid phase below the moving interface will expand by climb due to diffusion of vacancies to the loop. Loops formed at a large angle to the interface may eventually intersect the interface and the two points of emergence formed in this way will act as a site for the propagation of dislocations into the new crystal.

8.3 Homogeneous Nucleation of Dislocations

When a dislocation is created in a region of the crystal that is free from any defects the nucleation is referred to as *homogeneous*. This occurs only under extreme conditions because a very large stress is required. A method of estimating the stress has been described by Cottrell (1953). Imagine that in a crystal under an applied resolved shear stress τ, slip is nucleated by the creation on the slip plane of a small dislocation loop of radius r and Burgers vector $\mathbf{b}$. The increase in elastic energy of the crystal is given by equation (5.11) and the work done by the applied stress is $\pi r^2 \tau b$. The increase in energy associated with loop creation is therefore

$$E = \frac{1}{2} G b^2 r \ln\left(\frac{2r}{r_0}\right) - \pi r^2 \tau b \tag{8.1}$$

where ν has been taken as zero for simplicity. The energy increases for small r, reaching a maximum when $\mathrm{d}E/\mathrm{d}r = 0$ at the critical radius r_c, and then decreases for increasing r. From differentiation of equation (8.1), r_c satisfies the relation

$$r_c = \frac{Gb}{4\pi\tau}\left[\ln\left(\frac{2r_c}{r_0}\right) + 1\right] \tag{8.2}$$

and the maximum energy is

$$E_c = \frac{1}{4} G b^2 r_c \left[\ln\left(\frac{2r_c}{r_0}\right) - 1\right] \tag{8.3}$$

If the loop has at least the critical radius r_c, it will be a stable nucleus and will grow under the applied stress: E_c is the activation energy for the process.

In the absence of thermal fluctuations of energy, nucleation can only occur when $E_c = 0$, i.e. $\ln(2r_c/r_0) = 1$ or $r_c = 1.36 r_0$, and from equation (8.2) this requires a stress

$$\tau = Gb/2\pi r_c \simeq G/10 \tag{8.4}$$

in agreement with the theoretical shear stress estimate (equation (1.5)). At the more realistic upper limit for the yield stress of $\tau \simeq G/1000$, equation (8.2) gives $r_c \simeq 500b$ with r_0 taken as $2b$. For a critical nucleus of this size, equation (8.3) gives $E_c \simeq 650\ Gb^3$, which is about 3 keV for a typical metal. Since the probability of energy E_c being provided by thermal fluctuations is proportional to $\exp(-E_c/kT)$ and $kT = 1/40\,\mathrm{eV}$ at room temperature, it is clear that homogeneous nucleation of dislocations cannot occur at the yield stress. Although stress concentrations in inhomogeneous materials many raise the stress to the level required locally (section 8.4), plastic flow occurs generally by the movement and multiplication of *pre-existing* dislocations.

8.4 Nucleation of Dislocations at Stress Concentrators

A well-known example of the nucleation of dislocations at a stress concentration is illustrated in Fig. 8.1. Spherical glass inclusions were introduced into a crystal of silver chloride which was subsequently given a treatment to reduce the dislocation density to a low value. The crystal was held at 370 °C to homogenise the temperature, and remove any internal strains associated with the inclusion, and then cooled to 20 °C. The dislocations were revealed by a decoration technique (section 2.3). The photograph shows a row of prismatic dislocation loops, viewed edge on, which have been *punched out* from around the glass inclusion during cooling. The axis of the loops is parallel to a $\langle 110 \rangle$ direction which is the principal slip direction in this structure. The nucleation of the dislocations results from the stress produced around the sphere by the differential contraction of the crystal and the glass inclusion during cooling.

Suppose that at 370 °C the inclusion has unit radius; it will be resting in a hole in the silver chloride also of unit radius. If the coefficient of expansion of glass and silver chloride are α_1 and α_2 respectively, and $\alpha_1 < \alpha_2$, then on cooling to 20 °C the natural radius of the inclusion will be $1 - 350\alpha_1$ and the natural radius of the hole will be $1 - 350\alpha_2$. If the inclusion is unyielding this will result in a spherically symmetrical strain field in the surrounding matrix which can be estimated by analogy with

Figure 8.1 System of prismatic dislocation loops produced in a recrystallised, dislocation-free, crystal of silver chloride to relax the strain field around the small glass sphere caused by differential contraction which occurs during cooling. (From Mitchell (1958), *Growth and Perfection of Crystals*, p. 386, Wiley.)

the strain field around a spherical hole with an internal pressure. Thus, consider a particle of radius $r_1(1+\varepsilon)$ in a hole of natural radius r_1 in an infinite, isotropic elastic medium. The displacement in the matrix is purely radial and decreases with radial distance r as r^{-2}. Since the radial displacement u_r equals εr_1 at $r = r_1$, the displacement at r ($>r_1$) is

$$u_r = \frac{\varepsilon r_1^3}{r^2} \tag{8.5}$$

The displacement components in rectangular Cartesian coordinates (see section 4.2) are therefore

$$u_x = \varepsilon r_1^3 \frac{x}{r^3} \quad u_y = \varepsilon r_1^3 \frac{y}{r^3} \quad u_z = \varepsilon r_1^3 \frac{z}{r^3} \tag{8.6}$$

where $r^2 = x^2 + y^2 + z^2$. It is easy to show from relations (4.2)–(4.4) that the strain field is one of pure shear, and that the maximum shear strain acting in a radial direction on a cylindrical surface with a radial axis occurs at the interface between the inclusion and matrix and on a cylinder of diameter $\sqrt{2}r_1$. This is illustrated in Fig. 8.2. The magnitude of the corresponding shear stress is

$$\tau_{\max} = 3\varepsilon G \tag{8.7}$$

Taking $\alpha_1 = 34 \times 10^{-7}\,\mathrm{deg}^{-1}$ for the glass particle and $\alpha_2 = 345 \times 10^{-7}\,\mathrm{deg}^{-1}$ for the silver chloride crystal gives $\varepsilon \simeq 0.01$ after the crystal and inclusion have cooled to 20 °C and

$$\tau_{\max} = \frac{G}{33} \tag{8.8}$$

This is close to the stress required to nucleate dislocations. The dislocations formed in this way at the interface will move away under the influence of the strain field of the inclusion.

The mechanism for formation of the loops is as follows. Under the action of $\tau_{\max}$ in the slip direction a small half loop forms on the surface

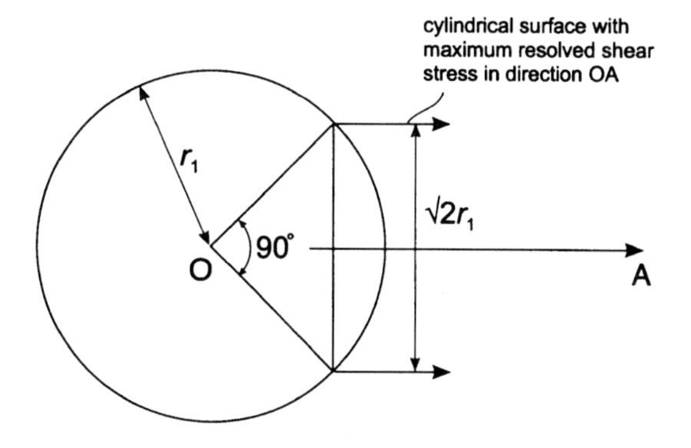

Figure 8.2 Section of spherical, misfitting particle.

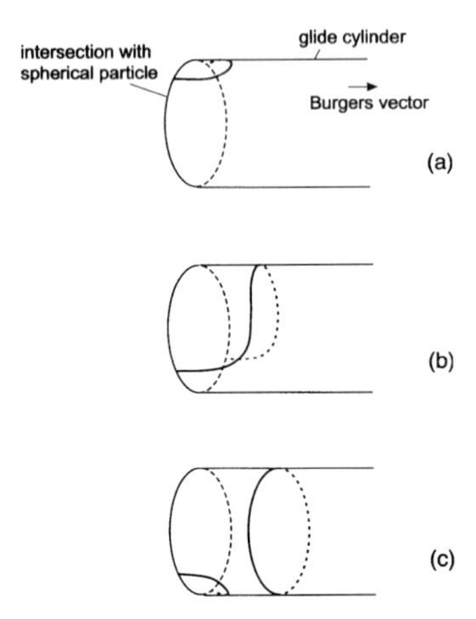

Figure 8.3 Mechanism of formation of prismatic loops around an inclusion. (a) Small loop at the surface of inclusion. (b) Loop expands around glide cylinder. (c) Prismatic loop. Burgers vector is parallel to the cylinder axis.

of the glide cylinder indicated in Fig. 8.3. The forward edge component glides away from the interface under the influence of a stress field which decreases in intensity with increasing distance from the interface. The screw components of the loop, which are parallel to the axis of the glide cylinder, experience a tangential force which causes them to glide in opposite directions around the surface of the cylinder. Dislocation movement is restricted to the glide cylinder because any other motion requires climb. When the two ends of the dislocation intersect they annihilate and a dislocation loop will be produced. This is a positive prismatic dislocation loop for $\varepsilon > 0$ and as it moves away along the slip axis it is effectively moving material away from the inclusion and so relaxing the strain field. The process can be repeated to produce a series of loops. More complex patterns are obtained when glide cylinders of different orientation exist at one particle.

Although this is a somewhat ideal model, dislocation generation at local regions of stress concentration is common. When inclusions and precipitates have a complex shape the associated strain field and resultant dislocation distribution is correspondingly more complex and tangled arrays of dislocations are produced. Dislocation generation also occurs in alloys when precipitates lose coherency (see section 9.7) provided the misfit is sufficiently large ($\varepsilon \simeq 0.01$–0.05) for spontaneous nucleation, i.e. zero activation energy. Other stress concentrators such as surface irregularities, cracks, etc., have a similar effect.

8.5 Multiplication of Dislocations by Frank–Read Sources

To account for the large plastic strain that can be produced in crystals, it is necessary to have *regenerative multiplication* of dislocations. Two mechanisms are important. One is the *Frank–Read type sources* to be described in this section and the other is *multiple cross glide* described in section 8.6.

The first model proposed by Frank and Read resembles the model proposed to account for the role of dislocations in crystal growth. In an irregular array some dislocations lie partly in their slip planes and partly in other planes. This is illustrated in Fig. 8.4(a). The length *BC* of the edge dislocation *ABC* lies in the slip plane *CEF* and can move freely in this plane. The length *AB* is not in the slip plane and is treated as sessile. Thus *BC* will be anchored at one end and can only move by rotating about *B*. The dislocation will tend to wind up into a spiral as illustrated in Fig. 8.4(b). Two things are noted about this mechanism:

(a) Each revolution around *B* produces a displacement of the crystal above the slip plane relative to that below by one atom spacing *b*; the process is *regenerative* since the process can repeat itself so that *n* revolutions will produce a displacement *nb*. A large *slip step* will be produced at the surface of the crystal.
(b) The spiralling around *B* results in an increase in the total length of dislocation line. This mechanism is similar to the single ended sources mentioned in section 7.4.

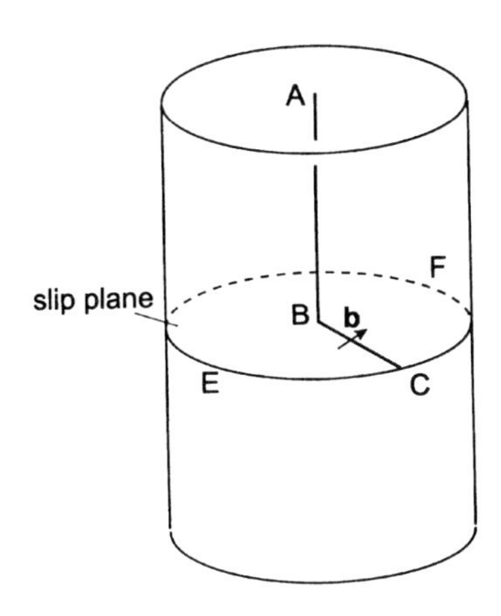

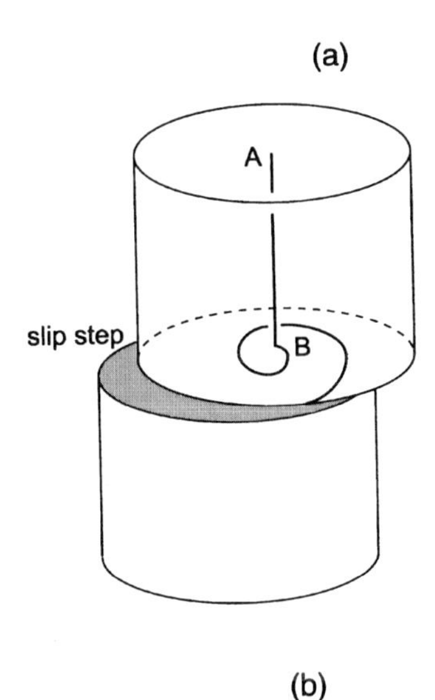

Figure 8.4 Single ended Frank–Read source. (a) Dislocation lying partly in slip plane *CEF*. (b) Formation of a slip step and spiral dislocation by rotation of *BC* about *B*.

The well-known *Frank–Read source* is an extension of the above mechanism to a dislocation line held at each end as illustrated in Fig. 8.5. The segment *AB* has the slip plane indicated in Fig. 8.5(a), i.e. its Burgers vector lies in this plane, and is held at both ends by an unspecified barrier, which may be dislocation intersections or nodes, composite jogs, precipitates, etc. An applied resolved shear stress τ exerts a force τb per unit length of line and tends to make the dislocation bow out as described in section 4.4: the radius of curvature R depends on the stress according to equation (4.30). Thus, as τ increases, R decreases and the line bows out until the minimum value of R is reached at the position illustrated in Fig. 8.5(c), where now the slip plane is represented by the plane of the paper. Here, R equals $L/2$, where L is the length of AB, and, with $\alpha = 0.5$, the stress is

$$\tau_{\text{max}} = Gb/L \tag{8.9}$$

As the line continues to expand at this stress, R increases and so the dislocation becomes unstable, for equation (4.30) cannot be satisfied unless τ is reduced. The subsequent events are shown in Fig. 8.5(d)–(f). The dislocation forms a large kidney-shaped loop, and the segments m and n annihilate on meeting. This occurs because m and n, which move in opposite directions under the same stress, have the same Burgers vector but opposite line sense. The result is a large outer loop, which continues to expand, and a regenerated dislocation AB, which repeats the process.

Figure 8.5 and equation (4.30) neglect the orientation of the Burgers vector in the slip plane: as explained in section 4.5, the bowing line usually adopts in practice a shape elongated in the Burgers vector direction. An excellent example of a Frank–Read source is illustrated in Fig. 8.6. The dislocation is held at each end by other parts of the dislocation network. These are not in the $\{111\}$ plane of the loops and are consequently out of focus. The dislocation lines tend to lie along the $\langle 110 \rangle$ directions in which they have a minimum energy, indicating the strong anisotropy of this material.

The expression for τ_{max} shows that dislocation segments of length $L \simeq 10^4 b$ can act as Frank–Read sources at applied stresses close to the yield stress. Considerable multiplication probably occurs by the Frank–Read mechanism, but additional processes must occur to account for the experimental observations. For example, the Frank–Read source does not explain the broadening of slip bands which is a common feature of the early stages of deformation of some crystals, e.g. iron and lithium fluoride, and is described in the following.

8.6 Multiplication by Multiple Cross Glide

An example of the multiplication of dislocations associated with the initiation and broadening of a slip band is shown in Fig. 8.7, from work by Gilman and Johnston on lithium fluoride. The dislocations are revealed by the etch pit technique (section 2.2); a succession of etching

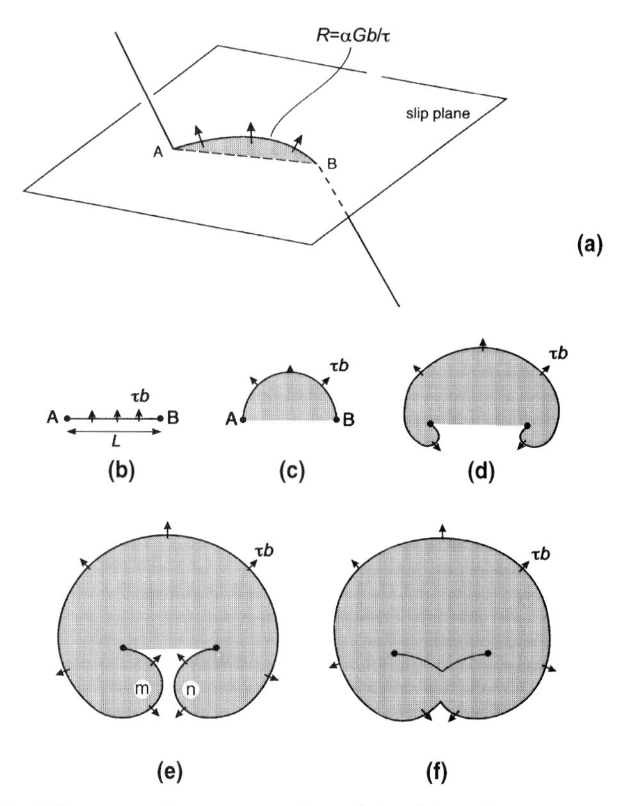

Figure 8.5 Diagrammatic representation of the dislocation movement in the Frank–Read source. Unit slip has occurred in the shaded area. (After Read (1953), *Dislocations in Crystals*, McGraw-Hill.)

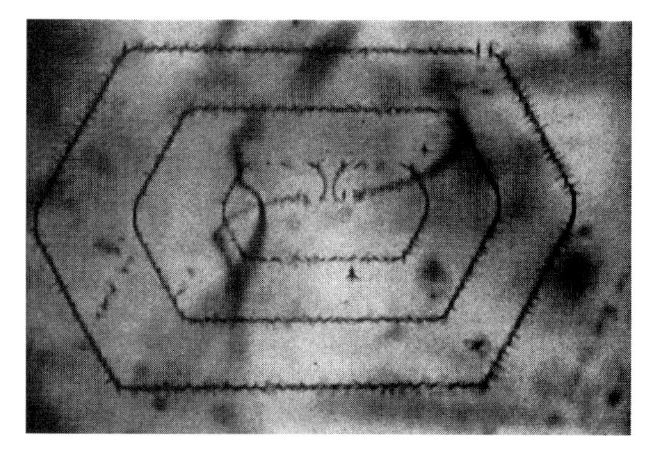

Figure 8.6 Frank–Read source in a silicon crystal. The dislocations have been revealed by the decoration technique described in section 2.3. (From Dash (1957), *Dislocation and Mechanical Properties of Crystals*, Wiley.)

and deformation treatments revealed the growth of the slip band. Initially (Fig. 8.7(a)), the slip band started as a single loop of dislocation and the two points of emergence are revealed by the two large pits. Deformation resulted in the formation of dislocations on each side of the original loop. Since the band of dislocation etch pits has a finite width, it follows that the dislocations do not lie on the same glide plane, but on a set of parallel glide planes. Widening of a glide band is shown in Fig. 8.7(b); after the first deformation the band was W_1 wide and after the second deformation W_2 wide. The dislocation density is approximately uniform throughout the band.

The widening can be accounted for by a process called *multiple cross glide*. In principle, it is the same as the process illustrated in Fig. 3.9. Thus, a screw dislocation lying along AB can cross glide onto position CD on a parallel glide plane. If the stress is greater on the primary plane the long jogs AC and BD are relatively immobile. However the segments lying in the primary slip planes will be free to expand and can each operate as a Frank–Read source. A similar mechanism can result from cross-slipped segments in particle strengthened alloys (Fig. 7.12). When cross glide can occur readily, the Frank–Read sources may never complete a cycle and there will be one continuous dislocation line lying on each of many parallel glide planes, connected by jogs. Thus, it is possible for a single dislocation loop to expand and multiply in such a way that the slip spreads from plane to plane, so producing a wide slip band. Multiple cross glide is a more effective mechanism than the simple Frank–Read source since it results in a more rapid multiplication.

8.7 Multiplication by Climb

Two mechanisms, involving climb, which increase the total dislocation length, have been described already; (a) the expansion of a prismatic loop and (b) the spiralling of a dislocation with a predominantly screw character (section 3.7). A *regenerative multiplication* known as the *Bardeen–Herring source* can occur by climb in a similar way to the Frank–Read mechanism illustrated in Fig. 8.5. Suppose that the dislocation line AB in Fig. 8.5 is an edge dislocation with Burgers vector perpendicular to the plane of the paper in Fig. 8.5(b)–(f), i.e. the plane shown in Fig. 8.5(a) is not the slip plane of AB but contains its extra half-plane. If the dislocation line is held at A and B, the presence of an excess concentration of vacancies will cause the dislocation to climb, the force causing climb being given by equation (4.40). An additional condition to the normal Frank–Read source must be satisfied if this process is to be regenerative; the anchor points D and D' must end on dislocations with screw character. If this were not so, one cycle of the source, Fig. 8.5(e), would remove the extra half-plane without creating a new dislocation along AB. Bardeen–Herring sources have been observed experimentally, as illustrated in Fig. 8.8 for an aluminium–3.5 per cent magnesium alloy quenched from 550 °C into silicone oil. The large concentration of excess vacancies has resulted in the formation of four concentric loops. Each loop represents

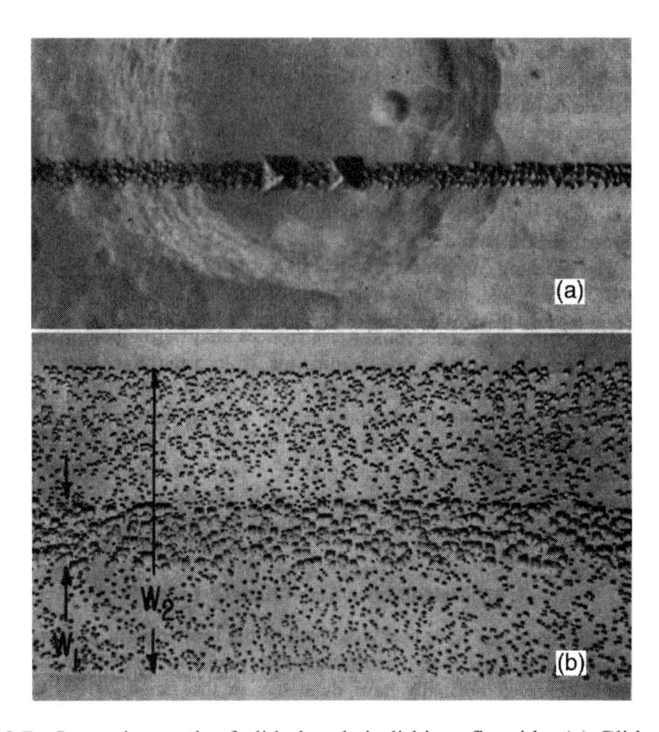

Figure 8.7 Lateral growth of glide bands in lithium fluoride. (a) Glide band formed at a single loop. Large pits show the position of the loop. Small pits show that new dislocations lie on both sides of the glide plane of the original loop. (b) Widening of a glide band. (From Gilman and Johnston, *Solid State Physics* **13**, 147, 1962.)

the removal of a plane of atoms. More complicated source arrangements have also been observed in this and other alloys (see 'Further Reading').

For a dislocation line bowing by slip, the work done by the applied shear stress balances in equilibrium the increase in line energy (section 4.5). Similarly for bowing by climb, the increase in line energy is balanced by the gain in energy due to the loss or creation of point defects (section 4.7). Hence, from equations (4.30) and (4.40), the chemical force per unit length which produces a radius of curvature R by climb is

$$f = \frac{bkT}{\Omega} \ln\left(\frac{c}{c_0}\right) = \frac{\alpha G b^2}{R} \tag{8.10}$$

By analogy with the analysis for the Frank–Read source, the critical vacancy supersaturation required to operate a Bardeen–Herring source of length L is therefore

$$\ln\left(\frac{c}{c_0}\right) = \frac{2\alpha G b \Omega}{LkT} \tag{8.11}$$

where $\alpha \simeq 0.5$. Consider a typical metal with $\Omega = b^3$ and $Gb^3 = 5\,\text{eV}$. At $T = 600\,\text{K}$, $kT \simeq 0.05\,\text{eV}$ and equation (8.11) gives $\ln(c/c_0) \simeq 100\,b/L$.

Thus, for $L = 10^3 b$, $c/c_0 = 1.11$, and for $L = 10^4 b$, $c/c_0 = 1.01$. These are small supersaturations in comparison with those met in rapid quenches, for which c/c_0 can be as large as $\sim 10^4$, and Bardeen–Herring sources can probably operate throughout the period of fast cooling.

8.8 Grain Boundary Sources

An important source of dislocations during plastic flow in poly-crystalline materials is the boundary region between the grains. Many investigations have shown that dislocations can be emitted from grain boundaries, and it is probable that they subsequently multiply within grains by the multiple cross-slip process (section 8.6). Several mechanisms may be involved in the emission. In low-angle boundaries (Chapter 9), the segments of dislocation networks forming the boundary can act as Frank–Read (or Bardeen–Herring) sources. Another possible source is an adsorbed lattice dislocation which is stabilised at the boundary in the stress-free state by the boundary structure. The resulting boundary ledge can provide sites for dislocation nucleation under stress, for the reduction in boundary area at nucleation reduces the activation energy. The dislocations generated by sources within a grain produce large stress concentrations when piled up at the grain boundary (section 10.9): these can trigger boundary sources at relatively low applied stress. It has also been established that migrating grain boundaries generate dislocations in the lattice they pass through. Gleiter *et al.* (1980) have proposed that these dislocations are nucleated by accidental mispacking of atoms at the boundary when one grain grows at the expense of another. This process provides a source for dislocations in crystals grown in the solid state, such as those formed by recrystallisation and phase transformations.

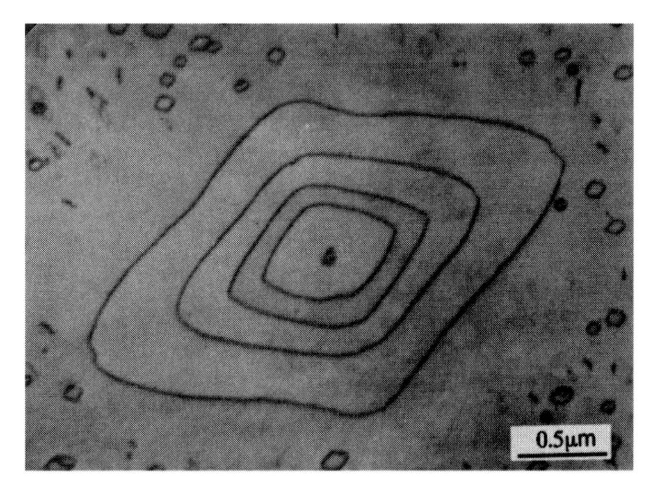

Figure 8.8 Transmission electron micrograph of concentric loops formed at a climb source in aluminium–3.5 per cent magnesium alloy quenched from 550 °C. (From Westmacott, Barnes and Smallman, *Phil. Mag.* **7**, 1585, 1962.)

Further Reading

Ashby, M. F. and Johnson, L. (1969) 'On the generation of dislocations at misfitting particles in a ductile matrix', *Phil. Mag.* **20**, 1009.

Boyd, J. D. and Edington, J. W. (1971) 'Dislocation climb sources and vacancy loops in quenched Al-2.5%Cu', *Phil. Mag.* **23**, 633.

Brown, L. M. and Woolhouse, G. R. (1970) 'The loss of coherency of precipitates and the generation of dislocations', *Phil. Mag.* **21**, 329.

Cottrell, A. H. (1953) *Dislocations and Plastic Flow in Crystals*, Oxford University Press.

Elbaum, C. (1959) 'Substructure in crystals grown from the melt', *Progress in Metal Physics*, vol. 8, p. 203, Pergamon, London.

Frank, F. C. (1952) 'Crystal growth and dislocations', *Adv. Physics*, **1**, 91.

Gilman, J. J. and Johnston, W. G. (1957) 'Origin and growth of glide bands', *Dislocations and Mechanical Properties of Crystals*, p. 116, Wiley.

Gleiter, H., Mahajan, S. and Bachman, K. J. (1980) 'The generation of lattice dislocations by migrating boundaries', *Acta Met.* **28**, 1603.

Hirth, J. P. and Lothe, J. (1982) *Theory of Dislocations*, Wiley.

Jackson, P. J. (1983) 'The role of cross-slip in the plastic deformation of crystals', *Mater. Sci. Eng.* **57**, 39.

Kuhlmann-Wilsdorf, D. (1958) 'Origin of dislocations', *Phil. Mag.* **3**, 125.

Li, J. C. M. (1981) 'Dislocation sources', *Dislocation Modelling of Physical Systems*, p. 498, Pergamon.

Mutaftschiev, B. (1980) 'Crystal Growth and Dislocations', *Dislocations in Solids*, vol. 5, p. 57, North-Holland.

Pamplin, B. R. (Ed.) (1975) *Crystal Growth*, Pergamon.

9 Dislocation Arrays and Crystal Boundaries

9.1 Plastic Deformation, Recovery and Recrystallisation

Plastic deformation of crystalline materials leads to the formation of three-dimensional arrays or distributions of dislocations which are characteristic of (1) crystal structure of the material being deformed, (2) temperature of deformation, (3) strain and (4) strain rate. Additionally, such features as grain boundaries, precipitates, and stacking fault energy affect the distribution of the dislocations. Two distributions are illustrated in Fig. 9.1, which shows the effect of deforming pure iron specimens at 20 °C and −135 °C, respectively. Deformation at 20 °C has resulted in the formation of dense *tangles of dislocations* arranged in *walls* surrounding regions or *cells* almost free from dislocations. The *cell size* reaches a limit in the early stages of deformation and changes only slightly thereafter. The cell walls tend to orient themselves in certain crystallographic directions. Deformation at −135 °C produces a much more homogeneous distribution of dislocations.

The hardening of crystals during plastic deformation is due to the increase in dislocation density and the mutual interaction between dislocations (Chapter 10). Most of the work done by the external load

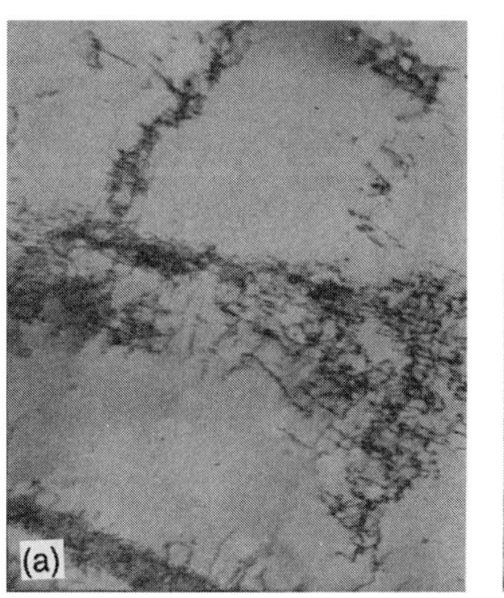

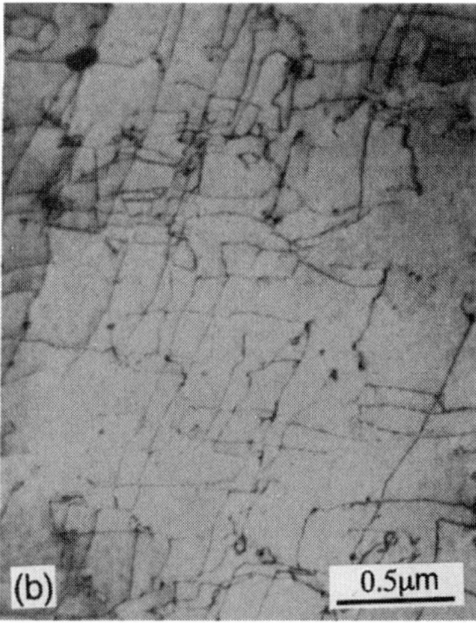

Figure 9.1 Dislocation arrangements produced by plastic deformation of iron. (a) Dense tangles and dislocation free cells formed after 9 per cent strain at 20 °C. (b) Uniform array of straight dislocations formed after 7 per cent strain at −135 °C. (From Keh and Weissmann (1963), *Electron Microscopy and Strength of Crystals*, p. 231, Interscience.)

during plastic deformation is dissipated as heat, but a small proportion is retained in the material as *stored energy*. This is accomplished by an increase in elastic strain energy resulting from the increase in dislocation density. An additional, small amount of energy is stored when point defects are produced during plastic deformation. The energy only remains stored within a crystal if the temperature is sufficiently low for the atoms to be effectively immobile, i.e. $T \lesssim 0.3T_m$: the plastic deformation that meets this requirement is known as *cold-work*. The stored energy can be released if the dislocations rearrange themselves into configurations of lower energy. These are called low-angle boundaries and can be represented by a uniform array of one, two, three or more sets of dislocations (sections 9.2 and 9.3). They separate regions of the crystal which differ in orientation by $\lesssim 5°$. A considerable amount of energy is released by the local rearrangement of the dislocations in the *tangles* and further release of energy occurs when low-angle boundaries are formed. Both these processes involve climb of the dislocations and will occur, therefore, only when there is sufficient thermal activation to allow local and long-range diffusion of point defects, i.e. $T \gtrsim 0.3T_m$. These changes are accompanied by a pronounced softening of the dislocation-hardened crystal. The process is called *recovery* and it occurs when a plastically deformed crystal is heated to moderate temperatures. The later stages of the recovery process in which low-angle boundaries are formed is called *polygonisation*.

When a heavily cold-worked metal is heated above a critical temperature new grains relatively free from dislocations are produced in the 'recovered' structure, resulting in a process called *recrystallisation*. *Large-angle grain boundaries* with a misorientation $\gtrsim 10°$ are produced. The sequence of photographs in Fig. 9.2 illustrates the change from a heavily deformed structure with a uniform distribution of tangled dislocations to a recrystallised structure containing large and small angle grain boundaries. The grain structure is small immediately after recrystallisation but grows progressively with longer annealing times and higher temperatures. This is called *grain growth* and results in a small reduction in energy because the total area of grain boundary is reduced. Some recent advances in our knowledge of the dislocation content of high-angle grain boundaries are described in section 9.6.

The basic process in the formation of low-angle boundaries is illustrated in its most simple form in Fig. 9.3. Consider a crystal which is bent about its z-axis: the resulting dislocations will be distributed randomly on the glide planes and there will be an excess of edge dislocations of one sign, as illustrated in Fig. 9.3(a). The energy of the crystal can be reduced by rearranging the dislocations into a vertical wall to form a symmetrical tilt boundary, Fig. 9.3(b) (section 9.2). Alternatively, imagine that a thin wedge-shaped section, symmetrical about the plane $x = 0$, is cut from the perfect crystal $ABCD$, Fig. 9.4(a), and that the two cut faces are then placed together as in Fig. 9.4(b). This is geometrically the same as the tilt boundary in Fig. 9.3(b) and illustrates an

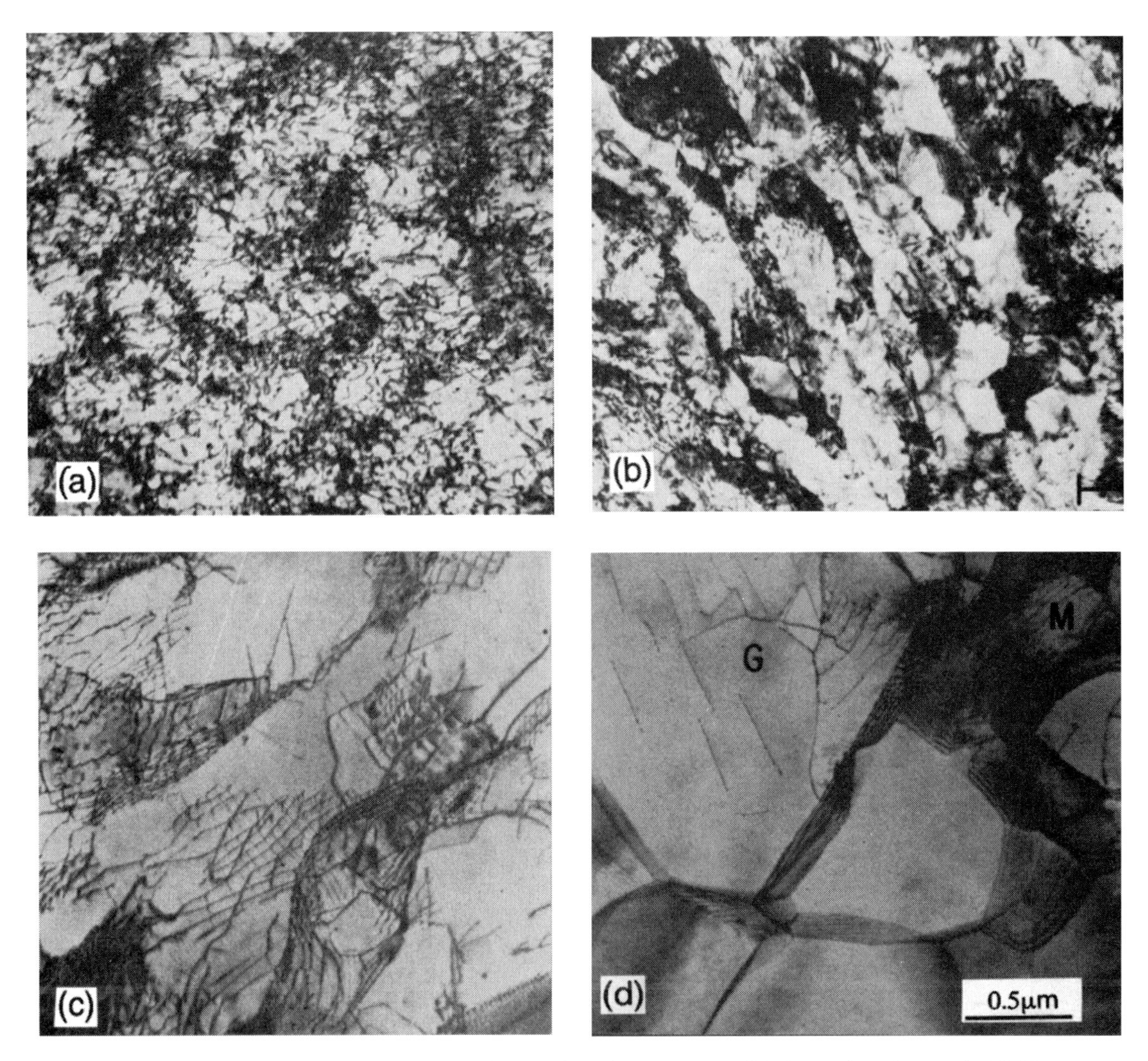

Figure 9.2 Transmission electron micrographs illustrating the structure of deformed and annealed 3.25 per cent silicon iron. (a) Approximately uniform distribution of dislocations in a crystal rolled 20 per cent. (b) Formation of small sub-grains in rolled material annealed 15 min at 500 °C. (c) As for (b) annealed 15 min at 600 °C. (d) As for (b) annealed 30 min at 600 °C. (From Hu, *Trans. Met. Soc. AIME* **230**, 572, 1964.)

important feature of small angle boundaries, namely that they have *no long range stress field*. The strain is localised in the region around the dislocations.

The example illustrated in Fig. 9.4 is a boundary with *one degree of freedom* since the axis of relative rotation is fixed in the same crystallographic direction in both crystals and the boundary is a mirror plane of the bicrystal. In the general case illustrated in Fig. 9.5, the boundary is formed by cutting a single crystal on two arbitrary planes, removing the wedge of material between the cuts, and placing the grains together

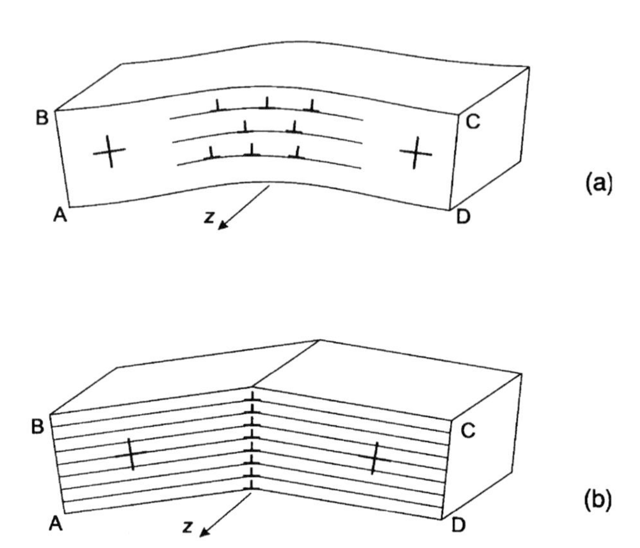

Figure 9.3 Formation of a low-angle boundary. The orientation of the slip planes is denoted by a cross. (a) Bent crystal with random dislocations. (b) Rearrangement of dislocations to form symmetrical tilt boundary. Both climb and glide are required to produce this boundary.

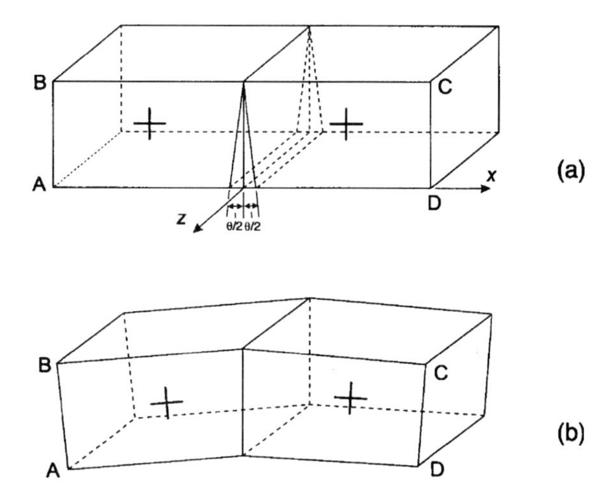

Figure 9.4 Geometry of the formation of a symmetrical tilt boundary. The relative rotation of the two crystals is produced by cutting a wedge from the perfect crystal.

with a twist. The boundary has *five* degrees of freedom, three corresponding to the rotation of one grain with respect to the other about three perpendicular axes, and two due to the rotation of the boundary with respect to the two grains individually.

The requirement that there are no long-range stress fields places severe conditions on the dislocation geometry of low-angle boundaries and this is discussed in sections 9.3 and 9.4. The regular dislocation arrays observed after recovery are true low-angle boundaries. They can adopt the form typified in Fig. 9.3 because there are no external constraints on the two grains. These boundaries are unlikely to form during plastic deformation, however, because the material in regions away from the boundary will be unable to adopt, in general, the orientation required: long-range strain fields then result. The tangled arrangements in Fig. 9.1 may well be of this form.

9.2 Simple Dislocation Boundaries

The simplest boundary is the *symmetrical tilt boundary* (Figs 2.3(b), 9.3 and 9.4). The atomic mismatch at the boundary is accommodated by regions of good and bad fit: the latter are dislocations. In a simple cubic lattice with edge dislocations $\mathbf{b} = [100]$, the boundary will consist of a sheet of equally spaced dislocations lying parallel to the z-axis; the plane of the sheet will be the symmetry plane $x = 0$, i.e. (100). The extra half-planes of the dislocations terminate at the boundary from the left- and right-hand sides alternately. The crystals on each side of the boundary are rotated by equal and opposite amounts about the z-axis and differ in orientation by the angle θ (Fig. 2.3(b)). If the spacing of the dislocations is D, then:

$$\frac{b}{2D} = \sin\frac{\theta}{2} \tag{9.1}$$

and for small values of θ (in radians)

$$\frac{b}{D} \sim \theta \tag{9.2}$$

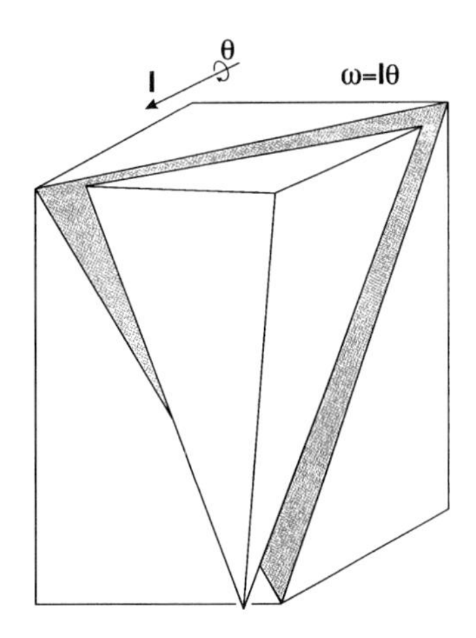

Figure 9.5 Formation of a general grain boundary with five degrees of freedom. Three degrees arise from the rotation about **l** of one grain with respect to the other: vector **l** has three components.

If $\theta = 1^\circ$ and $b = 0.25\,\text{nm}$ the spacing between dislocations will be 14 nm. When the spacing is less than a few lattice spacings, say five, $\theta \sim 10^\circ$ and the individual identity of the dislocations is in doubt, the boundary is called a *large-angle boundary*. Experimental evidence for tilt boundaries has been presented in Fig. 2.3. Many such observations of boundaries have been reported using decoration and thin film transmission electron microscope techniques.

A *more general tilt boundary* is illustrated in Fig. 9.6. By introducing a second degree of freedom the boundary plane can be rotated about the z-axis. The boundary plane is no longer a symmetry plane for the two crystals but forms an angle ϕ with the mean [100] direction of the two grains. Two sets of atom planes end on the boundary and the geometrical conditions require the boundary to consist of two sets of uniformly spaced edge dislocations with mutually perpendicular Burgers vectors, $\mathbf{b}_1$ and $\mathbf{b}_2$. Since $EC = AC\cos(\phi - \theta/2)$ and $AB = AC\cos(\phi + \theta/2)$, the number of (100) planes intersecting EC is $(AC/b_1)\cos(\phi - \theta/2)$ and the number intersecting AB is $(AC/b_1)\cos(\phi + \theta/2)$. The number terminating at the boundary is therefore

$$(AC/b_1)[\cos(\phi - \theta/2) - \cos(\phi + \theta/2)] \simeq (AC/b_1)\theta\sin\phi$$

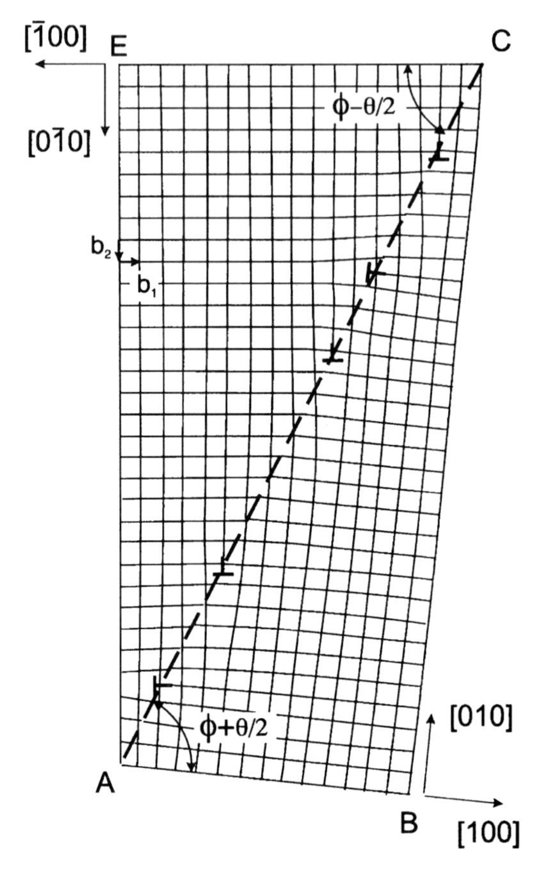

Figure 9.6 Tilt boundary with two degrees of freedom. This is the same boundary as in Fig. 9.3 except that the plane of the boundary makes an arbitrary angle ϕ with the mean of the (010) planes in the two grains. The atom planes end on the boundary in two sets of evenly spaced edge dislocations. (After Read (1953), *Dislocations in Crystals*, McGraw-Hill.)

if $\theta \ll 1$. Similarly, the number of terminating (010) planes is $(AC/b_2)\theta \cos\phi$. The spacings along the boundary of the two sets of dislocations are therefore

$$D_1 = \frac{b_1}{\theta \sin\phi} \quad D_2 = \frac{b_2}{\theta \cos\phi} \tag{9.3}$$

A simple boundary formed from a *cross grid of pure screw dislocations* is illustrated in Fig. 9.7. A single set of screw dislocations has a long-range stress field and is therefore unstable but the stress field is cancelled by the second set of screw dislocations. The two sets of equally spaced parallel dislocations lie in the boundary which also lies in the plane of the diagram. They produce a rotation about an axis normal to the boundary of one-half of the crystal with respect to the other. Such

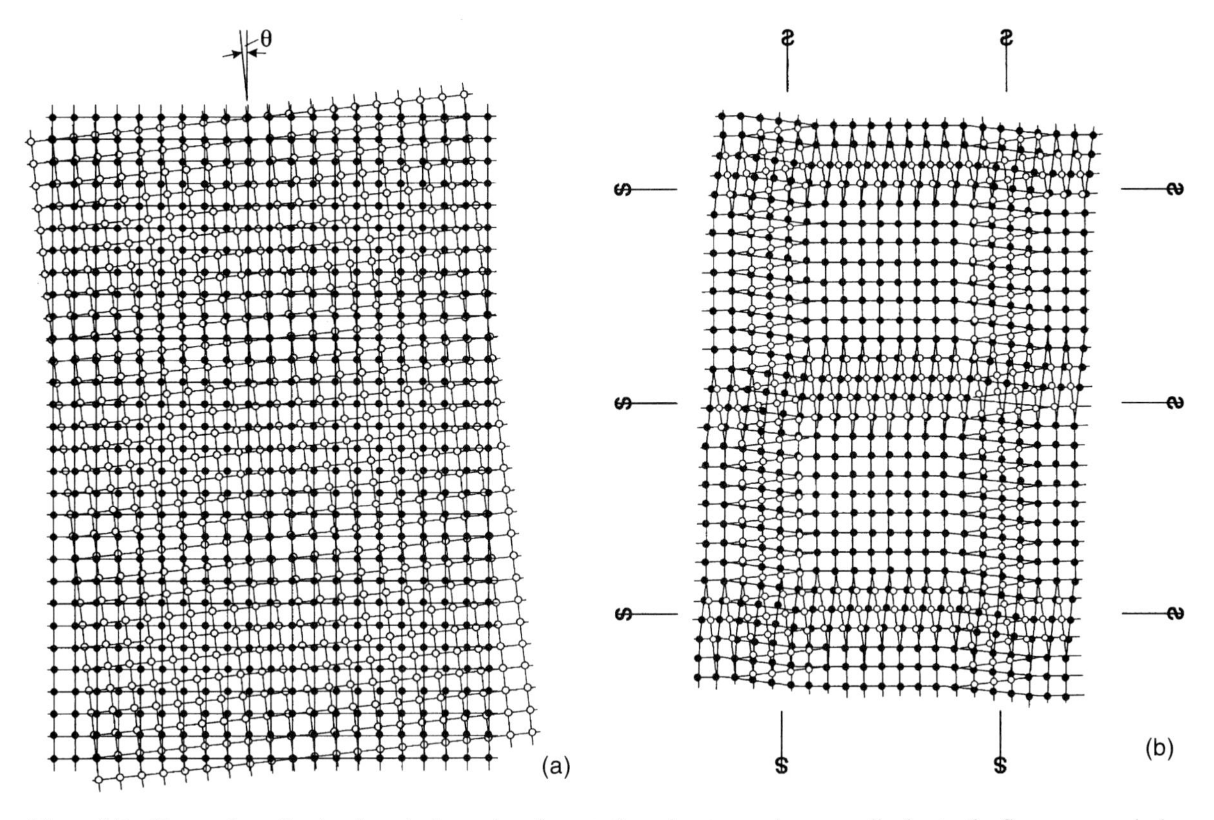

Figure 9.7 Formation of a simple twist boundary by rotation about an axis perpendicular to the figure: open circles represent atoms just above the boundary and solid circles those just below. (a) Atom positions resulting from rigid rotation through angle θ. (b) Accommodation of mismatch in (a) by localised distortions, seen to be two sets of parallel screw dislocations labelled *S-S*. (After Read (1953) *Dislocations in Crystals*, McGraw-Hill.)

a boundary is called a *twist boundary*. The spacing between dislocations in each set is:

$$D = \frac{b}{\theta}$$

9.3 General Low-angle Boundaries

It was explained in section 9.1 that the general planar boundary has five degrees of freedom. Three correspond to the difference in orientation of the grains, for if $\mathbf{l}$ is a unit vector parallel to the axis of relative rotation of the grains (Fig. 9.5), the rotation may be represented by the vector $\boldsymbol{\omega} = \mathbf{l}\theta$, which has three independent components. The orientation of the boundary may be specified by a unit vector $\mathbf{n}$ normal to the boundary plane. This introduces only two degrees of freedom, one for each crystal, for rotation of the boundary about $\mathbf{n}$ is not significant. The tilt and twist boundaries can now be defined as follows: in a *pure tilt* boundary, $\mathbf{l}$ and $\mathbf{n}$ are at right-angles and in a *pure twist* boundary $\mathbf{l} = \mathbf{n}$. In the examples in section 9.2 the tilt boundaries are described by arrays of edge dislocations and the twist boundaries by screw dislocations, but these are special cases. For example, if the dislocations in the crystals of Fig. 2.3(b) have Burgers vectors $\frac{1}{2}\langle 111\rangle$ rather than $\langle 100\rangle$, the pure tilt boundary on (100) would consist of a wall of dislocations of mixed character, the edge components all having the same sign and the screw components alternating in sign; e.g. the Burgers vectors may alternate between $\frac{1}{2}[111]$ and $\frac{1}{2}[1\bar{1}\bar{1}]$. The screw components contribute to the energy of the tilt boundary but not to its misfit. Similarly, the pure twist boundary will, in general, contain dislocations with (alternating) edge character. The general boundary with **l** neither perpendicular nor parallel to **n** has *mixed* tilt and twist character.

The orientation of the crystals and the boundary formed in the way illustrated in Figs 9.4 and 9.5 can be defined by the parameters **l**, **n** and θ. Frank (1950) has derived a relation which may be used to determine the arrangements of dislocations which will produce such a boundary, or conversely the orientation of a boundary produced by a given set of dislocations or *dislocation network*. For a general boundary, Frank's relation is

$$\mathbf{d} = (\mathbf{r} \times \mathbf{l})2\sin\frac{\theta}{2}$$

or (9.4)

$$\mathbf{d} = (\mathbf{r} \times \mathbf{l})\theta$$

for small values of θ. The vector **r** represents an arbitrary vector lying in the plane of the boundary which contains the dislocation network, and **d** is the sum of the Burgers vectors of all the dislocations intersected by **r**, i.e.

$$\mathbf{d} = \sum_i N_i \mathbf{b}_i \tag{9.5}$$

where N_i is the number of dislocations of Burgers vectors $\mathbf{b}_i$ cut by **r**. Relation (9.4) is derived by application of the Burgers circuit construction (section 1.4). Consider a closed circuit starting at the end point of **r**, passing through one grain to the start point of **r** and returning back through the second grain. In a perfect reference crystal, the circuit starts at the end point of **r** and finishes at the end point of the vector $\mathbf{r}'$ obtained from **r** by the rotation $\boldsymbol{\omega} = \mathbf{l}\theta$. **d** equals the closure failure $\mathbf{r} - \mathbf{r}'$ of the circuit which, for small θ, equals $\mathbf{r} \times \boldsymbol{\omega} = (\mathbf{r} \times \mathbf{l})\theta$.

A number of points should be noted regarding this relation. (a) It applies only to boundaries which are essentially flat and have no long-range stress field, i.e. the elastic distortion is restricted to the region close to the dislocations. (b) The formula does not uniquely determine the dislocations present, or their pattern, for a given orientation of crystal and boundary. Thus, a variety of possibilities may arise and the most probable will be the one of lowest energy. (c) The density of a given set of dislocations in a boundary is directly proportional to θ (for small θ).

(d) Each set of dislocation lines will be straight, equally spaced and parallel even for a boundary containing several sets of dislocations with different Burgers vectors.

A general boundary requires three sets of dislocations with three non-coplanar Burgers vectors, so that the boundaries formed from one or two sets are restricted. Frank's formula can be applied to analyse all cases, either to determine the possible dislocation arrangements if **n**, **l** and θ are known or to find the orientation, etc. if the dislocation content is specified. This is illustrated here by simple examples. (Examples of more detailed cases may be found in Chap. 19 of Hirth and Lothe, 1982.)

If a boundary contains only *one* set of dislocations, each having Burgers vector **b**, equations (9.4) and (9.5) become

$$N\mathbf{b} = (\mathbf{r} \times \mathbf{l})\theta \tag{9.6}$$

Clearly **b** is perpendicular to **r** and **l**, and since the direction of **r** in the boundary is arbitrary, **b** is parallel to **n** and **l** is perpendicular to **n**. When **r** is chosen parallel to **l**, $\mathbf{r} \times \mathbf{l} = 0$ and no dislocations are intersected by **r**, showing that the dislocations are parallel to the rotation axis **l**. From these observations, the boundary is a pure tilt boundary consisting of pure edge dislocations. If **r** is taken perpendicular to **l** and (obviously) **n**, then $\mathbf{r} = (\mathbf{l} \times \mathbf{n})r$, where $r = |\mathbf{r}|$ is the length of **r**. Equation (9.6) then becomes

$$N\mathbf{b} = [(\mathbf{l} \times \mathbf{n}) \times \mathbf{l}]r\theta$$

i.e.

$$N\mathbf{b} = \mathbf{n}r\theta$$

Since $\mathbf{b} = b\mathbf{n}$, the spacing of the dislocations is

$$D = \frac{r}{N} = \frac{b}{\theta}$$

in agreement with equation (9.2). The simple tilt boundary of Figs 9.4 and 2.3(b) with $\mathbf{l} = [001]$ and $\mathbf{n} = [100]$ is of this type, for when **r** lies in the [001] direction, $\mathbf{r} \times \mathbf{l}$ is zero, demonstrating that the dislocations are all parallel to [001]. When **r** is in the [010] direction $\mathbf{r} \times \mathbf{l}$ equals $\mathbf{r}[100]$, showing that if the dislocations have Burgers vectors $\mathbf{b} = [100]$, their spacing is b/θ.

Consider next a boundary containing *two* sets of dislocations with Burgers vectors $\mathbf{b}_1$ and $\mathbf{b}_2$, which are not parallel. Frank's formula is

$$N_1\mathbf{b}_1 + N_2\mathbf{b}_2 = (\mathbf{r} \times \mathbf{l})\theta \tag{9.7}$$

Taking the scalar product of both sides with $\mathbf{b}_1 \times \mathbf{b}_2$ gives

$$(\mathbf{r} \times \mathbf{l}) \cdot (\mathbf{b}_1 \times \mathbf{b}_2) = 0$$

i.e.

$$\mathbf{r} \cdot [\mathbf{l} \times (\mathbf{b}_1 \times \mathbf{b}_2)] = 0 \tag{9.8}$$

Since the direction of **r** in the boundary is arbitrary, this condition is satisfied whenever $[\mathbf{l} \times (\mathbf{b}_1 \times \mathbf{b}_2)]$ is parallel to **n**; i.e. $(\mathbf{b}_1 \times \mathbf{b}_2)$ and **l** lie in

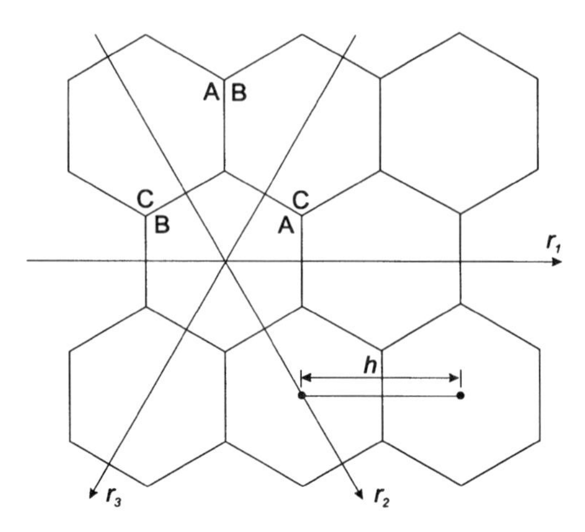

Figure 9.8 Hexagonal network of perfect dislocations ($\mathbf{b} = \frac{1}{2}\langle 110 \rangle$) on a $\{111\}$ plane in a face-centred cubic lattice.

the boundary plane. This defines a pure tilt boundary. With $\mathbf{r}$ chosen parallel to $\mathbf{l}$, the right-hand side of equation (9.7) is zero and so both sets of dislocations are parallel to $\mathbf{l}$. Their density is found by setting $\mathbf{r}$ perpendicular to $\mathbf{l}$, i.e. $\mathbf{r} = (\mathbf{l} \times \mathbf{n})r$, and equation (9.7) becomes

$$N_1\mathbf{b}_1 + N_2\mathbf{b}_2 = \mathbf{n}r\theta \tag{9.9}$$

The boundary of Fig. 9.6 has this form. From the figure, $\mathbf{b}_1 = [100]$, $\mathbf{b}_2 = [0\bar{1}0]$ and $\mathbf{n} = [\sin\phi, -\cos\phi, 0]$. Substitution into equation (9.9) gives $D_1 = b/\theta \sin\phi$ and $D_2 = b/\theta \cos\phi$, where b is the lattice parameter, in agreement with the result (9.3).

Condition (9.8) is also satisfied when $\mathbf{l}$ is parallel to $\mathbf{b}_1 \times \mathbf{b}_2$, i.e. $\mathbf{l} = (\mathbf{b}_1 \times \mathbf{b}_2)/|\mathbf{b}_1 \times \mathbf{b}_2|$. Equation (9.7) then becomes

$$N_1\mathbf{b}_1 + N_2\mathbf{b}_2 = [\mathbf{b}_1(\mathbf{r} \cdot \mathbf{b}_2) - \mathbf{b}_2(\mathbf{r} \cdot \mathbf{b}_1)]\theta/|\mathbf{b}_1 \times \mathbf{b}_2| \tag{9.10}$$

It may be shown that the two sets of dislocations are not parallel. For the pure twist boundary, $\mathbf{n}$ equals $\mathbf{l}$ and the spacings simplify to $\mathbf{b}_1/\theta$ and $\mathbf{b}_2/\theta$ respectively. In the special case when $\mathbf{b}_1$ is perpendicular to $\mathbf{b}_2$, the line directions are $\mathbf{b}_1$ and $\mathbf{b}_2$, respectively, so that the dislocations are pure screw. This corresponds to the boundary of Fig. 9.7.

Finally, consider the following boundary in a face-centred cubic lattice. The dislocation network is an example of that defined by equation (9.10), except that at the point where a dislocation of one set intersects one from the other, they react to form a third. This results in an hexagonal net of lower line length (and energy), as illustrated in Fig. 9.8. The Burgers vectors are described using the Thompson notation (section 5.4). They lie in the plane of the net, e.g. $\mathbf{AB} = \frac{1}{2}[\bar{1}10]$, $\mathbf{BC} = \frac{1}{2}[10\bar{1}]$, $\mathbf{CA} = \frac{1}{2}[0\bar{1}1]$ in plane (111), and one results from the addition of the other two; e.g. $\mathbf{AB} + \mathbf{BC}$ equals $\mathbf{AC}$. It is important that the sequence of lettering around any node is consistent. The Burgers vector of each dislocation is represented by two letters written on either

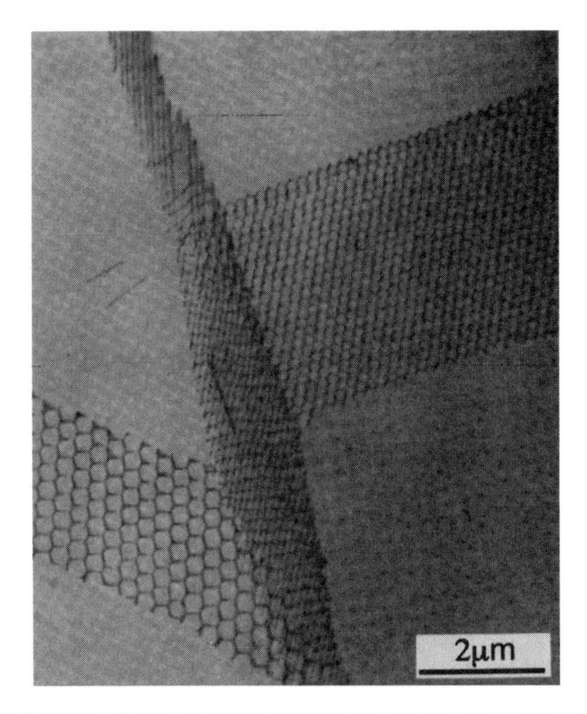

Figure 9.9 Transmission electron micrograph of extensive dislocation networks in body-centred cubic iron. Each network consists of three sets of dislocations, Burgers vectors $\frac{1}{2}\langle 111\rangle, \frac{1}{2}\langle 1\bar{1}\bar{1}\rangle$ and $\langle 100\rangle$. The plane of the networks is almost parallel to the plane of the foil. (Courtesy Dadian and Talbot-Besnard.)

side of it. The lettering should be such that any track around the node makes a continuous path, for example, *AB, BC, CA*, or *AB, CA, BC*, but not *BC, CA, BA*, or *AB, BC, AC*. Given that h is the distance between the centres of the mesh, the orientation of the plane of the net and the individual dislocations in the net can now be determined. If vector $\mathbf{r}$ is placed horizontally on the net in Fig. 9.8, i.e. $\mathbf{r}_1$, it crosses r_1/h dislocations of Burgers vector **AB**. Frank's formula becomes

$$\mathbf{AB}r_1/h = (\mathbf{r}_1 \times \mathbf{l})\theta$$

Similarly for vectors $\mathbf{r}_2$ and $\mathbf{r}_3$ shown in the figure:

$$\mathbf{BC}r_2/h = (\mathbf{r}_2 \times \mathbf{l})\theta$$
$$\mathbf{CA}r_3/h = (\mathbf{r}_3 \times \mathbf{l})\theta$$

It follows that $\mathbf{l}$ is perpendicular to the net, which therefore forms a pure twist boundary. Furthermore, since **AB** is perpendicular to $\mathbf{r}_1$ and $\mathbf{l}$, the dislocation segments perpendicular to $\mathbf{r}_1$ are pure screw in character. So, too, are those perpendicular to $\mathbf{r}_2$ and $\mathbf{r}_3$. All the dislocations are pure screws, and θ, the angle of rotation of the boundary, is b/h for small θ, where $b = |\mathbf{AB}|$. Such a net of regular hexagons cannot lie in any other plane without producing long-range stress fields in the crystal. Any

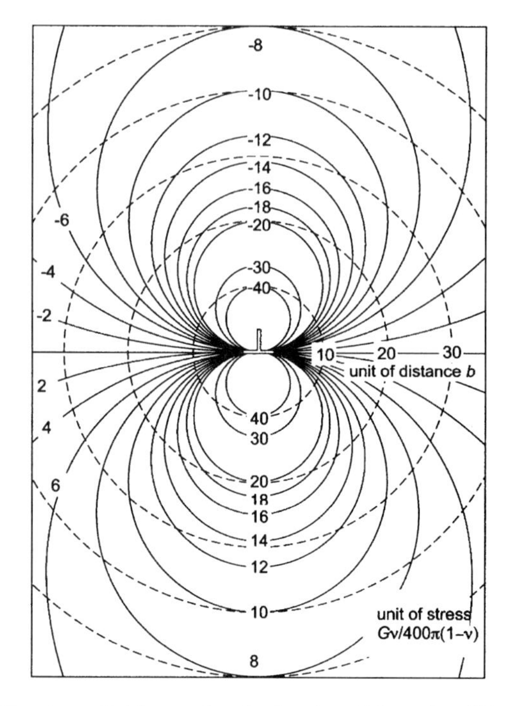

Figure 9.10 The stress field σ_{zz} of a single edge dislocation (From Li (1963), *Electron Microscopy and Strength of Crystals*, p. 713. Interscience.)

deviation from the (111) boundary plane must be accompanied by a modification of the regular hexagon structure if long-range stresses are to be avoided. The dislocations can dissociate to form extended nodes (section 7.8) whilst still satisfying Frank's formula. Similar stable networks can also arise in other structures. The arrangement of dislocations in a typical boundary in a body-centred cubic crystal is shown in Fig. 9.9.

Although low-angle boundaries are formed primarily under conditions in which dislocations can climb freely, it is possible to produce them by slip, but the geometrical conditions are very restrictive. The orientation and character of the dislocations in the boundary must have common slip Burgers vectors and lie on slip planes, and also satisfy the Frank equation.

9.4 Stress Field of Dislocation Arrays

Dislocation nets may have long- and short-range stress fields. The distribution of stress is sensitive to the arrangement, orientation and Burgers vectors of the dislocations. Boundaries satisfying Frank's condition (equation (9.4)) do not produce long-range fields. A few examples of the most elementary boundaries are sufficient to illustrate these points.

The components of the stress field of single edge and screw dislocations have been given in equations (4.16) and (4.13), respectively. The total stress field of an array is obtained by a summation of the components of the stress field of the individual dislocations sited in the array. Thus, consider a wall of edge dislocations of Burgers vector **b** making up a symmetrical tilt boundary lying in the plane $x = 0$ with the dislocations parallel to the z-axis. With one dislocation lying along the z-axis, and N others above and N below, the stress field of the boundary is given by

$$\sigma_{xx} = \frac{-Gb}{2\pi(1-\nu)} \sum_{n=-N}^{N} \frac{y_n(3x^2 + y_n^2)}{(x^2 + y_n^2)^2} \tag{9.11}$$

$$\sigma_{yy} = \frac{Gb}{2\pi(1-\nu)} \sum_{n=-N}^{N} \frac{y_n(x^2 - y_n^2)}{(x^2 + y_n^2)^2} \tag{9.12}$$

$$\sigma_{xy} = \frac{Gb}{2\pi(1-\nu)} \sum_{n=-N}^{N} \frac{x(x^2 - y_n^2)}{(x^2 + y_n^2)^2} \tag{9.13}$$

where $y_n = y - nD$ and D is the dislocation spacing. The term with $n = 0$ is the contribution of the dislocation at $y = 0$ (see equations (4.16)). The effect of the summation is for the individual contributions to tend to cancel within a distance $\sim ND$ of the origin. For example, it is readily shown from the above equations that the stresses, σ_{xx}, σ_{yy} and $\sigma_{zz} = -\nu(\sigma_{xx} + \sigma_{yy})$ at $y = D/2$ due to dislocation $n = 0$ are cancelled by those due to dislocation $n = 1$. This mutual cancellation occurs over a large part of the region $0 < y < D$, as demonstrated explicitly for σ_{zz} by comparing Figs 9.10 and 9.11. Addition of dislocations $n = -1$,

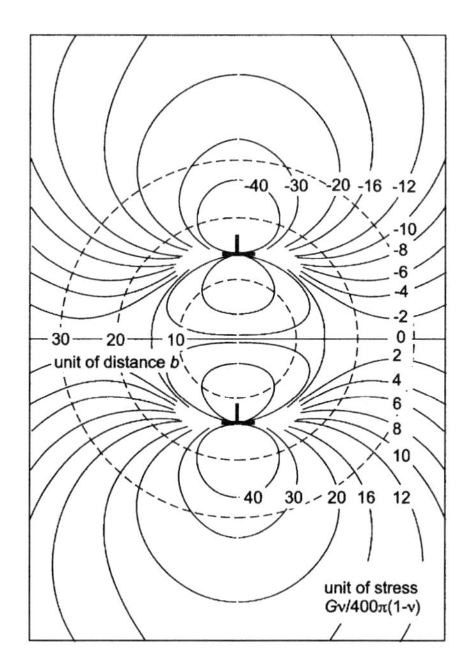

Figure 9.11 The stress field σ_{zz} of a vertical wall of two edge dislocations spaced $28b$ apart.

$n = 2$, and then $n = -2$, $n = 3$, etc., extends the region of cancellation. Exact summation of equations (9.11)–(9.13) is possible for infinite arrays, i.e. $N \to \infty$. The solution for σ_{xy} is shown in Fig. 9.12. For $|x| \gtrsim D/2\pi$, σ_{xy} is approximated by

$$\sigma_{xy} \simeq \frac{2\pi Gbx}{(1-\nu)D^2}\cos\left(\frac{2\pi y}{D}\right)\exp\left(\frac{-2\pi|x|}{D}\right) \tag{9.14}$$

so that away from the boundary, σ_{xy} decreases exponentially with x. Near the wall ($|x| \lesssim D/2\pi$), the stress is dominated by the nearest one to three dislocations (Fig. 9.12). Because the stress fields are localised at the boundary the strain is small and therefore the boundary represents a stable configuration with respect to slip. However, the dislocations in the boundary can climb, resulting in an increased separation of the dislocations and a reduction in θ. Although the infinite wall of edge dislocations has no long-range stress field, a finite wall (N not infinite) has, as implied by Fig. 9.11.

If the dislocations in the vertical wall were to be moved by slip so that the boundary made an angle ϕ with the original low-energy position, it can be shown that the wall will have a long-range stress field and a higher energy because the stress field of the individual dislocations no longer cancel each other so effectively. Such a boundary is most likely to be formed during plastic deformation. The long-range stresses can be removed by the boundary combining with a second wall of dislocations with Burgers vectors normal to the original ones as shown in the stable tilt boundary in Fig. 9.6. Similar arguments can be used to show that an infinite wall of parallel screw dislocations always has a long-range stress field, which is removed by the introduction of a second set of screw dislocations to form a cross grid as in Fig. 9.7.

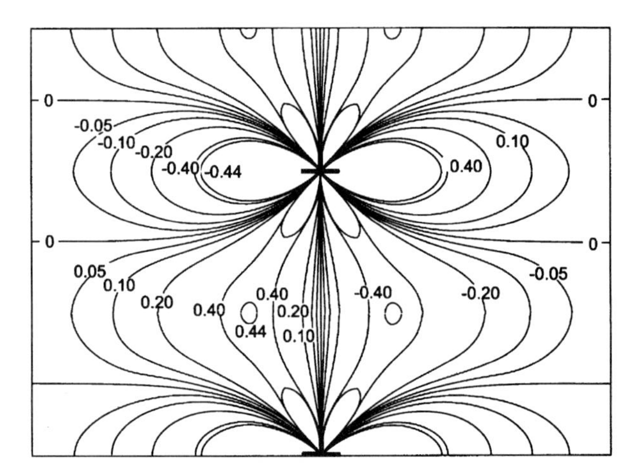

Figure 9.12 The shear stress field σ_{xy} of an infinite array of edge dislocations. Unit of stress $Gb/2(1-\nu)D$. (From Li, *Acta Metall.* **8**, 296, 1960.)

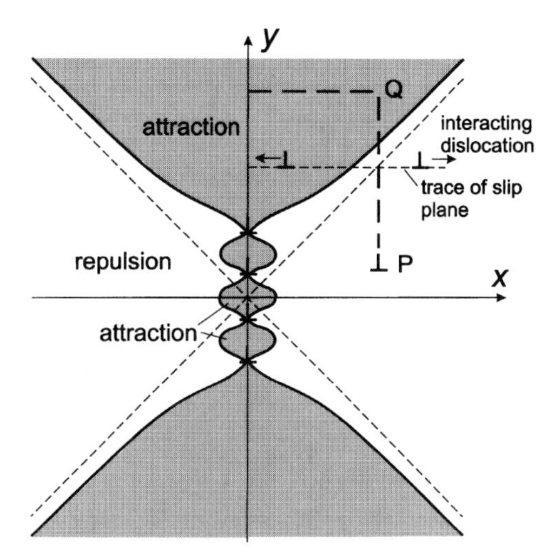

Figure 9.13 Dislocation wall containing four edge dislocations of the same sign. Similar dislocations of the same sign in parallel glide planes are attracted or repelled depending on their position.

As in the case of a single dislocation, work will be done when a dislocation is introduced into the stress field of a dislocation wall. The interaction force can be determined from equations such as (4.35) and (4.37) by using the appropriate stress summations. The problem may be complicated by the movement of the boundary dislocations during the interaction. Some qualitative features are worth noting and, for simplicity, the intersection of an edge dislocation with a wall of edge dislocations of the same sign is considered, as depicted in Fig. 9.13. Near the wall, the dislocation experiences short-range forces similar to those for the infinite boundary (Fig. 9.12). In the attractive regions, this may result in a small displacement of an adjacent wall dislocation, so that the two 'share' one wall position. Alternatively, if the temperature is high enough, the wall dislocations may climb to accommodate the extra dislocation within the wall. The regions of attraction and repulsion are shown in Fig. 9.13. It is seen that at distances $\gtrsim D$ from the wall, attraction only occurs in the shaded regions delineated approximately by the lines $y = \pm x$. In these regions, the stress fields of the dislocations in the wall tend to reinforce each other rather than cancel: i.e. the dislocation in the matrix in Fig. 9.13 effectively experiences at large distances a force due to a single dislocation at the origin with Burgers vector four times that of the individual wall dislocations.

Figure 9.13 illustrates also a probable way in which a low-angle boundary develops during the *recovery process*. Dislocations in the unshaded regions will tend to be repelled by the boundary, but can climb by vacancy diffusion processes. If a dislocation at P climbs into the shaded region Q it will then experience an attractive force tending to align the dislocation in the low-energy configuration at the top of the existing wall. The path taken is indicated by the dotted line.

9.5 Strain Energy of Dislocation Arrays

The energy per unit area of a low-angle boundary is found by multiplying the energy per unit length per dislocation by the number of dislocations per unit area, which is $D^{-1} = \theta/b$ for an array of parallel dislocations. The result, first published by Read and Shockley (see Read, 1953), is

$$E = E_0\theta(A - \ln\theta) \tag{9.15}$$

where constant E_0 is a function of the elastic properties of the material and A is a constant which depends on the core energy of an individual dislocation. An approximate derivation of this remarkably simple formula for the symmetrical tilt boundary of edge dislocations is as follows.

Consider the work done in creating one dislocation of the complete wall of identical dislocations. By analogy with the derivation of equation (4.22) in section 4.4, the contribution to the strain energy per unit length for the dislocation at the origin $y = 0$, for example, is

$$E(\text{per dislocation}) = \frac{1}{2}b\int_{r_0}^{\infty}\sigma_{xy}\,\mathrm{d}x \tag{9.16}$$

where σ_{xy} is given by equation (9.13) evaluated on $y = 0$, and r_0 is the dislocation core radius. It was shown in the preceding section that σ_{xy} for small x is approximately the stress due to a single dislocation at $y = 0$, i.e. $Gb/2\pi(1 - \nu)x$, and that for large x it has the approximate exponential form of equation (9.14). The radius of the transition between these two expressions is proportional to D and is approximately $x = D/2\pi$. The contributions the two parts make to the energy are readily found from equation (9.16) to be

$$\left.\begin{aligned}\frac{1}{2}b\int_{r_0}^{D/2\pi}\sigma_{xy}\,\mathrm{d}x &\simeq E_0 b\,\ln(D/2\pi r_0)\\ \frac{1}{2}b\int_{D/2\pi}^{\infty}\sigma_{xy}\,\mathrm{d}x &= E_0 b2/e\end{aligned}\right\} \tag{9.17}$$

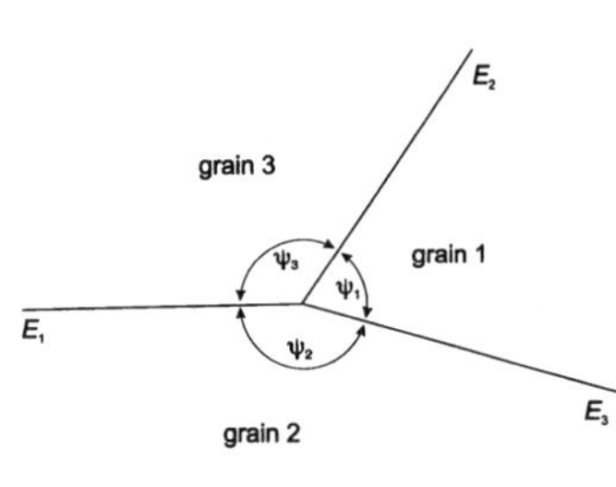

Figure 9.14 Three grain boundaries that meet along a line (normal to the figure). Each boundary has an energy, E, per unit area which is a function of its misorientation angle, θ, and its crystallographic orientation.

where $E_0 = Gb/4\pi(1 - \nu)$. Multiplying the sum of these two terms by θ/b and replacing D by b/θ gives the Read–Shockley formula (equation (9.15)) with $A = \ln(b/2\pi r_0) + 2/e$. The term $2/e$ in A is replaced by 1 if the more accurate form of σ_{xy} is used.

This derivation emphasises the fact that the $\ln(\theta)$ term arises from the *elastic* energy per dislocation. It *decreases* as θ *increases* because the stress fields of the dislocations overlap and cancel more fully as D decreases. If the non-linear dislocation core energy is added to equation (9.15), parameter A is modified in value but remains independent of θ.

The Read–Shockley formula shows that it is energetically favourable for two low-angle boundaries with misorientation angles θ_1 and θ_2 to combine to form a single boundary with angle $(\theta_1 + \theta_2)$. Furthermore, it predicts that E has a maximum at $\theta = \exp(A - 1)$, as can be verified by differentiation of equation (9.15).

Experimental verification of equation (9.15) was one of the early successes of dislocation theory. The relative energy of a dislocation boundary was measured using the simple principle illustrated in Fig. 9.14. This shows three boundaries looking along their common axis. If E_1, E_2 and E_3 are the energies of the grain boundaries per unit area, each boundary will have an effective surface tension equal to its energy. The situation will approximate, under equilibrium conditions, to a triangle of forces, and for the equilibrium of these forces acting at a point

$$\frac{E_1}{\sin \psi_1} = \frac{E_2}{\sin \psi_2} = \frac{E_3}{\sin \psi_3}$$

By measuring ψ_1, ψ_2 and ψ_3 and the misorientation across the boundaries it is possible to determine the relative energies.

Early measurements of E as a function of θ showed good agreement with the Read–Shockley formula, even for angles as large as $\sim 40°$. This was somewhat surprising, because for large-angle boundaries ($\theta \gtrsim 5–10°$), the dislocation cores overlap to such an extent that the derivation of equation (9.15) based on linear-elastic energy considerations ceases to be valid. Subsequent research on the energy of low-angle boundaries found that equation (9.15) does indeed give a good fit to the data for small θ, and that the apparent fit for high angles is fortuitous: it is only obtained by changing the constants E_0 and A from the values fitted at low angles. This is demonstrated by the data for tilt boundaries in copper in Fig. 9.15.

9.6 Dislocations and Steps in Interfaces

Admissible Defects

Line defects in interfaces between crystals can play an important role in transformations involving the growth of one crystal at the expense of another, e.g. twinning, martensitic transformations and precipitation. In the simplest picture, the defect is akin to a step and its motion over the interface is the mechanism that transforms atoms of one crystal to sites of the other. The word 'step' is misleading, however, because line defects in interfaces have specific character, which may or may not be step-like, and need to be defined with care. A brief description of this topic is presented here and more rigorous treatments are referenced at the end of the Chapter.

The topological properties, e.g. the Burgers vector **b** and step height h, of interfacial defects that can arise when two crystals are joined to form a bicrystal are determined by the symmetry and orientation of the crystals. Specifically, they are defined by the crystal symmetries that are broken.

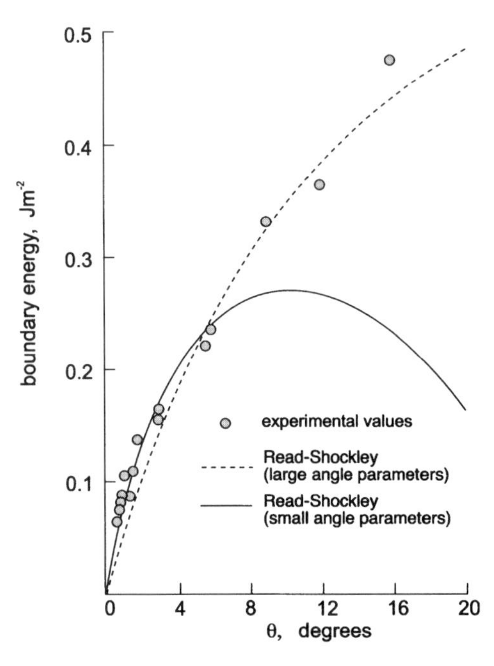

Figure 9.15 Relation between boundary energy and misorientation angle θ for [001] tilt boundaries in copper. Theoretical fits of the Read–Shockley equation using either high-angle or low-angle data are shown. (After Gjostein and Rhines, *Acta Metall.* **7**, 319 (1959).)

An *interfacial dislocation* arises when either the translation or point group symmetry operations are broken. Consider the two crystal labelled μ and λ (black and white) in Fig. 9.16. If a bicrystal is created by bringing the upper planar surface of μ, with outward normal vector $\mathbf{n}$, into contact with the lower surface of λ, as in Fig. 9.16(a), a defect-free interface can result. If a surface contains a step and the surface structure is identical on either side of it, the step must be related to the crystal symmetry: in the simplest case, it is characterised by a translation vector $\mathbf{t}$ of the lattice, as indicated by the steps on the surfaces drawn in Figs 9.16(b)–(c). The height of such a step is $h = \mathbf{n} \cdot \mathbf{t}$ and can be signed positive or negative. For the situation in Fig. 9.16(b), the surfaces have steps that are neither parallel nor of the same height ($\mathbf{t}_\lambda \neq \mathbf{t}_\mu$, $h_\lambda \neq h_\mu$), so that when they are brought together to form a bicrystal with no atoms removed and no spaces left along the stepped interface, the material in the vicinity of the resulting overlap step will have to be distorted to fit, as in the 'cutting and rebonding' process for crystal dislocations presented in section 3.2. Thus, a dislocation with Burgers vector

$$\mathbf{b} = \mathbf{t}_\lambda - \mathbf{t}_\mu \qquad (9.18)$$

is created, separating identical interfaces on either side. (Note that the sign convention used for $\mathbf{b}$ assumes that the positive line sense of the defect is out of the paper in Fig. 9.16.) Interfacial defects with non-zero **b** and step height h have been termed *disconnections*. If the two surfaces have steps of equal height but opposite sign, as in Fig. 9.16(c), the resultant interface contains a defect that is a dislocation with no step character. Finally, if the two crystals have symmetry in common, i.e.

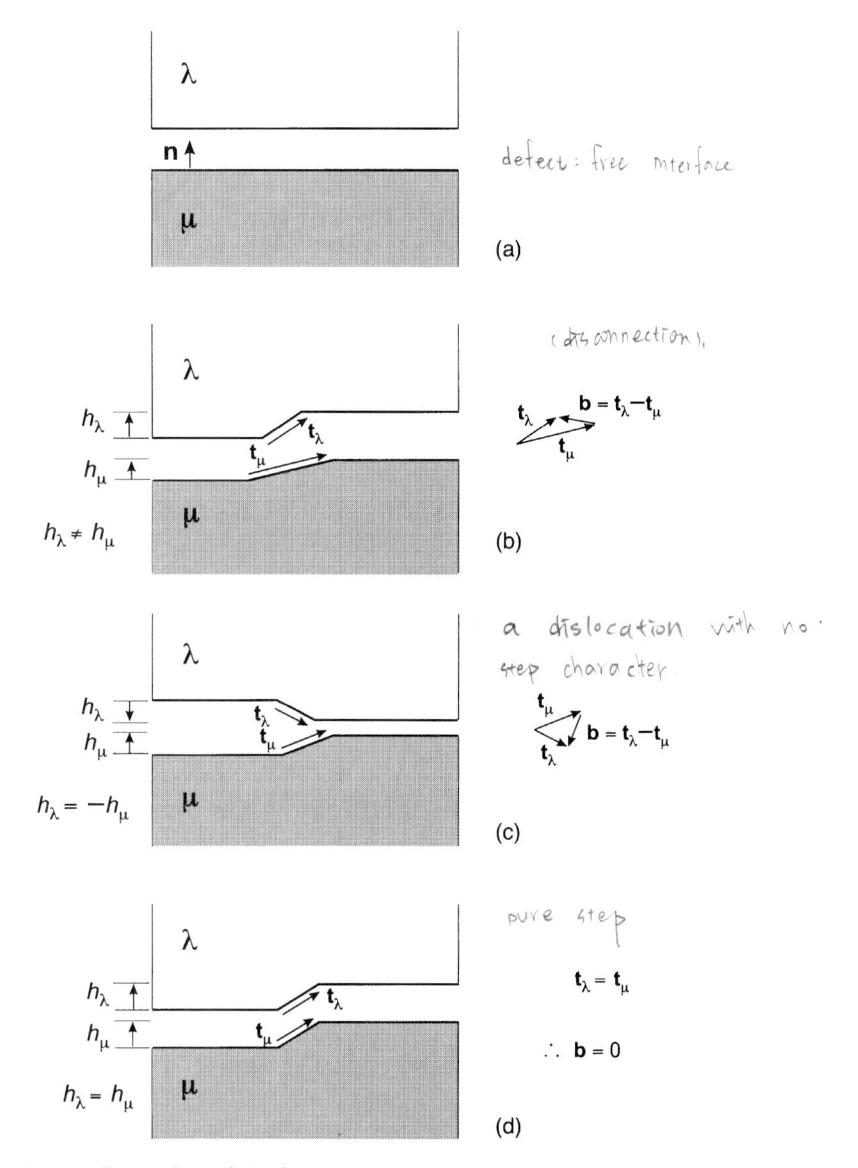

Figure 9.16 Illustration of the formation of a bicrystal by joining surfaces of crystals λ and μ. (a) Reference space has no interfacial defects. (b) Interfacial dislocation with step character and (c) one without. (d) Interface step with no dislocation character. The interface structures are identical on either side of the defect in (b)–(d).

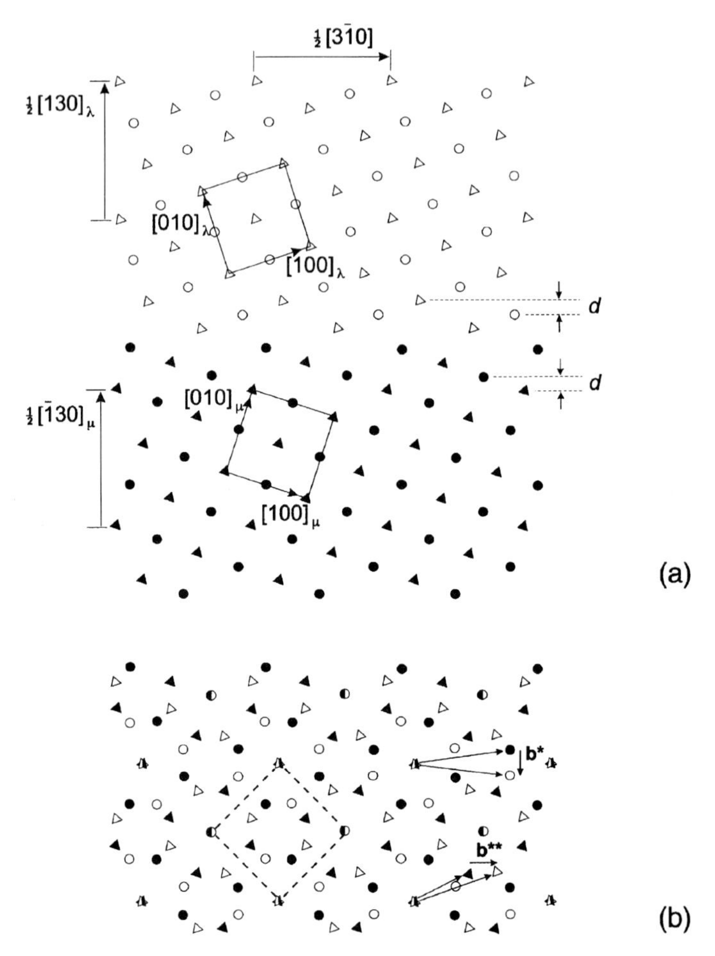

Figure 9.17 (a) Atom positions in two adjacent (001) planes in two face-centred cubic crystals separated by a symmetric $\{130\}/\langle 001\rangle$ 37°-tilt boundary. The unit cells are shown in outline and the atoms in different planes are plotted as circles and triangles. (b) Two lattices superimposed to show coincident sites, marked as half-filled symbols. The unit cell of this CSL with $\Sigma = 5$ is indicated by the dashed lines.

$\mathbf{t}_\lambda = \mathbf{t}_\mu$, then it is possible to create a defect as a *pure step* without dislocation character, as in Fig. 9.16(d).

To illustrate these features in a specific case, consider a bicrystal formed by joining two face-centred cubic crystals of the same metal along a $\{130\}$ plane of each after one has been rotated with respect to the other by an angle of $2\tan^{-1}(1/3) = 36.9°$ about a common $\langle 001\rangle$ axis. The [001] projection of atoms that lie in two adjacent (001) planes is shown in Fig. 9.17(a). (The misorientation angle of this symmetric tilt boundary is too large for the interface structure to be described as an array of distinct dislocations, as in sections 2.2 and 9.2). Now imagine the two lattices to interpenetrate and bring a lattice point of each into coincidence by translation. This creates the *dichromatic pattern* of Fig. 9.17(b). It can be seen that for this particular misorientation of the crystals many sites (one in five) are in coincidence, shown by half-filled symbols. When such a situation occurs a *coincident-site lattice (CSL)* exists. The reciprocal of the ratio of CSL sites to lattice sites of one of the crystals is denoted by Σ, which is 5 in this case. CSLs arise most commonly for rotations about low-index directions in cubic crystals, but cannot be created so readily for lower symmetry systems or when crystals of different structure are joined, and so their general significance for models of interfaces should not be over-emphasized. Nevertheless, the example of Fig. 9.17 serves a useful purpose for illustrating the interfacial defects discussed above.

Since some of the translation vectors of each lattice are coincident – these are the vectors of the CSL – it is possible to form a pure step, as in Fig. 9.16(d). This is demonstrated for the (130) interface in Fig. 9.18(a), where $\mathbf{t}_\lambda$ and $\mathbf{t}_\mu$ are $\frac{1}{2}[130]_\lambda$ and $\frac{1}{2}[\bar{1}30]_\mu$, respectively. Both h_λ and h_μ equal $10d$, where d is the interplanar spacing of the $\{130\}$ planes drawn in Fig. 9.17(a). Although the indices of $\mathbf{t}_\lambda$ and $\mathbf{t}_\mu$ are different when expressed in the coordinate frame of their own lattice, they are identical when referred to a common frame, and so $\mathbf{b} = \mathbf{t}_\lambda - \mathbf{t}_\mu$ is zero. The defect is a step of height $10d$.

An interfacial dislocation with no step character, as in Fig. 9.16(c), can be formed by joining the λ and μ surfaces shown in Fig. 9.18(b). Here, $\mathbf{t}_\lambda$ is $\frac{1}{2}[2\bar{1}\bar{1}]_\lambda$ with height $h_\lambda = -d$ and $\mathbf{t}_\mu$ is $\frac{1}{2}[211]_\mu$ with $h_\mu = d$. The resultant defect has zero step height since $h_\lambda = -h_\mu$, but $\mathbf{b} = \mathbf{t}_\lambda - \mathbf{t}_\mu$ is not zero. It is a vector of the dichromatic pattern and is denoted as $\mathbf{b}^*$ in Fig. 9.17(b). In the coordinate frame of the white crystal, it is $\mathbf{b} = \frac{1}{10}[\bar{1}\bar{3}0]_\lambda$. An interfacial dislocation with step character, i.e. a disconnection as in Fig. 9.16(b), where the step has a small height can be formed as depicted in Fig. 9.18(c). In this case, $\mathbf{t}_\lambda$ is $[100]_\lambda$ with $h_\lambda = 2d$ and $\mathbf{t}_\mu$ is $\frac{1}{2}[110]_\mu$ with $h_\mu = 2d$. The resultant defect is a disconnection with step height $2d$ and $\mathbf{b} = \frac{1}{10}[3\bar{1}0]_\lambda$, which is denoted as $\mathbf{b}^{**}$ in Fig. 9.17(b). The dashed lines in Figs 9.18(b) and (c) indicate the extra half-planes and distortions associated with these edge dislocations.

It has been noted that both $\mathbf{b}^*$ and $\mathbf{b}^{**}$ are vectors from black to white sites of the dichromatic pattern. This is a general condition for the Burgers vector of dislocations that can exist in an interface of a bicrystal where translation symmetry is broken and the interface structure is identical on either side of the defect. If an interface can have more than one stable form, e.g. by rigid-body translation of one crystal with respect to the other, then an interfacial dislocation could separate structures that are not identical and its Burgers vector would not be a black-to-white vector of the dichromatic pattern. As noted above, CSLs arise only in special cases and are not a requirement for the approach to defect characterisation presented here. The twin interface in a hexagonal-close-packed metal described in section 9.7 is an example where the reference space is not a three-dimensional lattice.

Identification of Interfacial Dislocations

The descriptions of the preceding section provide a means of defining the topological features of line defects that can arise in interfaces. Suppose, however, that a defect is observed in an interface by, say, high resolution electron microscopy: how can it be characterised? As with the Burgers circuit construction for a crystal dislocation (section 1.4), a circuit map can be used. A closed circuit around a feature to be identified is first made between crystallographically-equivalent white sites and then black sites, crossing the interface at two places. It is then mapped on to a reference space, but unlike the Burgers circuit for a crystal dislocation, there are two reference crystals, black and white. The presence of a closure failure ('finish-to-start') in this reference space identifies the Burgers vector content enclosed by the circuit in the real bicrystal.

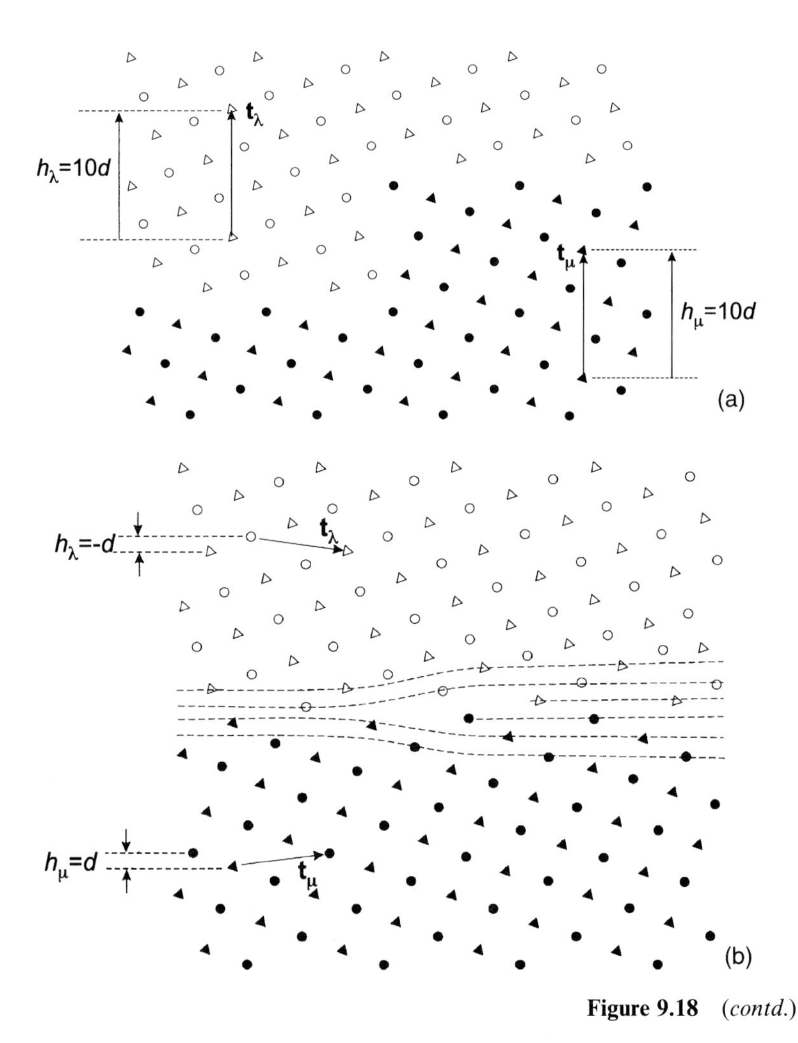

Figure 9.18 (*contd.*)

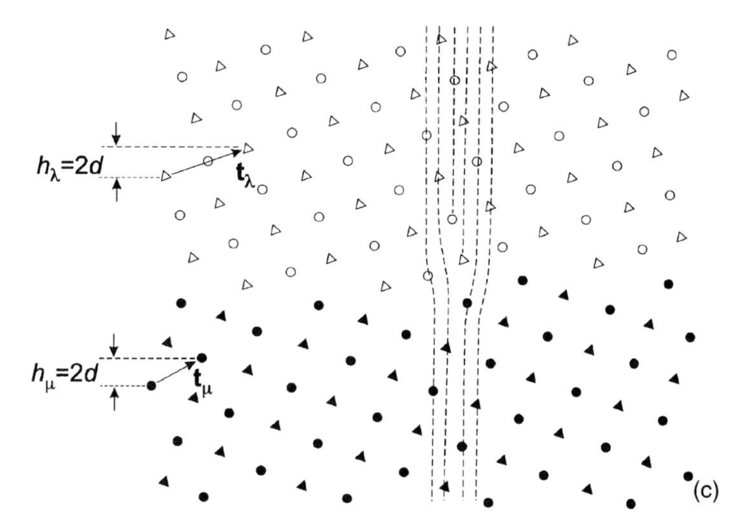

Figure 9.18 Some interfacial defects in a symmetric $\{130\}/\langle 001\rangle$ 37°-tilt boundary between face-centred cubic crystals: (a) pure step, (b) edge dislocation with no step and (c) edge dislocation with a step. Lattice vectors $\mathbf{t}_\lambda$ and $\mathbf{t}_\mu$ define the defect and its Burgers vector $\mathbf{b} = (\mathbf{t}_\lambda - \mathbf{t}_\mu)$. The horizontal and vertical dashed lines in (b) and (c) indicate the distortion associated with the dislocations.

Consider the disconnection line defect of Fig. 9.18(c). It is reproduced in Fig. 9.19(a) and, assuming that the defect has a positive sense out of the paper, a possible right-handed circuit starting and finishing at S is drawn on the figure. The steps XY and ZS cross the interface in equivalent ways and are not part of either crystal. The reference spaces taken from Fig. 9.18 are plotted in Fig. 9.19(b) and the circuit in (a) is mapped onto them. The presence of the closure failure confirms that the defect does indeed have dislocation character with $\mathbf{b} = \mathbf{FS}$, which is the same as $\mathbf{b}^{**}$ in Fig. 9.17(b).

Dislocations in Epitaxial Interfaces

The structure and defect content of the interface between a substrate and material grown on it by epitaxy is of considerable importance, particularly in the technology of solid-state electronic devices, and the methods of the previous two sections are useful for describing them. Consider for simplicity two cubic crystals λ and μ with their unit cells in parallel orientations, but let their lattice parameters a_λ and a_μ be different. If, for instance, $8a_\lambda = 7a_\mu$, as depicted in Fig. 9.20(a), the (001) interface is periodic because the arrangement shown repeats periodically along the interface. The admissible interfacial defects arising from the broken translation symmetry have $\mathbf{b} = (\mathbf{t}_\lambda - \mathbf{t}_\mu)$. The shortest lattice vectors are $\mathbf{t}_\lambda = [100]_\lambda$ and $\mathbf{t}_\mu = [100]_\mu$, and so $\mathbf{b}_{\min} = \frac{1}{7}[\bar{1}00]_\lambda$ with $h = 0$.

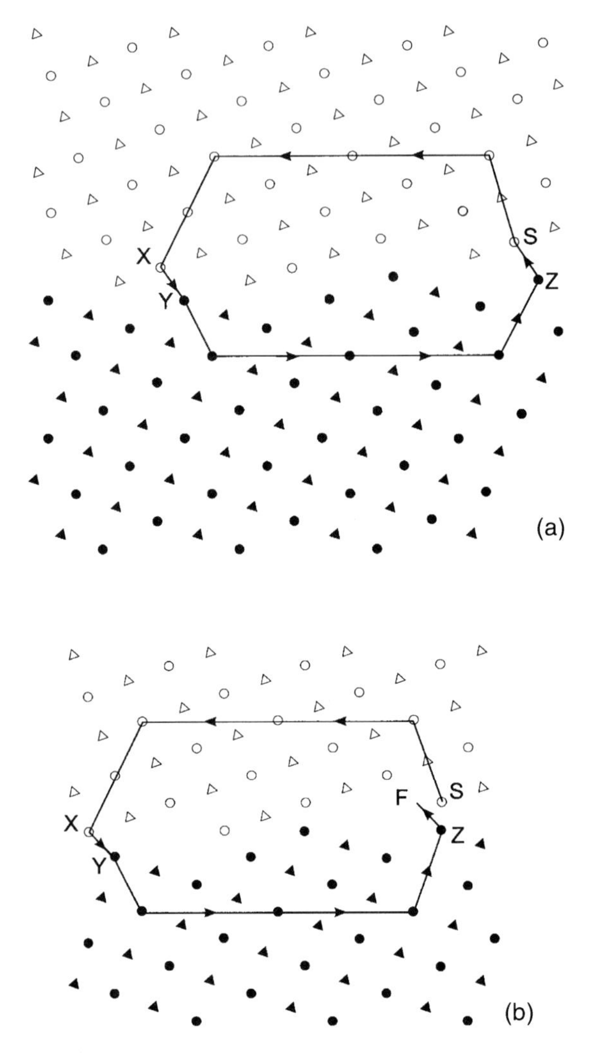

Figure 9.19 (a) Closed circuit drawn around the interfacial defect of Fig. 9.18(c). (b) Circuit of (a) mapped onto the reference space of the two crystals, showing that the defect has a Burgers vector **FS** (= $\mathbf{b}^{**}$ in Fig. 9.17(b)).

Now consider deforming the two crystals until they are identical with $a_\lambda = a_\mu$, e.g. expand the λ crystal until the bicrystal in Fig. 9.20(b) can be formed. This is a *commensurate* interface, sometimes referred to as *coherent*. In dislocation terms, we can construct a closed circuit *SABCDE* as in Fig. 9.20(b) and map it on to the λ and μ reference frames to test for a closure failure. This is shown in Fig. 9.20(c), and demonstrates that the circuit in (b) does indeed enclose one or more dislocations with a total Burgers vector $\mathbf{b} = \mathbf{FS} = [\bar{1}00]_\lambda$. The uniformity of the interface structure in Fig. 9.20(b) implies that this dislocation content may be considered to be distributed along the interface in the

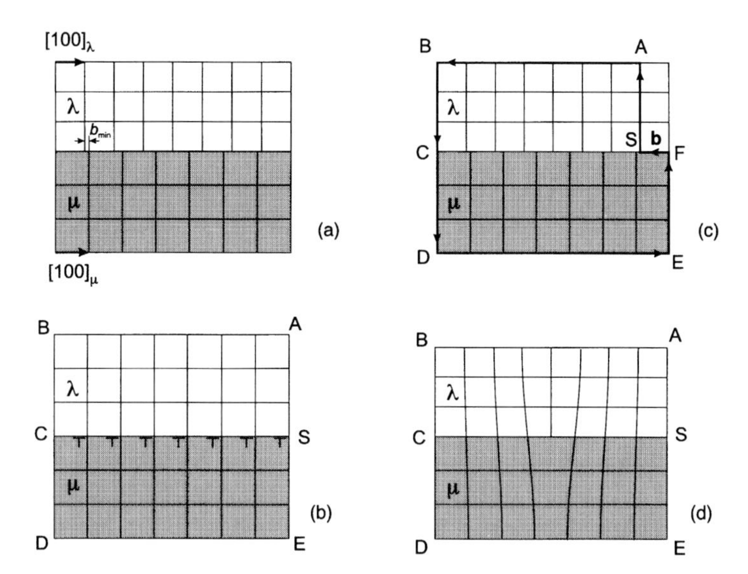

Figure 9.20 Illustration of the dislocation content of epitaxial interfaces: (a) reference λ and μ crystals; (b) crystal λ strained to produce a coherent interface which, with respect to (a) and as demonstrated in (c), contains the distribution of dislocations indicated by the usual symbols with $\mathbf{b} = \mathbf{b}_{\text{min}}$; (d) interface containing a 'misfit dislocation', but the dislocation content is actually zero. (After Pond and Hirth, Fig. 2.6 of 'Defects at surfaces and interfaces', *Solid State Physics*, Vol. 47, p. 287 (1994)).

form of seven dislocations with $\mathbf{b} = \frac{1}{7}[\bar{1}00]_\lambda = \mathbf{b}_{\text{min}}$. Their presence is indicated schematically in Fig. 9.20(b).

When a strained layer grows with a coherent interface, as in Fig. 9.20(b), the elastic strain energy increases in proportion to the volume, but can be relieved by the introduction of crystal dislocations (with $\mathbf{b} = \mathbf{t}_\lambda$ or $\mathbf{t}_\mu$) into the interface. They are known as *misfit dislocations* and enable the lattice parameter to return to the stress-free, equilibrium value. This is drawn schematically in Fig. 9.20(d) and an example obtained by high resolution electron microscopy for the interface between gallium arsenide and silicon is presented in Fig. 9.21. Note, however, that if the closed circuit *SABCDE* of Fig. 9.20(d) is mapped onto the reference space of Fig. 9.20(c), no closure failure results. In other words, the structure of Fig. 9.20(d) has zero dislocation content because the long-range stress field of the misfit dislocation is annulled by the uniform distribution of dislocations of opposite sign (Fig. 9.20(b)), resulting in a total **b** of zero.

9.7 Movement of Boundaries

For the movement of a boundary containing interfacial dislocations, several conditions must be satisfied. Firstly, if slip is involved the dislocation must be free to glide on its slip plane, i.e. the plane that contains

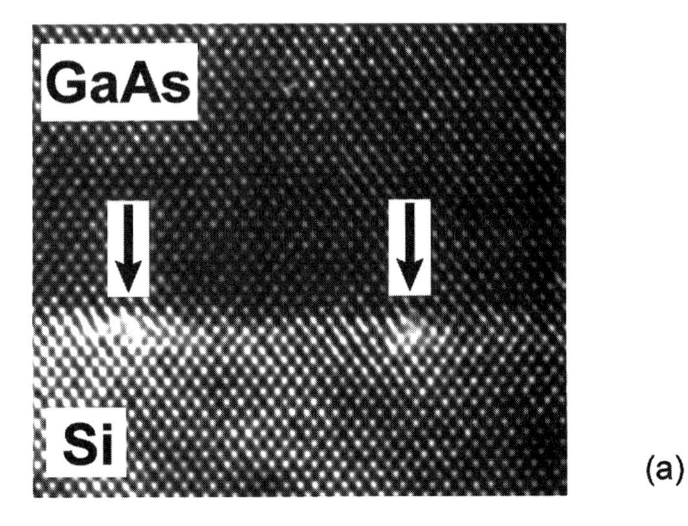

Figure 9.21 (a) High resolution electron microscope image revealing the atomic structure of the interface between gallium arsenide and silicon which contains misfit dislocations (arrowed). (b) Tilted and transformed image showing the presence of the misfit dislocations more clearly. (Courtesy R. Beanland and A. F. Calder.)

the line and its Burgers vector. Secondly, matter must be conserved if a diffusional flux of atoms is required for motion of interfacial defects. Thirdly, the thermodynamic condition that movement results in either a reduction in the energy of the boundary or, in the case of movement induced by an externally applied stress, that the stress does work when the boundary moves. Fourthly, the driving force due, for example, to the externally applied stress or the excess vacancy concentration must be sufficient to produce dislocation movement. Some aspects of the stress required to move dislocations are considered in the next chapter.

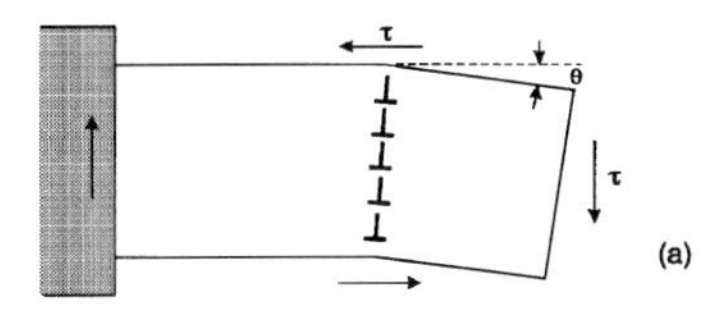

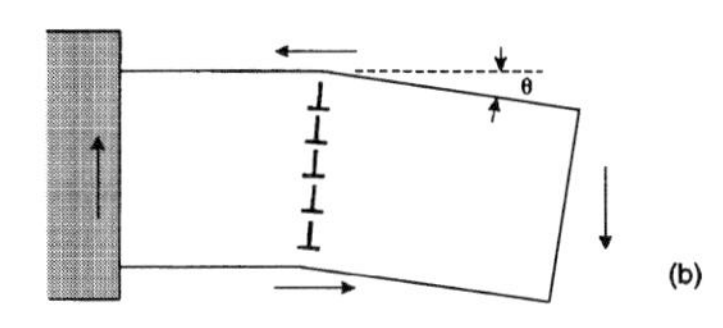

Figure 9.22 Stress-induced movement of a symmetrical pure tilt boundary.

Arrays of Crystal Dislocations

Consider first the low-angle boundaries formed by dislocations whose Burgers vectors are lattice vectors (sections 9.2 and 9.3). The condition for slip is particularly restrictive on the movement of such boundaries and can be illustrated with reference to the motion of low-angle tilt boundaries entirely by slip: the dislocations must have parallel glide planes. Consider the symmetrical tilt boundary illustrated in Fig. 9.22(a). A shear stress can be applied to the boundary by adding a weight to one end of the crystal as illustrated. For a shear stress τ, acting on the slip planes in the slip direction, the force on every dislocation will be $b\tau$ per unit length, and since there are θ/b dislocation lines per unit boundary height, the force per unit area on the boundary is

$$F = \theta\tau \tag{9.19}$$

If this is sufficient to move the dislocations the boundary will move to the left as illustrated in Fig. 9.22(b). Since every dislocation remains in the same position relative to the others in the boundary the geometry of the boundary is conserved. The work done by τ is $\tau\theta$ per unit volume swept out by the boundary. The movement of such a boundary was observed directly by Washburn and Parker (1952) in single crystals of zinc. Tilt boundaries were introduced by bending and annealing to produce polygonisation as in Fig. 9.3. The boundary was moved backwards and forwards by reversing the direction of the applied stress.

Now consider the tilt boundary illustrated in Fig. 9.6, in which the Burgers vectors of the component edge dislocations are at right-angles. There are three possibilities for the movement of such a boundary under an applied stress. (a) The dislocations move by glide and remain in the

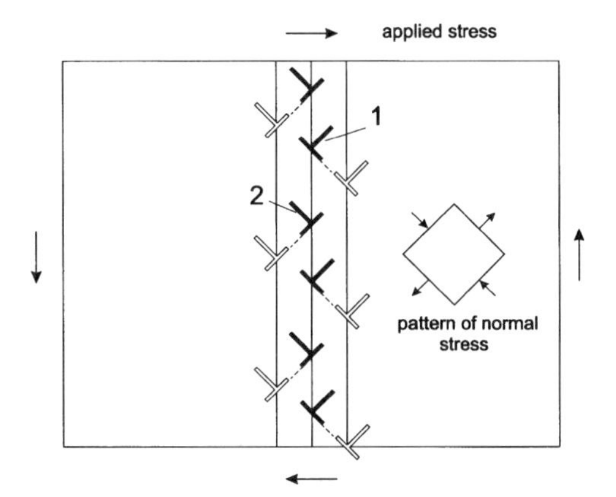

Figure 9.23 Stress-induced movement of a tilt boundary containing edge dislocations with mutually perpendicular glide planes. (After Read (1953), *Dislocations in Crystals*, McGraw-Hill.

same plane parallel to the original boundary. Since the slip planes are orthogonal the resolved shear stresses on the two systems are identical for a particular uniaxial stress, but induce the two sets of dislocations to move in opposite directions, as illustrated in Fig. 9.23. Thus, if all the dislocations were to move together, the net work performed would be zero and there would be no tendency for this to occur. (b) The dislocations move by glide but in opposite directions (Fig. 9.23). If both sets of dislocations move in such a way that work is done, the boundary will tend to split in the way illustrated, whereas the forces between dislocations will tend to keep the two boundaries together in the minimum energy configuration. Extra energy will be required and the process will be difficult. (c) The boundary moves uniformly as a whole, such that the dislocation arrangement is conserved. This can occur only by a combination of glide and climb. The stress field indicated on Fig. 9.23 shows that the set of dislocations (1) is under a compressive stress and will tend to climb up and to the right by the addition of vacancies. The set of dislocations (2) is under a tensile stress and will tend to climb down and to the right by emission of vacancies. Thus by a combination of climb and glide the boundary can move as a whole and the vacancies created at one set of dislocations are absorbed by the other. Only short-range diffusion will be required and the process can occur at high enough temperature.

Apart from the tilt boundary discussed above, and illustrated in Fig. 9.22, the only low-angle boundary that can move entirely by glide is a cross grid of screw dislocations and in this case it is essential that the junctions do not dissociate (see Fig. 7.19). In general, for stress induced boundary movement some diffusion is required, otherwise the boundary tends to break up, such as when a polygonised structure is deformed at low temperature.

Glide of Interfacial Defects

The discussion above is concerned with crystal dislocations. Now consider boundaries containing more general interfacial defects (section 9.6). The glide plane of a dislocation whose Burgers vector lies in the interface is the plane of the interface itself. A disconnection (Fig. 9.16(b)) with $h_\lambda = h_\mu$, as in Fig. 9.18(c), meets this condition and can glide if the resolved shear stress is high enough. Glide of the step allows one crystal to grow at the expense of the other: for example, glide of the defect in Fig. 9.18(c) to the right would allow black atoms to transfer to sites in the white crystal as the step passes. The simplest, yet very important, example of this occurs in *deformation twinning*.

Deformation twinning on the $\langle 111\rangle\{11\bar{2}\}$ system is common in the body-centred cubic metals, as noted in section 6.3. The direction and magnitude of the twinning shear is consistent with the glide of dislocations with $\mathbf{b} = \frac{1}{6}\langle 111\rangle$ on every successive $\{11\bar{2}\}$ atomic plane. This can be understood in terms of the description of admissible interfacial defects given in section 9.6. The defect-free interface is plotted in [110] projection in Fig. 9.24(a). (Visualisation of the atomic positions shown

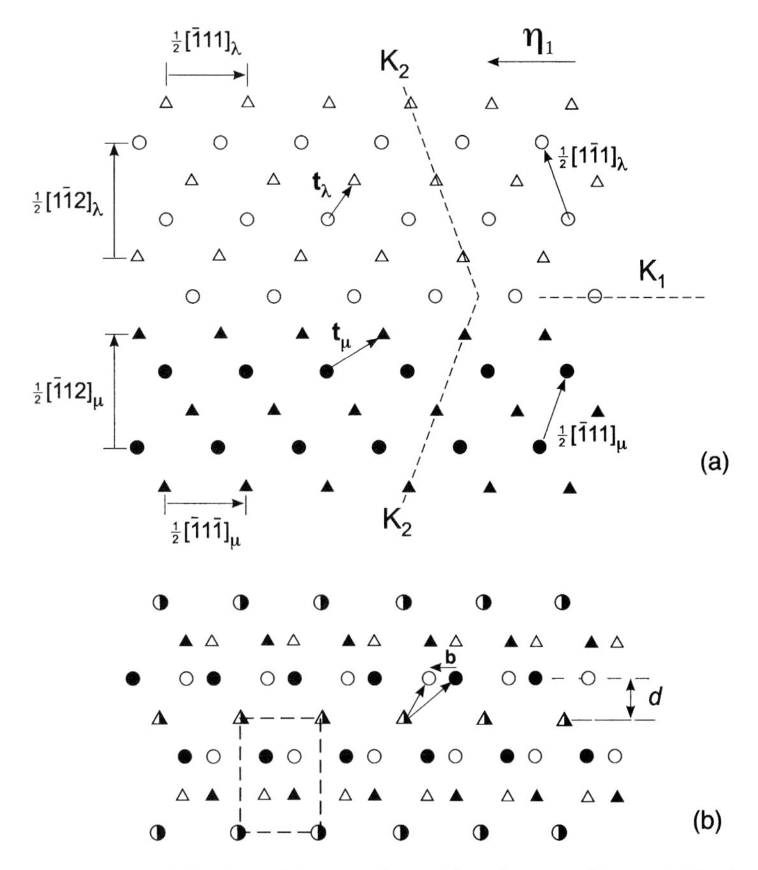

Figure 9.24 (a) Projection of the atomic positions in two adjacent (110) planes in a bicrystal formed by $\langle 111\rangle\{112\}$ twinning in a body-centred cubic metal. The twin habit plane is K_1, the direction of the twinning shear is η_1 and [110] is out of the paper. (b) Dichromatic pattern associated with (a). $\mathbf{b} = (\mathbf{t}_\lambda - \mathbf{t}_\mu)$ is the Burgers vector of a twinning dislocation formed by joining steps $\mathbf{t}_\lambda$ and $\mathbf{t}_\mu$ on the two crystals, as indicated.

is assisted by comparison with Fig. 1.6.) The twin habit plane is K_1 and the twinning shear in the direction η_1 reorientates the complementary twinning plane K_2 as indicated. The dichromatic pattern for the bicrystal is plotted in Fig. 9.24(b), which is seen to be a CSL with $\Sigma = 3$. The interfacial defect with a step up (from left to right) and height d, the spacing of the $\{112\}$ planes, and the shortest Burgers vector lying parallel to the interface, is created by joining the steps defined by $\mathbf{t}_\lambda = \frac{1}{2}[\bar{1}\bar{1}1]_\lambda$ and $\mathbf{t}_\mu = \frac{1}{2}[010]_\mu$ as indicated. This disconnection has $\mathbf{b} = \frac{1}{6}[1\bar{1}\bar{1}]_\lambda$ and is a twinning dislocation, as described in section 6.3, and for the step sign used is a negative edge dislocation. Thus, the atomic displacements that give rise to the macroscopic twinning shear occur by the glide of these defects on successive planes, as illustrated schematically in Fig. 9.25. An experimental observation of dislocations

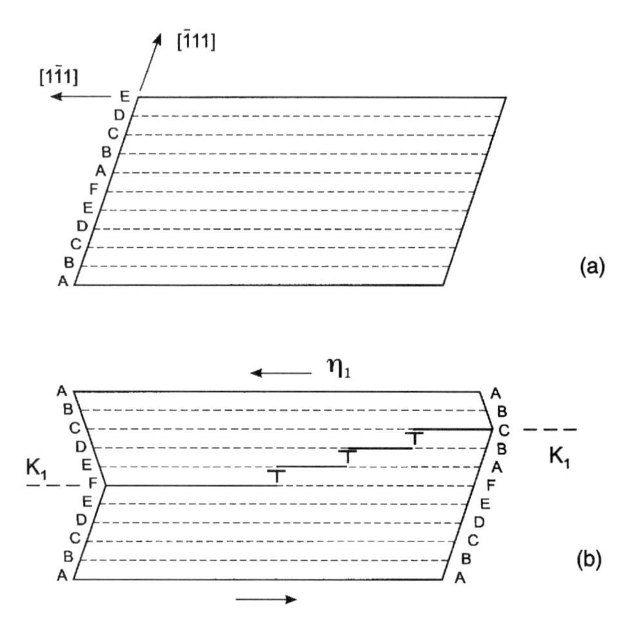

Figure 9.25 Schematic illustration of twinning in a body-centred cubic metal. The projection is the same as that in Fig. 9.24 and the dislocations are the same as the one defined there. (a) Untwinned crystal in [110] projection showing the stacking sequence *A, B,*... of the ($\bar{1}$12) planes. (b) Twinning dislocations with $\mathbf{b} = \frac{1}{6}[1\bar{1}1]_\mu$ glide to the right on successive ($\bar{1}$12) planes under the applied shear stress indicated to produce a twin-orientated region.

of this type is presented in Fig. 9.26. Fig. 9.26(a) illustrates the cross-section shape of the small twin seen in the transmission electron microscope image in (b) and an explanation for the contrast from the individual twinning dislocations is sketched in (c).

Twinning is an important mode of deformation for many crystalline solids, particularly if the number of independent slip systems associated with the glide of crystal dislocations is restricted (section 10.9). In all cases, the **b** and *h* characteristics of the twinning dislocations can be analysed using treatments similar to those presented here. (Further details can be found in references at the end of the Chapter.) The glide motion of twinning dislocations can occur under low resolved shear stress in many materials and is therefore relatively easy. Note, however, that if the crystal structure has more than one atom per lattice site, then the simple shear associated with the passage of such dislocations may restore the lattice in the twinned orientation but not all the atoms. Shuffles of these atoms will be required. This is illustrated by the {10$\bar{1}$2} twin in an hexagonal-close-packed metal, which has two atoms per lattice site. The structure of the twinning dislocation found by computer simulation (section 2.7) is shown in Fig. 9.27(a). The positive edge dislocation defined by $\mathbf{t}_\lambda - \mathbf{t}_\mu$ results in a step down (from left to right) of height $2d$, where d is the spacing of the {10$\bar{1}$2} lattice planes.

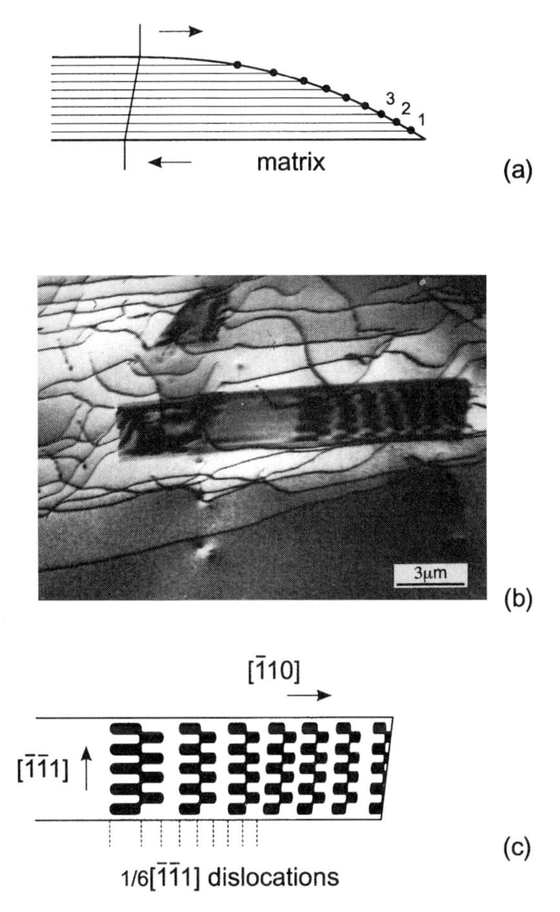

Figure 9.26 Experimental observation of a small deformation twin in a molybdenum – 35% rhenium alloy by transmission electron microscopy. (a) Illustration of the shape of the small twin shown in (b); the dislocations are represented by dots. (b) Diffraction contrast produced by twin which lies at an angle of 20° to the plane of the thin foil. (c) Diagrammatic illustration of the diffraction contrast observed in (b). Each change in the fringe sequence is due to a twinning dislocation. (From Hull (1962), *Proc. 5th Int. Conf. Electron Microscopy*, p. B9, Academic Press.)

This structure has been verified experimentally, as demonstrated by the high resolution transmission electron microscopy image in Fig. 9.27(b). The black dots indicate the match between the positions of atoms near the interface in this image and those in Fig. 9.27(a). It can be seen from Fig. 9.27(a) that the atoms in the two atomic planes labelled S that traverse the step have to shuffle as the dislocation glides along the boundary, because atoms such as 1 and 2 are closer than 2 and 3 on the left whereas 2 and 3 are closer than 1 and 2 on the right. The shuffles are short and easily achieved in this particular case, and so the core of the dislocation spreads along the interface. Computer simulation shows

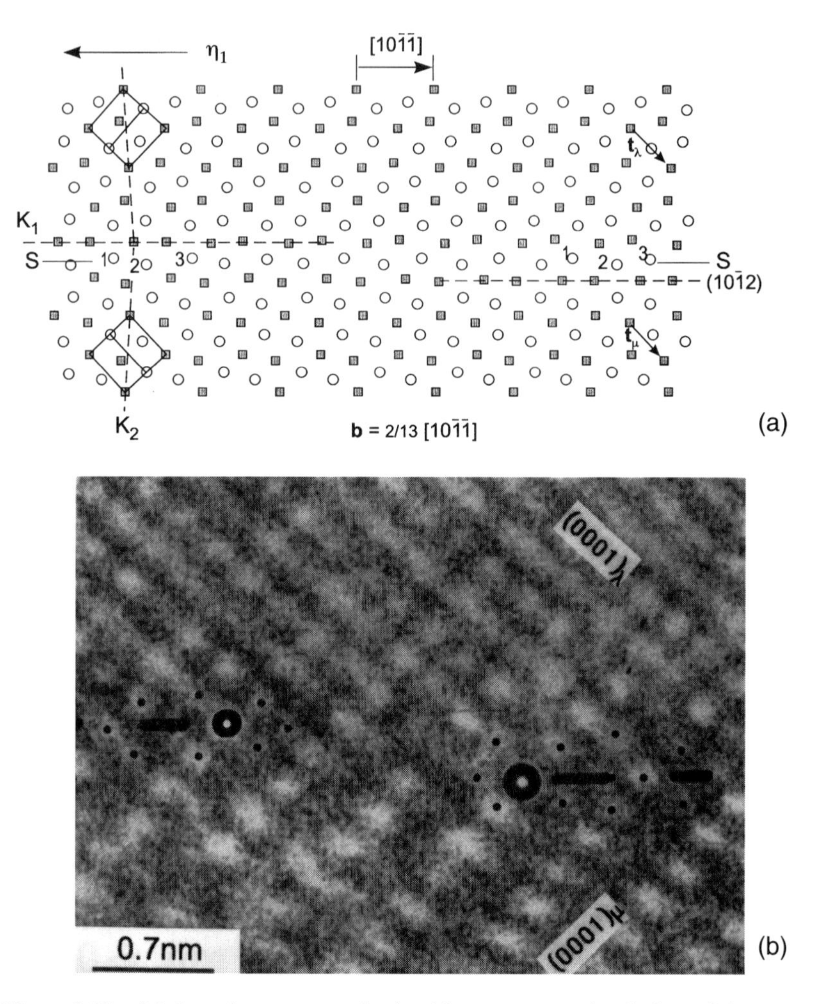

Figure 9.27 (a) Atomic structure obtained by computer simulation of the structure of a twinning dislocation in a $(10\bar{1}2)$ twin boundary in titanium, an hexagonal-close-packed metal. Unit cells are shown in outline and the position of the boundary is indicated by a dashed line. The twinning dislocation, defined by the lattice vectors $\mathbf{t}_\lambda$ and $\mathbf{t}_\mu$, has a very small Burgers vector, but requires shuffling of atoms in the layers labelled *S*. (After Bacon and Serra (1994), *Twinning in Advanced Materials*, eds. M. H. Yoo and M. Wuttig, p. 83. The Minerals, Metals and Materials Society (TMS).) (b) Experimental HRTEM image of a boundary in titanium containing a twinning dislocation. The dashed lines show the location of the interface and the dots indicate the positions of some atoms near the interface. (From Braisaz, Nouet, Serra, Komninou, Kehagias and Karakostas, *Phil. Mag. Letters* **74**, 331 (1996), with permission from Taylor and Francis Ltd (www.tandf.co.uk/journals)

that this twinning dislocation moves easily. For boundaries where complex shuffles are necessary, the glide of twinning dislocations can require relatively high stress and the assistance of elevated temperature.

Diffusion-Assisted Motion of Interfacial Defects

The slip plane of the interfacial dislocations described in the preceding section is the plane of the interface itself and so they glide under the action of a resolved shear stress. If the dislocation has a Burgers vector with a component perpendicular to the interface, however, it can only move along the boundary by climb, irrespective of the nature of the two crystals, and so requires the diffusion of atoms to or from it. Furthermore, when either the atomic density or the chemical composition of the two crystals is different, i.e. the interface is an *interphase boundary*, a defect with step character can only move along the interface if atomic transport can occur. These two statements indicate that analysis of the atomic flux required for motion of interfacial defects has to treat both the dislocation and step components of the defect. This is demonstrated by the following simple example.

For the step drawn in Fig. 9.28, $\mathbf{t}_\lambda$ and $\mathbf{t}_\mu$ are parallel but of different length, and so $h_\lambda \neq h_\mu$ and the dislocation has a Burgers vector component b_n perpendicular to the interface. Consider the step part first (Fig. 9.28(b)). If it moves a distance x to the right, the volume change of the λ crystal is

$$V = h_\lambda x \tag{9.20}$$

per unit length of step. Let λ have atoms of species A, B, C, ... with X_λ^A, X_λ^B, X_λ^C, ... atoms per unit volume, and similarly for crystal μ. The

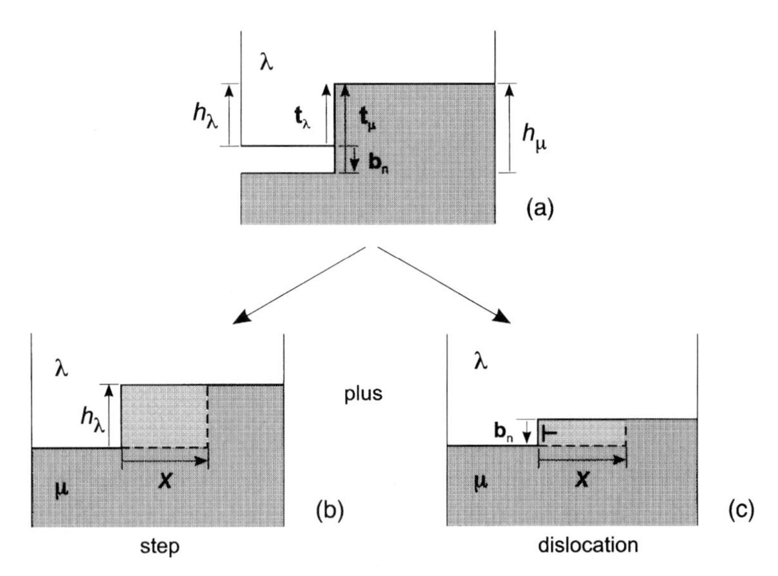

Figure 9.28 Step and dislocation parts of the interfacial defect in the bicrystal in (a) are shown in (b) and (c). As the defect moves to the right along the interface the volume of the λ crystal increases at the expense of the μ crystal because of the step component and the volume of the μ crystal decreases due to the climb of the dislocation component.

change in the number of A atoms in the volume swept by the step is then

$$\Delta N^A = h_\lambda x(X_\lambda^A - X_\mu^A) \tag{9.21}$$

For the step to move with a velocity ν, ΔN^A atoms will have to arrive or leave in the time x/ν, and so the diffusional flux (atoms per second per unit step length) will have to be

$$J^A = h_\lambda \nu(X_\lambda^A - X_\mu^A) \tag{9.22}$$

Now consider the dislocation part of the defect (Fig. 9.28(c)). As shown by the description of climb in section 3.6, the volume change of the μ crystal is

$$V = b_n x \tag{9.23}$$

per unit length of dislocation. The number of A atoms involved is

$$\Delta N^A = b_n x X_\mu^A \tag{9.24}$$

and the diffusive flux to the dislocation is

$$J^A = b_n \nu X_\mu^A \tag{9.25}$$

Thus, from equations (9.22) and (9.25) the net flux of A atoms to the complete defect is

$$J^A = \nu\left[h_\lambda(X_\lambda^A - X_\mu^A) + b_n X_\mu^A\right] \tag{9.26}$$

and similarly for $B, C, \ldots$ (Note that V, ΔN^A and J^A are negative ($b_n < 0$) in equations (9.23–9.25) because the dislocation climb shown removes μ crystal. It does not create λ crystal: that is done by motion of the step. Care must be exercised in choosing an unambiguous sign convention for the step height and b_n for use in the equations above.)

For interfacial defect motion along *grain boundaries*, of which the twin interface is a special case, the two crystals have identical chemical composition and atomic density and so $X_\lambda^A = X_\mu^A$. Equation (9.26) shows that motion is controlled solely by the normal component b_n of the Burgers vector, irrespective of whether a step exists or not. For motion along an *interphase boundary*, $(X_\lambda^A - X_\mu^A)$ is not zero in general because even if the chemical composition of the two phases is the same, a difference in crystal structure usually results in a difference in the number of atoms per unit volume. Treatments based on equation (9.26) for some specific material systems are presented in references at the end of the chapter.

9.8 Dislocation Pile-ups

The dislocation pile-up is a completely different kind of dislocation array that forms during plastic deformation. Thus, consider a dislocation

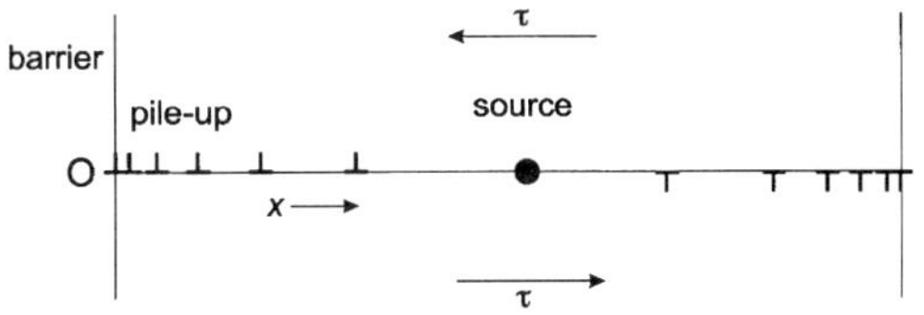

Figure 9.29 Linear arrays of edge dislocations piled-up against barriers under an applied shear stress τ.

source which emits a series of dislocations all lying in the same slip plane (Fig. 9.29). Eventually the leading dislocation will meet a barrier such as a grain boundary or sessile dislocation configuration and further expansion of the loop is prevented. The dislocations then pile-up behind the leading dislocation, but, being of the same sign, do not combine. They interact elastically and their spacing, which depends on the applied shear stress and the type of dislocation, decreases towards the front of the pile-up.

The stress τ_1 experienced by the leading dislocation of a pile-up can be deduced as follows. Suppose there are n dislocations. The leading dislocation experiences a forward force due to the applied stress τ and the other $(n-1)$ dislocations, and a backward force due to the internal stress τ_0 produced by the obstacle. If the leading dislocation moves forward by a small distance δx, so do the others, and the applied stress does work per unit length of dislocation equal to $nb\tau\delta x$. The increase in the interaction energy between the leading dislocation and τ_0 is $b\tau_0\delta x$. In equilibrium, these energies are equal and $\tau_1 = \tau_0$, so that

$$\tau_1 = n\tau \tag{9.27}$$

Thus, the stress at the head of the pile-up is magnified to n times the applied stress. The pile-up exerts a back-stress τ_b on the source, which can only continue to generate dislocations provided $(\tau - \tau_b)$ is greater than the critical stress for source operation. Eshelby *et al.* (1951) first calculated the spatial distribution of dislocations in a pile-up. They found that for a single-ended pile-up spread over the region $0 \leq x \leq L$, the number of dislocations in the pile-up is

$$n = \frac{L\tau}{A} \tag{9.28}$$

where A is Gb/π for screw dislocations and $Gb/\pi(1-\nu)$ for edges. The spacing of the first two dislocations is $0.92L/n^2$. The shear stress outside the pile-up well away from the first and last dislocations is approximately the same as that produced by a single superdislocation of Burgers vector nb at the centre of gravity of the pile-up $(x = 3L/4)$.

Thus, unlike the Read–Shockley arrays studied in the first part of this chapter, dislocation pile-ups produce large, long-range stress. At grain boundaries, this can nucleate either yielding in the adjacent grains or boundary cracks. It can also assist in the cross slip of screw dislocations held up at obstacles such as precipitates and dislocation locks. Dislocation pile-ups have been observed many times using transmission electron microscopy. Most pile-ups will be made up of dislocations with an edge component to the Burgers vector since screw dislocations can cross slip out of the slip plane. It should also be noted that certain crack configurations can be modelled in elasticity theory by pile-ups.

Further Reading

Christian, J. W. (1975) *The Theory of Transformations in Metals and Alloys*, Pergamon Press.

Christian, J. W. (1982) 'Deformation by moving interfaces', *Metall. Trans.* **A13**, 509.

Christian, J. W. and Crocker, A. G. (1980) 'Dislocations and lattice transformations', *Dislocations in Solids*, vol. 3, p. 165 (ed. F. R. N. Nabarro), North-Holland.

Eshelby, J. D., Frank, F. C. and Nabarro, F. R. N. (1951) 'The equilibrium of linear arrays of dislocations', *Phil. Mag.* **42**, 351.

Frank, F. C. (1950) *Conf. on Plastic Deformation of Crystalline Solids*, p. 150, Carnegie Inst. of Tech. and Office of Naval Research.

Hirth, J. P. and Lothe, J. (1982) *Theory of Dislocations*, Wiley.

Matthews, J. W. (1979) 'Misfit dislocations', *Dislocations in Solids*, vol. 2, p. 461 (ed. F. R. N. Nabarro), North-Holland.

Pond, R. C. (1989) 'Line defects and interfaces', *Dislocations in Solids*, vol. 8, p. 1 (ed. F. R. N. Nabarro), North-Holland.

Pond, R. C. and Hirth, J. P. (1994) 'Defects at surfaces and interfaces', *Solid State Physics*, vol. 47, p. 287.

Read, W. T. (1953) *Dislocations in Crystals*, McGraw-Hill, New York.

Read, W. T. and Shockley, W. (1952) 'Dislocation models of grain boundaries', *Imperfections in Nearly Perfect Crystals*, p. 352, Wiley.

Sutton, A. P. and Balluffi, R. W. (1995) *Interfaces in Crystalline Materials*, Oxford University Press.

10 Strength of Crystalline Solids

10.1 Introduction

In Chapter 1 it was shown that the theoretical shear strength of a perfect lattice is many orders of magnitude greater than the observed critical shear strength of real crystals which contain dislocations. In this chapter the factors affecting the strength of crystals are considered as an introduction to the way that an understanding of the properties of dislocations can be used to interpret the behaviour of crystalline solids.

Apart from effects associated with diffusion at high temperature, plastic deformation occurs by the glide of dislocations and hence the critical shear stress for the onset of plastic deformation is the stress required to move dislocations. This is measured usually by a tensile test in which the specimen is elongated at a *constant rate* and the load on the specimen is measured simultaneously with the extension. Representations of typical stress–strain curves are shown in Fig. 10.1. Curves (a) and (b) are typical of many solids: (a) represents a *ductile* material which undergoes extensive plastic deformation before fracture at F, and (b) represents a *brittle* material which exhibits little plasticity. The dislocations in (b) are either too low in density or too immobile to allow the specimen strain to match the elongation imposed by the testing machine. The curves show a linear region OE in which the specimen deforms elastically, i.e. the stress is proportional to strain according to Hooke's Law, followed by *yielding* at E and subsequent *strain* (or *work*) *hardening* up to F. In the latter process, the *flow stress* required to maintain plastic flow increases with increasing strain. For most materials, the change from elastic to plastic behaviour is not abrupt and the *yield stress* σ_y is not unique. This is because some non-linear *microplasticity* occurs in the preyield region OE due to limited dislocation motion. Then, as shown in Fig. 10.1(a), yielding is defined to occur when the plastic strain reaches a prescribed value, say 0.1 per cent: the corresponding *proof stress* is taken as the yield stress. Fig. 10.1(c) is typical of many body-centred cubic polycrystalline metals which do not yield uniformly. An example for iron is shown in Fig. 10.17(b). The curve can be divided into four regions, namely: (OE) elastic and preyield microplastic deformation, (EC) yield drop, (CD) yield propagation, and (DF) uniform hardening. The deformation between E and D is not homogeneous, for plastic flow occurs in only part of the specimen. This *Lüders band*, within which dislocations have rapidly multiplied, extends to occupy the entire length at D. In tests on single crystals, it

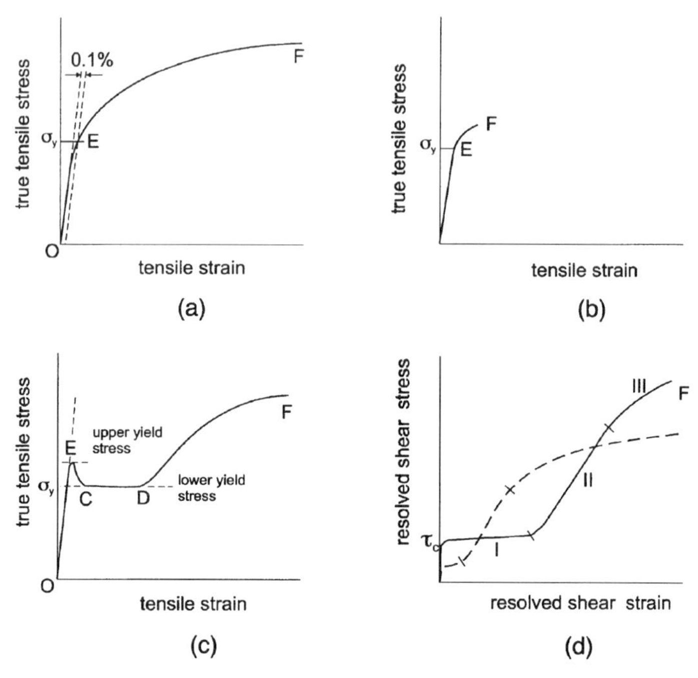

Figure 10.1 Typical stress–strain curves for (a), (b), (c) polycrystals and (d) single crystals.

is usual to resolve the stress and strain onto the plane and direction for which slip occurs first. The resulting stress–strain curve often has the form shown in Fig. 10.1(d). An actual example for copper is reproduced in Fig. 10.17(a). Above the *critical resolved shear stress* τ_c (section 3.1), the curve has three parts: stages I, II and III. The extent of these stages depends on the crystal orientation of the load axis, as indicated by the two curves. Polycrystalline specimens of the same metal (Fig. 10.1(a)) do not show stage I but deform in a manner equivalent to stages II and III.

It should be noted that the form of all the curves in Fig. 10.1 is dependent on test variables such as temperature and applied strain rate. Material parameters such as crystal structure, alloy composition, dislocation arrangement and grain size also affect the yield and flow stresses. It is therefore possible to modify materials in order to improve their performance. The understanding of these effects has been one of the successes of dislocation theory and is the subject of this chapter. In section 10.2 the manner in which the strength of barriers to dislocation motion affects the dependence of the flow stress on temperature and strain rate is considered. The nature of these barriers and the ways in which they are introduced are treated in the following sections. The presentation is, in part, necessarily simplified, and phenomena such as creep, fatigue and fracture are not touched upon at all. Reviews and texts with comprehensive descriptions are given in the references.

10.2 Temperature- and Strain-rate-dependence of the Flow Stress

The energy that has to be provided for dislocations to overcome the barriers they encounter during slip determines the dependence of the flow stress on temperature and applied strain rate. If the energy barriers are sufficiently small for thermal energy ($\sim kT$) to be significant, thermal vibrations of the crystal atoms may assist the dislocations to overcome obstacles at lower values of applied stress than that required at 0 K. Under such conditions, an increase in temperature, or a reduction in applied strain rate, will reduce the flow stress. The principles involved can be demonstrated as follows.

Consider a dislocation gliding in the x direction under an applied resolved shear stress τ^*, which produces a force $\tau^* b$ per unit length on the line. Suppose the dislocation encounters obstacles, each of which produces a resisting force K, as shown schematically in Fig. 10.2(a). Let the spacing of the obstacles along the line be l, so that the applied forward force on the line per obstacle is $\tau^* bl$. At the temperature 0 K, glide will cease if $\tau^* bl$ is less than $K_{\max}$, and the line will stop at position x_1. To overcome the barrier, the line must move to x_2. The isothermal energy change required is the area under the K versus x curve between x_1 and x_2, i.e. the Helmholtz free energy change,

$$\Delta F^* = \int_{x_1}^{x_2} K \, dx \tag{10.1}$$

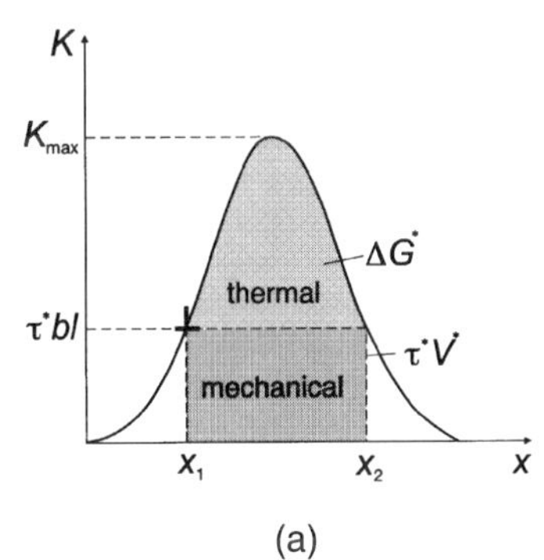

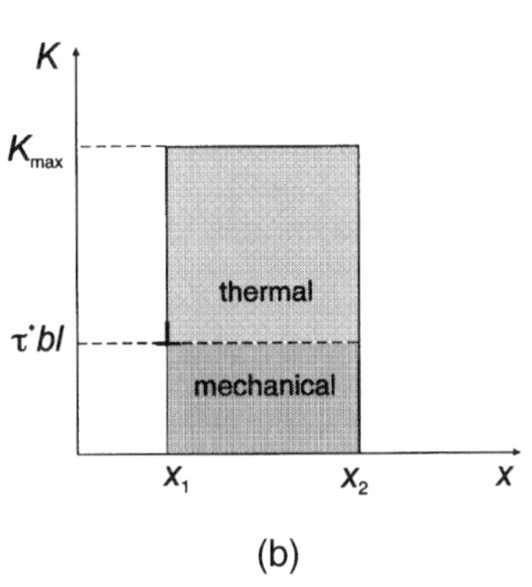

Figure 10.2 Profiles of resistance force K versus distance x for barriers opposing dislocation motion.

Part of this energy can be provided in the form of mechanical work done by the applied load, and is $\tau^* bl(x_2 - x_1)$, as shown in Fig. 10.2(a). It is customary to rewrite this mechanical contribution as $\tau^* V^*$, where V^* is known as the *activation volume* for the process. The remainder of the energy required is labelled 'thermal' in Fig. 10.2(a). It is the *free energy of activation*:

$$\Delta G^* = \Delta F^* - \tau^* V^* \tag{10.2}$$

and is the Gibbs free energy change between the two states x_1 and x_2 at the same temperature and applied stress. The probability of energy ΔG^* occurring by thermal fluctuations at temperature T (such that $\Delta G^* \gg kT$) is given by the Boltzmann factor $\exp(-\Delta G^*/kT)$, so that if the dislocation is effectively vibrating at a frequency ν ($\lesssim$ atomic vibration frequency), it successfully overcomes barriers at a rate of $\nu \exp(-\Delta G^*/kT)$ per second. The dislocation velocity is therefore

$$\bar{v} = d\nu \exp(-\Delta G^*/kT) \tag{10.3}$$

where d is the distance moved for each obstacle overcome. From equations (3.10) and (10.3), the macroscopic plastic strain rate is

$$\dot{\varepsilon} = \rho_m A \exp(-\Delta G^*/kT) \tag{10.4}$$

where ρ_m is the mobile dislocation density and $A = bd\nu$. The stress-dependence of $\dot{\varepsilon}$ arises from the stress-dependence of ΔG^*. The resulting relation between flow stress and temperature can be derived simply for the following case.

Suppose that the obstacles form a regular array and that each one is a region of constant resisting force, as shown in Fig. 10.2(b). Then

$$\Delta G^* = \Delta F\left[1 - \frac{\tau^*(T)}{\tau^*(0)}\right] \tag{10.5}$$

where ΔF is the total area under the K versus x curve, i.e. the energy required to overcome the obstacle when $\tau^* = 0$, and $\tau^*(0)$ is the flow stress at 0 K, i.e. the stress required to overcome the obstacle when no thermal energy is available. Substituting equation (10.5) into (10.4) gives for the flow stress

$$\frac{\tau^*(T)}{\tau^*(0)} = \frac{kT}{\Delta F}\ln\left(\frac{\dot{\varepsilon}}{\rho_m A}\right) + 1 \tag{10.6}$$

At temperatures above a certain value, say T_c, there will be sufficient thermal energy for the barriers to be overcome by thermal activation alone. Then $\tau^*(\tau_c) = 0$ and from equation (10.6)

$$T_c = \frac{-\Delta F}{k\ \ln(\dot{\varepsilon}/\rho_m A)} \tag{10.7}$$

Substituting this into equation (10.6) gives

$$\frac{\tau^*(T)}{\tau^*(0)} = \left(1 - \frac{T}{T_c}\right) \tag{10.8}$$

Hence, as shown in Fig. 10.3, τ^* decreases from $\tau^*(0)$ to zero as T increases from 0 K to T_c.

Generally, the motion of a dislocation is opposed by short-range barriers, which can be overcome by thermal activation as described here, and long-range forces from, say, other dislocations. The latter produce barriers too large for thermal activation to be significant. The flow stress then consists of *two* contributions. One is the *thermal component* τ^*: the other is the *athermal component* τ_G, which is almost independent of temperature apart from the small variation of the shear modulus G with temperature. Thus

$$\tau = \tau^* + \tau_G \tag{10.9}$$

The result of combining equations (10.8) and (10.9) is shown in Fig. 10.3. The temperature-dependence of τ for $T < T_c$ depends on the size of $\tau^*(0)$, which is proportional to $K_{\max}$, and thus on the nature of the obstacles. Most of the barriers encountered by dislocations do not have a simple square profile, as considered in Fig. 10.2(b), and do not form regular arrays. In more general situations, equation (10.5) is replaced by

$$\Delta G^* = \Delta F\left\{1 - \left[\frac{\tau^*(T)}{\tau^*(0)}\right]^p\right\}^q \tag{10.10}$$

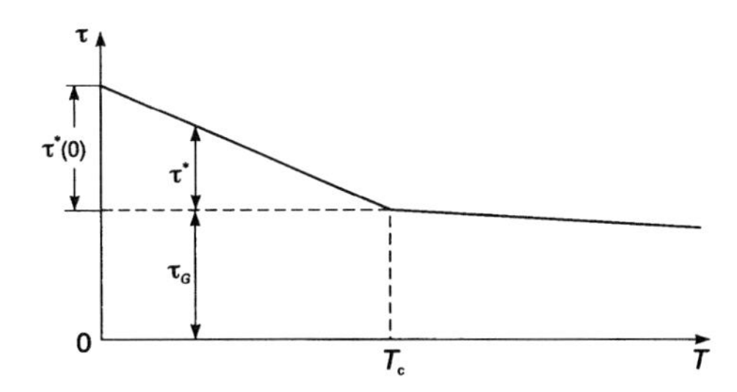

Figure 10.3 Variation of the flow stress with temperature for obstacles represented by Fig. 10.2(b).

with $0 \leq p \leq 1$ and $1 \leq q \leq 2$. This generally results in a curved rather than straight-line relationship between τ^* and T. The energy ΔF lies between about $0.05\,Gb^3$ and $2\,Gb^3$. This encompasses the range from small barriers such as solute atoms and the Peierls stress up to large barriers such as strong precipitates.

The approach of this section illustrates the features which control the flow stress at low to moderate temperatures. In particular, an increase in temperature or a decrease in applied strain rate provides an increase in the probability of thermal activation and therefore results in a reduction in flow stress. It should also be noted, however, that when ΔG^* is very small ($\simeq kT$), the Boltzmann distribution is no longer valid. Dislocation velocity is then determined not by thermally-activated release from obstacles, but by lattice dynamical effects, as discussed in section 3.5.

10.3 The Peierls Stress and Lattice Resistance

The applied resolved shear stress required to make a dislocation glide in an otherwise perfect crystal is called the *Peierls* (or *Peierls–Nabarro*) *stress*. It arises as a direct consequence of the periodic structure of the crystal lattice and depends sensitively on the form of the force–distance relation between individual atoms, i.e. on the nature of the interatomic bonding. It is a function of the core structure and, for this reason, a unique analytical expression for the Peierls stress cannot be derived. Some important features do emerge, however, from a qualitative assessment.

Dislocation Core Structure

Consider for simplicity a simple cubic crystal. When an extra half-plane of atoms is created by the presence of an edge dislocation of Burgers vector **b**, the atoms in planes above (A) and below (B) the slip plane are displaced by u, as illustrated in Fig. 10.4(a). To accommodate the dislocation, there is a *disregistry* of atomic coordination across the slip plane. It is defined as the *displacement difference* Δu between two atoms on adjacent sites above and below the slip plane i.e. $\Delta u = u(B) - u(A)$. The form of Δu, in units of b, versus x is shown in Fig. 10.4(b). The *width of the dislocation* w is defined as the distance over which the magnitude of the disregistry is greater than one-half of its maximum value, i.e. over which $-b/4 \leq \Delta u \leq b/4$. This parameter is shown in Fig. 10.4(b). The width provides a measure of the size of the dislocation core, i.e. the region within which the displacements and strains are unlikely to be close to the values of elasticity theory.

Typical forms of Δu curves for an edge dislocation in a simple cubic crystal are shown schematically in Fig. 10.5: in these cases, Δu has been increased by b when negative in order to produce continuous curves. The core widths of undissociated dislocations (Figs 10.5(a) and (b)) are usually found by computer simulation to be between b and $5b$, and to depend on the interatomic potential and crystal structure. Dissociation (Fig. 10.5(c)) can only occur when a stable stacking fault exists on the glide plane. The spacing d of the partials may be large and is determined principally by their elastic interaction and the stacking fault energy (equation (5.6)), whereas the width w of the partial cores is determined by inelastic atom–atom interactions. Another useful representation of core structure can be obtained simply from the derivative of the disregistry curve:

$$f(x) = \mathrm{d}(\Delta u)/\mathrm{d}x \tag{10.11}$$

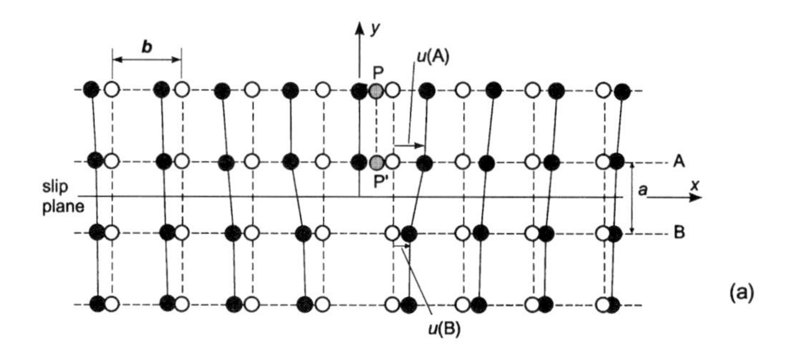

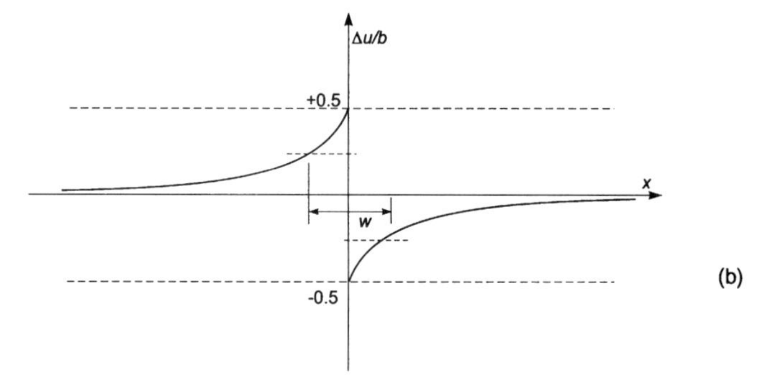

Figure 10.4 (a) Displacement of atoms at an edge dislocation. Open and full circles represent the atom positions before and after the extra half-plane is created. (b) Displacement difference Δu across the slip plane for the dislocation in (a). The width of the core in the slip plane is w.

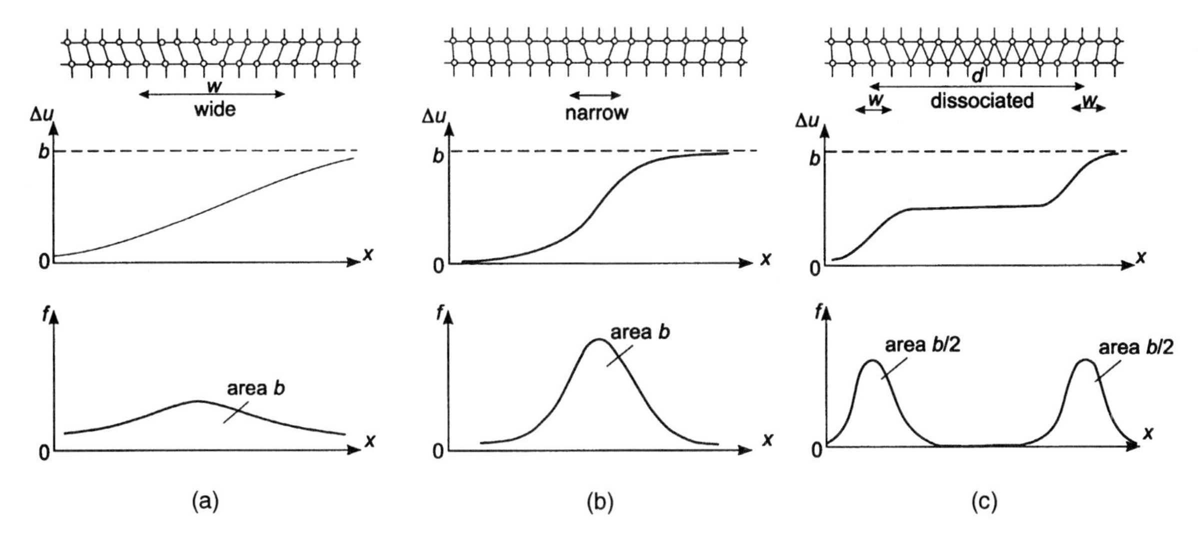

Figure 10.5 Atomic positions, disregistry Δu, and Burgers vector distribution f for (a) wide, (b) narrow and (c) dissociated edge dislocations in a simple cubic crystal.

The form of $f(x)$ for the three core structures of Fig. 10.5 is shown by the lower curve in each case. This function is known as the *distribution of Burgers vector* because the area under an $f(x)$ curve equals b. As can be seen from Fig. 10.5, the distribution curve shows clearly where the disregistry is concentrated: it is particularly useful when a dislocation consists of two or more closely-spaced partials. For non-planar cores in which the disregistry and Burgers vector are not distributed mainly on one plane – a situation sometimes found for screw dislocations – a two-dimensional plot in which the displacement difference is shown by the length of arrows can be illuminating. This procedure is illustrated for the core of the $\langle 111 \rangle$ screw dislocation in a body-centred cubic metal in Fig. 6.7. The displacement difference and Burgers vector distribution functions have proved to be particularly useful for presentation of details of the atomic structure of dislocation cores obtained by computer simulation (section 2.7).

The Peierls Barrier

The core disregistry imparts upon a dislocation a core energy and resistance to movement which are functions of the forces between atoms in the core region. In the first estimates of the lattice resistance (Peierls, 1940; Nabarro, 1947), it was assumed that the atom planes A and B (Fig. 10.4) interact by a simple sinusoidal force relation analogous to equation (1.4), and that in equilibrium the resulting disregistry forces on A and B are balanced by the elastic stresses from the two half-crystals above and below these planes. This condition provided an analytical solution for Δu, from which w was found to be $a/(1 - \nu)$ for an edge dislocation and a for a screw dislocation, where a is the interplanar spacing and ν is Poisson's ratio: the core is therefore 'narrow'. The dislocation energy was also found by combining the disregistry energy, calculated from Δu and the sinusoidal forces, with the elastic energy stored in the two half-crystals. It is similar in form to equation (4.22), with r_0 replaced by approximately $w/3$. When the dislocation in Fig. 10.4 moves to the right to PP', say, the atoms in planes A and B cease to satisfy the equilibrium distribution of Δu, and the disregistry energy increases. Peierls and Nabarro calculated the dislocation energy per unit length as a function of dislocation position, and found that it oscillates with period $b/2$ and maximum fluctuation (known as the *Peierls energy*) given by

$$E_p = \frac{Gb^2}{\pi(1-\nu)}\exp\left(\frac{-2\pi w}{b}\right) \tag{10.12}$$

The maximum slope of the periodic energy function is the critical force per unit length required to move the dislocation through the crystal. Dividing this by b therefore gives the Peierls stress

$$\tau_p = \frac{2\pi}{b^2}E_p = \frac{2G}{(1-\nu)}\exp\left(\frac{-2\pi w}{b}\right) \tag{10.13}$$

This simple model successfully predicts that τ_p is orders of magnitude smaller than the theoretical shear strength (equation (1.5)).

Although the Peierls model has now been superseded by more realistic approaches using computer simulation, it has qualitative features generally recognised to be valid. Slip usually occurs on the most widely-spaced planes, and a wide, planar core tends to produce low values of τ_p. For this reason, edge dislocations are generally more mobile than screws (section 3.5). Also, the Peierls energy can be very anisotropic, and dislocations will then tend to lie along the most closely-packed directions, for which τ_p is a maximum, as in Fig. 8.8. It should be noted, however, that the magnitudes of E_p and τ_p depend sensitively on the nature of the interatomic forces. τ_p is low ($\lesssim 10^{-6}$ to $10^{-5}\,G$) for the face-centred cubic and basal-slip hexagonal metals, in which dislocations dissociate, but is high ($\sim 10^{-2}\,G$) for covalent crystals such as silicon and diamond, in which dislocations have a preference for the $\langle 110\rangle$ orientations either parallel or at 60° to their $\frac{1}{2}\langle 110\rangle$ Burgers vector. The body-centred cubic metals, for which the $\langle 111\rangle$ screw does not have the planar form of the Peierls model (see section 6.3), and the prism-slip hexagonal metals fall between these extremes.

The energy of a dislocation as a function of position in the slip plane is illustrated in Fig. 10.6 for a dislocation lying predominantly parallel to the z-direction, which is a direction corresponding to a low core energy, e.g. one densely packed with atoms. The energy E_0 per unit length is given approximately by equations (4.22) and (4.23), but, as discussed above, superimposed on this is the fluctuation E_p (usually $\ll E_0$) due to the Peierls energy: it has a period a given by the repeat distance of the

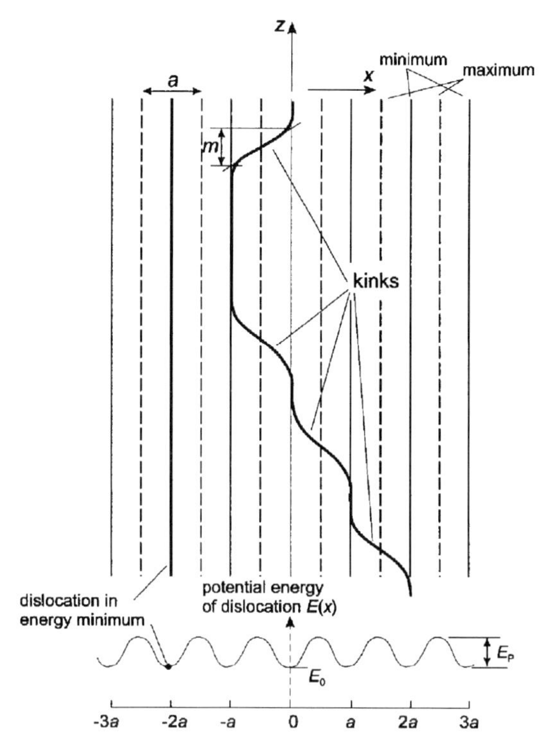

Figure 10.6 Schematic illustration of the potential energy surface of a dislocation line due to the Peierls stress τ_p. (After Seeger, Donth and Pfaff, *Disc. Faraday Soc.* **23**, 19, 1957.)

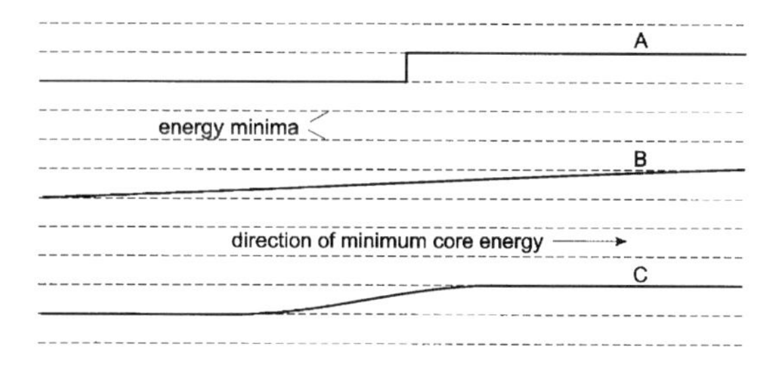

Figure 10.7 Shape of dislocations running almost parallel to a Peierls energy minimum. The energy per unit length of the dislocation is a minimum along the dashed lines and varies periodically at right angles to the lines. The shape of the dislocation (curve *C*) is somewhere between the extremes *A* and *B*.

lattice in the x-direction. If the dislocation is unable to lie entirely in one energy minimum, it contains kinks where it moves from one minimum to the next, as shown in Fig. 10.6. The shape and length m of a kink depend on the value of E_p, and result from the balance of two factors (Fig. 10.7): (i) the dislocation tends to lie as much as possible in the position of minimum energy – this factor alone gives $m = 0$ as for line *A*

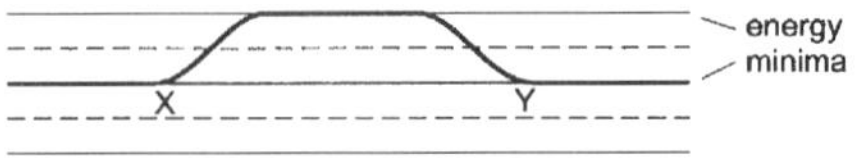

Figure 10.8 The process of double-kink nucleation.

– and (ii) the dislocation tends to reduce its energy by being as short as possible – this favours the straight line B with $m \gg a$. In practice, the shape falls between these extremes (line C), with high E_p producing low m and vice versa. The stress required to move a kink laterally along the line, and thus move the line from one energy minimum to the next, is less than τ_p, which is the stress required to move a long straight line rigidly over the energy hump E_p. Thus, at a low applied stress, pre-existing kinks can move laterally until reaching the nodes at the ends of the dislocation segments. The resulting plastic strain ('preyield microplasticity') is small and leaves long segments of line lying along energy minima. At 0 K, a stress of at least τ_p is therefore required for further (macroscopic) plastic flow. As the temperature is raised, however, there is an increasing probability that atomic vibrations resulting from thermal energy may enable the core to bulge from one minimum to the next over only part of the line (Fig. 10.8) and thus reduce the flow stress. This process of *double-kink nucleation* increases the line length and energy, however. Also, the two kinks X and Y are of opposite sign, i.e. they have the same **b** but approximately opposite line direction, and therefore tend to attract and annihilate each other. As a result, the double kink is not stable under an applied stress unless the length of the bulge, i.e. the spacing XY, is sufficiently large, typically $\simeq 20b$ and $\gg m$. Double-kink nucleation therefore has an activation energy which is a function of E_p.

The effect on the flow stress is as described in section 10.2. The flow stress for a given applied strain rate decreases with increasing temperature up to T_c as thermal activation becomes increasingly significant. Stress and strain rate at a given temperature are related by equation (10.4) with the Gibbs free energy of activation ΔG^* given by the empirical equation (10.10). The constants p and q in ΔG^* are found to be approximately 3/4 and 4/3 respectively (Frost and Ashby, 1982), and ΔF is the energy to form two well-separated kinks. It contains both elastic and core contributions, and it increases with E_p, but not in a simple way. It is of the order of $0.1\,Gb^3$ ($\simeq 1-2$ eV) for silicon, germanium and for $\langle 111 \rangle$ screw dislocations in the body-centred cubic transition metals, but is less than $0.05\,Gb^3$ ($\lesssim 0.2$ eV) for the face-centred cubic metals. The effect of lattice resistance on the flow stress in the latter case is therefore only significant at very low temperatures.

10.4 Interaction Between Point Defects and Dislocations

Point defects, i.e. vacancies, self-interstitials, and substitutional and interstitial impurities, interact with dislocations. The most important contribution to the interaction between a point defect and a dislocation

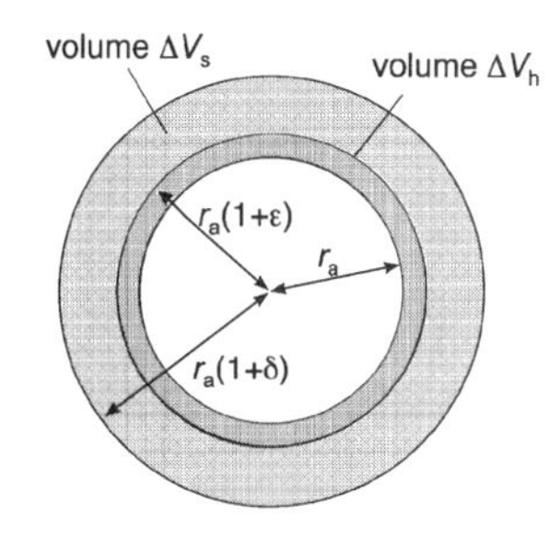

(a)

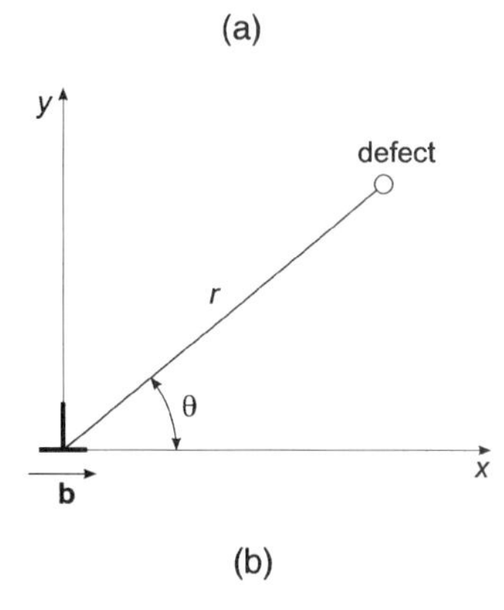

(b)

Figure 10.9 (a) Elastic model for a defect of natural radius $r_a(1+\delta)$ inserted in a hole of radius r_a. The final radius is $r_a(1+\varepsilon)$. (b) Geometry for the interaction of a defect with a dislocation lying along the z-axis.

is usually that due to the distortion the point defect produces in the surrounding crystal. The distortion may interact with the stress field of the dislocation to raise or lower the elastic strain energy of the crystal. This change is the *interaction energy* E_I. If the defect occupies a site where E_I is large and negative, work $|E_I|$ will be required to separate the dislocation from it. This situation may be met either when a dislocation glides through a crystal containing defects or by annealing at sufficiently high temperatures to enable defects to diffuse to such favoured positions. Whatever the cause, an increased stress will be required for slip and the crystal will be stronger, as discussed in sections 10.5 and 10.6. In this section, the form of E_I known as the Cottrell–Bilby formula is derived.

The simplest model of a point defect is an elastic sphere of natural radius $r_a(1+\delta)$ and volume V_s, which is inserted into a spherical hole of radius r_a and volume V_h in an elastic matrix (Fig. 10.9(a)). The sphere and matrix are isotropic with the same shear modulus G and Poisson's ratio ν. The difference between the defect and hole volumes is the *misfit volume* V_{mis}:

$$\begin{aligned} V_{\text{mis}} &= V_s - V_h \\ &\simeq 4\pi r_a^3 \delta \quad (\text{if } \delta \ll 1) \end{aligned} \tag{10.14}$$

The *misfit parameter* δ is positive for oversized defects and negative for undersized ones. On inserting the sphere in the hole, V_h changes by ΔV_h to leave a final defect radius $r_a(1+\varepsilon)$. The change ΔV_h is given by

$$\begin{aligned} \Delta V_h &= \frac{4}{3}\pi r_a^3 (1+\varepsilon)^3 - \frac{4}{3}\pi r_a^3 \\ &\simeq 4\pi r_a^3 \varepsilon \quad (\text{if } \varepsilon \ll 1) \end{aligned} \tag{10.15}$$

Parameter ε is determined by the condition that in the final state the inward and outward pressures developed on the sphere and hole surfaces are equal. The condition is

$$\Delta V_h = \frac{(1+\nu)}{3(1-\nu)} V_{\text{mis}}$$

i.e.

$$\varepsilon = \frac{(1+\nu)}{3(1-\nu)} \delta \tag{10.16}$$

The total volume change experienced by an infinite matrix is ΔV_h, for the strain in the matrix is pure shear with no dilatational part (section 8.4). In a finite body, however, the requirement that the outer surface be stress-free results in a total volume change given by

$$\Delta V = \frac{3(1-\nu)}{(1+\nu)} \Delta V_h \tag{10.17}$$

which, from equation (10.16), equals V_{mis}.

If the material is subjected to a pressure $p = -\frac{1}{3}(\sigma_{xx} + \sigma_{yy} + \sigma_{zz})$, the strain energy is changed by the presence of the point defect by

$$E_I = p\Delta V \tag{10.18}$$

For a dislocation, p is evaluated at the site of the defect. For a *screw* dislocation, $p = 0$ (section 4.3) and thus $E_I = 0$. However, for an *edge* dislocation lying along the z-axis (Fig. 10.9(b)), equations (4.17), (10.14) and (10.17) give

$$E_I = \frac{4(1+\nu)Gbr_a^3\delta}{3(1-\nu)} \frac{y}{(x^2+y^2)} \tag{10.19}$$

or, in cylindrical coordinates with $r = (x^2+y^2)^{1/2}$ and $\theta = \tan^{-1}(y/x)$,

$$E_I = \frac{4(1+\nu)Gbr_a^3\delta}{3(1-\nu)} \frac{\sin\theta}{r} \tag{10.20}$$

In terms of ε (equation (10.16)), E_I simplifies to

$$E_I = 4\,Gbr_a^3\varepsilon\frac{\sin\theta}{r} \tag{10.21}$$

but note that this equation would not be valid if the defect were, say, incompressible, for then equation (10.16) would be replaced by $\varepsilon = \delta$.

For an oversized defect ($\delta > 0$) E_I is positive for sites above the slip plane ($0 < \theta < \pi$) and negative below ($\pi < \theta < 2\pi$). This is because the edge dislocation produces compression in the region of the extra half-plane and tension below (section 4.3). The positions of attraction and repulsion are reversed for an undersized defect ($\delta < 0$). The value of r_a in equation (10.20) can be estimated from the fact that V_h is approximately equal to the volume per atom Ω for substitutional defects and vacancies, and is somewhat smaller for interstitial defects. A given species of defect produces strain in the lattice proportional to δ, and this parameter can be determined by measurement of the lattice parameter as a function of defect concentration. It is found to range from about -0.1 to 0 for vacancies, -0.15 to $+0.15$ for substitutional solutes, and 0.1 to 1.0 for interstitial atoms. The sites of minimum E_I, i.e. maximum *binding energy* of the defect to the dislocation, are at $\theta = \pi/2$ or $3\pi/2$, depending on the sign of δ. The strongest binding occurs within the dislocation core at $r \simeq b$, and from equation (10.20) the energy is approximately $G\Omega|\delta|$. It ranges from about $3|\delta|$ eV for the close-packed metals to $20|\delta|$ eV for silicon and germanium. Although the use of equation (4.17) for p within the dislocation core is questionable, these energy values may be considered upper limits and are within an order of magnitude of estimates obtained from experiment and computer simulation.

The interaction of a dislocation with a spherically symmetric defect is a special case of a more-general *size-effect* interaction. Many defects occupy sites having lower symmetry than the host crystal and thereby

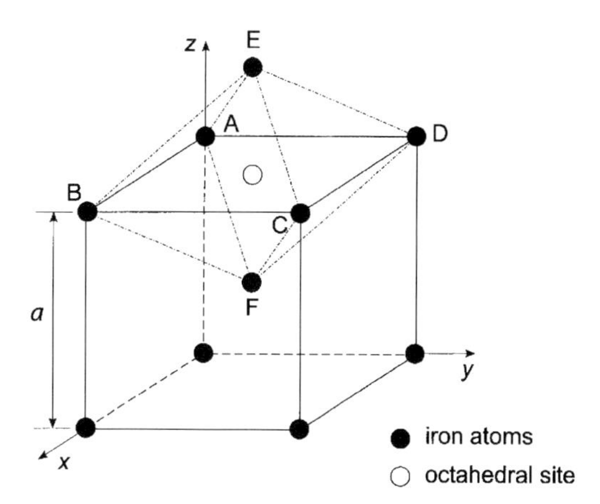

Figure 10.10 Octahedral interstitial site in a body-centred cubic cell.

produce asymmetric distortions. Unlike the spherical defect, the asymmetric defect interacts with both hydrostatic and shear stress fields, and therefore interacts with edge *and* screw dislocations. A well-known example of such a defect is the interstitial solute atom in a body-centred cubic metal, e.g. carbon in α-iron. The defect geometry is illustrated in Fig. 10.10. The most favourable site is the *octahedral* site at the centre of a cube face or, equivalently, at the centre of a cube edge. The interstitial has two nearest-neighbour atoms E and F a distance $a/2$ away and four second-nearest neighbours A, B, C and D a distance $a/\sqrt{2}$ away. An interstitial atom with a diameter larger than $(a - 2R)$, where R is the radius of an iron atom, produces a tetragonal distortion by displacing atoms E and F more strongly than the second-nearest neighbours. The misfits in the x, y, z-directions for carbon in iron are estimated to be

$$\delta_{xx} = \delta_{yy} = -0.05; \quad \delta_{zz} = +0.43 \tag{10.22}$$

The interaction energy with a dislocation is, by analogy with equation (10.18)

$$E_I = -(\sigma_{xx}\delta_{xx} + \sigma_{yy}\delta_{yy} + \sigma_{zz}\delta_{zz})\frac{4}{3}\pi r_a^3 \tag{10.23}$$

where σ_{xx}, σ_{yy} and σ_{zz} are the stresses produced by the dislocation. For the screw dislocation, which lies along $\langle 111 \rangle$, the magnitude of E_I is comparable with that for the edge. In both cases E_I is proportional to r^{-1}, as in equation (10.20), but has a more complicated orientation dependence than $\sin\theta$.

A second form of elastic interaction arises if the defect is considered to have different elastic constants from the surrounding matrix. Point defects can increase or decrease the elastic modulus. A vacancy, for example, is a soft region of zero modulus. Both hard and soft defects induce a change in the stress field of a dislocation, and this produces the *inhomogeneity* interaction energy. The interaction is always attractive for soft defects, because they are attracted to regions of high elastic energy density which they tend to reduce. (This is analogous to the attraction of dislocations to a free surface (section 4.8)). Hard defects are repelled. The dislocation strain energy over the defect volume is proportional to $1/r^2$ for both edges and screws (see equations (4.9), (4.15) and (4.16)), and the interaction energy therefore decreases as $1/r^2$. Although it is of second order in comparison with the $1/r$ size-effect interaction, it can be important for substitutional atoms and vacancies when the misfit parameter δ is small.

There are sources of interaction between point defects and dislocations which are not included in the elastic model. One is the *electrical interaction*. In *metals*, the conduction electron density tends to increase in the dilated region below the half-plane of the edge dislocation and to decrease in the compressed region. The resulting electric dipole could, in principle, interact with a solute atom of different valency from the solvent atoms, but free-electron screening is believed to reduce the interaction to negligible proportions in comparison with the size and inhomogeneity effects. In *ionic* crystals, lack of electron screening leads to a more significant interaction, but of more importance is the presence of charged jogs. The electric charge associated with jogs (section 6.4) produces a $1/r$ Coulomb electrostatic interaction with vacancies and charged impurity ions, and this probably dominates the total interaction energy in such materials. In *covalent* crystals, the relative importance of the variety of possible effects is less clearly understood. In addition to the elastic interaction and the electric-dipole interaction with impurities of different valency, an electrostatic-interaction can occur if conduction electrons are captured by the dangling bonds which may exist in some dislocation cores (section 6.6).

Finally, when a perfect dislocation dissociates into partial dislocations, the ribbon of stacking fault formed changes the crystal structure locally. For example, the faulted stacking sequence is hexagonal-close-packed in face-centred cubic metals (section 5.3) and face-centred cubic in the hexagonal metals (section 6.2). Thus, the solute solubility in the fault region may be different from that in the surrounding matrix. The resulting change in chemical potential will cause solute atoms to diffuse to the fault. This *chemical* (or *Suzuki*) *effect* is not a long-range interaction, but it may present a barrier to the motion of dissociated dislocations.

10.5 Solute Atmospheres and Yield Phenomena

Dislocation Locking

In a crystal containing a solution of point defects, the energy of a defect in the vicinity of a dislocation is changed by E_I, given by the sum of the terms discussed above. The equilibrium defect concentration c at a position (x, y) near the dislocation therefore changes from c_0, the value a long way from the dislocation, to

$$c(x, y) = c_0 \exp[-E_I(x, y)/kT] \qquad (10.24)$$

(This formula assumes the concentration c is weak, i.e. fewer than half the core sites are occupied, and that the defects do not interact with each other.) Defects therefore tend to congregate in core regions where E_I is large and negative, and dense atmospheres of solute atoms can form in otherwise weak solutions with small values of c_0. The condition for formation of these *Cottrell atmospheres* is that the temperature is sufficiently high for defect migration to occur but not so high that the entropy contribution to the free energy causes the atmosphere to 'evaporate' into the solvent matrix. It is readily shown from equation (10.24) that even in dilute solutions with, say, $c_0 = 0.001$, dense atmospheres with $c \geq 0.5$ may be expected at $T = 0.5T_m$, where T_m is the melting point, in regions where $-E_I \gtrsim 3kT_m$. This corresponds to a defect–dislocation binding energy of typically 0.2 to 0.5 eV for metals, which is fairly commonplace for interstitial solutes.

The kinetics of the migration of defects to dislocations are governed by the drift of defects under the influence of the interaction energy E_I superimposed on the random jumps associated with diffusion. The path taken by a defect depends on the precise form of the angular part of E_I. For example, a misfitting spherical defect interacts with a straight edge dislocation according to the form (equations (10.19) (10.20))

$$E_I = A\frac{y}{(x^2 + y^2)} = A\frac{\sin\theta}{r} \qquad (10.25)$$

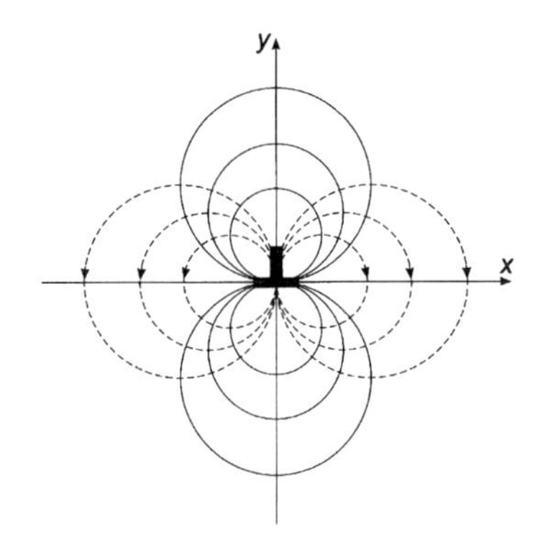

Figure 10.11 Equipotential contours (full lines) for the elastic interaction potential given by equation (10.25), between an edge dislocation along the z-axis with Burgers vector parallel to the x-axis and solute atoms. Broken lines are lines of flow for oversized solute atoms migrating to the dislocation.

so that lines of constant E_I are circles centred on the y-axis as shown in Fig. 10.11. A defect will tend to migrate into the core region by following a path which is everywhere perpendicular to a constant energy contour. Such a flow line is itself a circle which is centred on the x-axis, as indicated by the broken lines in Fig. 10.11. However, the rate of arrival of defects at the core depends on the exponent of r in E_I, rather than the shape of the flow lines. Solution of the appropriate diffusion equations shows that the rate varies with time t as $t^{2/3}$ for first-order interactions ($E_I \propto r^{-1}$) and $t^{1/2}$ for second-order ones ($E_I \propto r^{-2}$). The former dependence is that obtained by Cottrell and Bilby for carbon solutes in α-iron. Within the core itself, the arrival of defects can have several effects. Vacancies and self-interstitials can cause climb, as discussed in Chapter 3, or nucleate dislocation loops heterogeneously. Solute atoms in high local concentrations may interact to form precipitates along the core, such as those forming carbides in Fig. 10.12 and those discussed in the decoration technique in section 2.3. On the other hand, solutes may remain in solution as a dense atmosphere. Both effects can annul the long-range stress field of the dislocation and solute flow then ceases.

The importance of solute segregation in or near the core as far as mechanical properties are concerned, is that an extra stress is required to

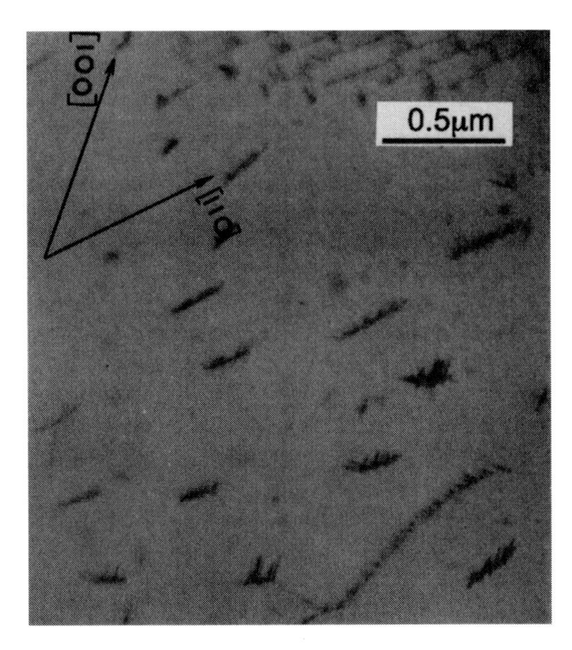

Figure 10.12 Transmission electron micrograph of carbide precipitate particles formed along dislocations in iron. The particles are in the form of platelets and are viewed edge-on. (From Hull and Mogford, *Phil. Mag.* **6**, 535, 1961.)

overcome the attractive dislocation–defect interactions and move the dislocation away from the concentrated solute region. The dislocation is said to be *locked* by the solutes. Once a sufficiently high stress has been applied to separate a dislocation from the defects, its movement is unaffected by locking. However, subsequent heat treatment which allows the defects to diffuse back to the dislocations re-establishes locking: this is the basis of *strain ageing*. Also, at sufficiently high deformation temperatures and low strain rates, point defect mobility may enable solutes to repeatedly lock dislocations during dislocation motion. This produces a repeated yielding process known as *dynamic strain ageing* (or *Portevin–Le Chatelier effect*), characterised by a serrated stress–strain curve.

The unlocking stress is treated here for the case of a row of misfitting spherical defects lying in the position of maximum binding along the core of a straight edge dislocation, as illustrated in Fig. 10.13. They represent, therefore, a small atmosphere of, say, carbon atoms (or possibly very small coherent precipitates). It is assumed that the core sites are saturated, i.e. that every defect site (of spacing b) along the core is occupied. (Note that even at saturation, the number of solutes per unit volume in such atmospheres $\approx \rho/b$, where ρ is the dislocation density, is only a small fraction of the total number in the crystal $\simeq c_0/b^3$. For example, taking $\rho = 10^6\,\text{mm}^{-2}$ and $b = 0.3\,\text{nm}$, the ratio $\rho b^2/c_0$ is only $10^{-7}/c_0$.) If the position of the solutes is r_0 ($\simeq b$) below the slip plane, as

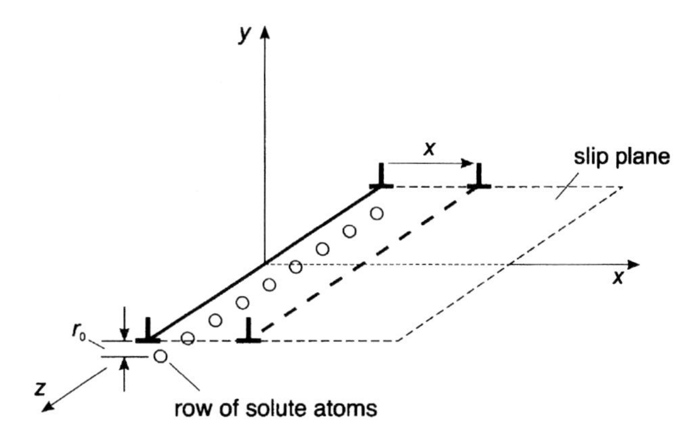

Figure 10.13 Diagrammatic representation of a row of solute atoms lying in the position of maximum binding at an edge dislocation. An applied shear stress will cause the dislocation to separate from the solute atoms by gliding in the slip plane to position x.

shown, the interaction energy of the dislocation with one defect when it has been displaced to position x in the slip plane is, from equation (10.25)

$$E_I(x) = \frac{A(-r_0)}{(x^2 + r_0^2)} \tag{10.26}$$

The force in the x-direction on the line in this position is

$$K(x) = \frac{-\mathrm{d}E_I(x)}{\mathrm{d}x} = \frac{-2Ar_0x}{(x^2 + r_0^2)^2} \tag{10.27}$$

The force due to the $1/b$ defects per unit length is $K(x)/b$. Thus the resolved shear stress necessary to displace the line to position x is

$$\tau = -\frac{1}{b^2}K(x) = \frac{2Ar_0x}{b^2(x^2 + r_0^2)^2} \tag{10.28}$$

It has a maximum τ_0 at $x = r_0/\sqrt{3}$ given by

$$\tau_0 = \frac{3\sqrt{3}A}{8b^2r_0^2} \tag{10.29}$$

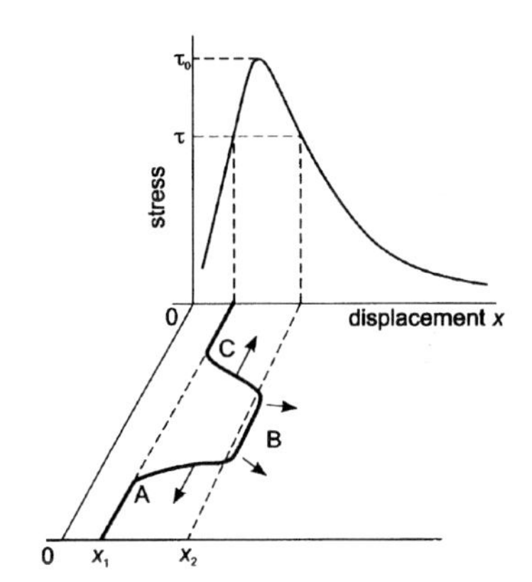

Figure 10.14 Separation of a dislocation from a row of condensed solute atoms. Initially the dislocation lies along the line $x = 0$. Under the applied stress τ it moves forward to the position of stable equilibrium x_1. To break away it must reach the position of unstable equilibrium x_2. *ABC* represents a loop of dislocation, formed by a thermal stress fluctuation, in the process of breaking away. (From Cottrell (1957), *Properties of Materials at High Rates of Strain*, Instn. Mech. Eng., London.)

The form of τ is shown in Fig. 10.14. From equation (10.20), $A \simeq Gb\Omega|\delta|$, so that taking $r_0 = b$, $\tau_0 \simeq 0.2G|\delta|$. This is undoubtedly an upper limit, since it is based on the application of elasticity within the dislocation core. Nevertheless, it demonstrates that a considerable increase in the yield strength may be expected from concentrated solute atmospheres.

The estimate for τ_0 is the yield stress at 0 K. At non-zero temperatures, thermal activation (section 10.2) may assist unlocking and reduce the applied stress required. The mechanism is illustrated in Fig. 10.14, where a straight dislocation AC under an applied stress τ ($<\tau_0$) is held in stable equilibrium at position x_1 until thermal fluctuations assist part of the line B to overcome the resisting force. The process is analogous to double-kink nucleation over the Peierls barrier (section 10.3), except that $(x_2 - x_1)$ depends on the atmosphere dimensions and is not simply the lattice spacing. The analysis of Cottrell and Bilby gives an activation energy in the form of equation (10.10) with $p = 1$ and $q = 1.5$. The applied stress for unlocking therefore decreases rapidly from τ_0 with increasing temperature. However, as can be deduced from the preceding analysis, the predicted values of barrier energy ΔF for dense atmospheres and core precipitates are frequently too large for thermally-activated unpinning to be responsible for the temperature dependence of the yield stress. In materials in which strong atmospheres form it is probable that only a small fraction of locked dislocations become free, as a result, perhaps, of local stress concentrations. The temperature-dependence of the yield stress is then dominated by other effects, such as the Peierls stress and solution strengthening. Nevertheless, locking may still have a significant influence on the actual yield behaviour, as discussed below.

Yield Drop

The presence of strong locking effects can lead in some alloys to a pronounced drop in the stress required for plastic deformation immediately after yielding. The stress–strain curve that results is shown in Fig. 10.1(c). If the deformed specimen is unloaded and reloaded immediately, a yield drop does not accompany the onset of plastic flow a second time. However, if the sample is given a suitable anneal prior to reloading, strain ageing causes the yield drop to return. The principles involved in the yield drop phenomenon are as follows.

Consider a tensile test, shown schematically in Fig. 10.15. The elasticity of the machine (including the load cell) and specimen is represented by an imaginary spring. The crosshead moves at a constant speed $s = \mathrm{d}l/\mathrm{d}t$ so that the cross head displacement at time t is st. The total elastic displacement of the spring is KF where F is the applied force and K the spring constant. The plastic elongation of the specimen is $\varepsilon_p l_0$ where ε_p is the plastic strain and l_0 is the original gauge length. Thus

$$st = KF + \varepsilon_p l_0 \tag{10.30}$$

The plastic strain of the specimen is therefore

$$\varepsilon_p = \frac{st - KF}{l_0} \tag{10.31}$$

and the plastic strain rate is

$$\dot{\varepsilon}_p = \frac{\mathrm{d}\varepsilon_p}{\mathrm{d}t} = \frac{s - K\dfrac{\mathrm{d}F}{\mathrm{d}t}}{l_0} \tag{10.32}$$

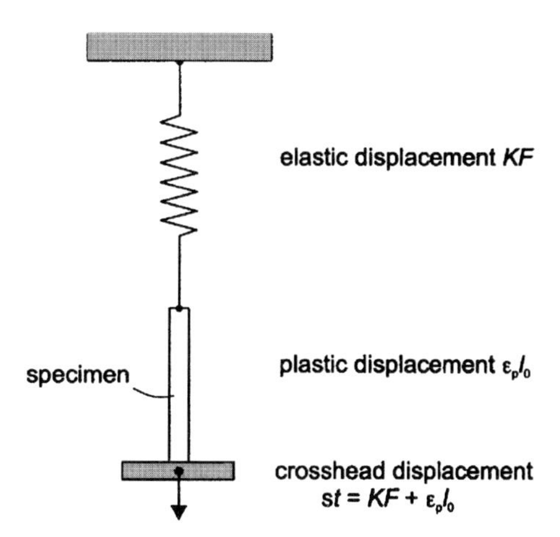

Figure 10.15 Schematic diagram of a tensile machine. The spring represents the elastic properties of the machine and specimen. (After Johnston and Gilman, *J. Appl. Phys.* **30**, 129, 1959.)

Taking the resolved shear stress τ as $F/2A_0$, where A_0 is the original cross-sectional area of the specimen, equation (10.30) also gives

$$\frac{\mathrm{d}\tau}{\mathrm{d}t} = \frac{s - \dot{\varepsilon}_p l_0}{2A_0 K} \tag{10.33}$$

Since $s = \mathrm{d}l/\mathrm{d}t$, the measured hardening rate

$$\frac{\mathrm{d}\tau}{\mathrm{d}l} = \frac{1}{2A_0 K}\left(1 - \frac{\dot{\varepsilon}_p l_0}{s}\right) \tag{10.34}$$

can be seen to be dependent on the elastic properties of the machine and specimen and the instantaneous plastic strain rate of the specimen.

The shape of the stress–strain curve given by equation (10.34) can be related to the dislocation behaviour through the plastic strain rate term $\dot{\varepsilon}_p$ using equation (3.10). However, since both ρ_m and $\bar{v}$ will vary with stress and strain, it is necessary to know the variation of both before any predictions can be made. This information is only available in isolated cases. Johnston (1962) reviewed the effects of all the machine and specimen dislocation variables on the stress–strain curve and some of his results are illustrated in Fig. 10.16. The calculations were based on data obtained from LiF crystals in which the measured dislocation density increases approximately linearly with strain in the early stages and the dislocation velocity can be related to the applied shear stress through a relation of the type (3.3a), namely

$$\bar{v} \propto \tau^m \tag{10.35}$$

where m was taken as 16.5 for LiF.

Figure 10.16(a) shows the effect of changing the *initial* density of mobile dislocations from 0 to $5 \times 10^{10}\,\mathrm{m}^{-2}(= 5 \times 10^6\,\mathrm{cm}^{-2})$. The line OA represents the situation where there is no dislocation movement and $\dot{\varepsilon}_p = 0$ in equation (10.34), i.e. completely elastic behaviour, $d\tau/dl = (2A_0K)^{-1}$. The curves show that a sharp yield drop is obtained for low initial values of ρ_m and that the yield drop decreases as ρ_m increases. The curves D, E, F in Fig. 10.16(a) have been displaced from the origin for convenience. Figure 10.16(b) shows the calculated stress–strain curves for different values of the parameter m in relation (10.35). The yielding shows a strong dependence on m. These effects are best understood by noting from equation (3.10) that at a constant plastic strain rate

$$\rho_{mu}\bar{v}_u = \rho_{ml}\bar{v}_l \tag{10.36}$$

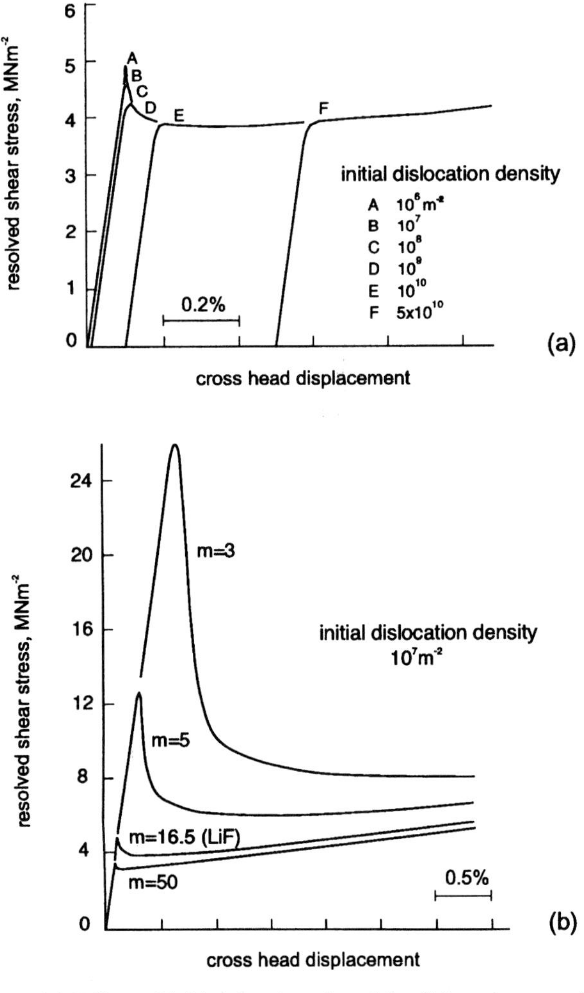

Figure 10.16 (a) Effect of initial density of mobile dislocations on the yield point. (b) Effect on the yield point of changing the stress dependence of dislocation velocity for an initial mobile dislocation density of 10^3 cm^{-2}. (After Johnston, *J. Appl. Phys.* **33**, 2716, 1962.)

where subscripts u and l refer to values at upper and lower yield points respectively (see Fig. 10.1(c)). From relations (10.35) and (10.36), the ratio of the upper to lower yield stresses is

$$\frac{\tau_u}{\tau_l} = \left[\frac{\rho_{ml}}{\rho_{mu}}\right]^{1/m} \qquad (10.37)$$

The ratio is therefore largest for small m and/or large (ρ_{ml}/ρ_{mu}). The value of m varies considerably from one material to another (section 3.5). In

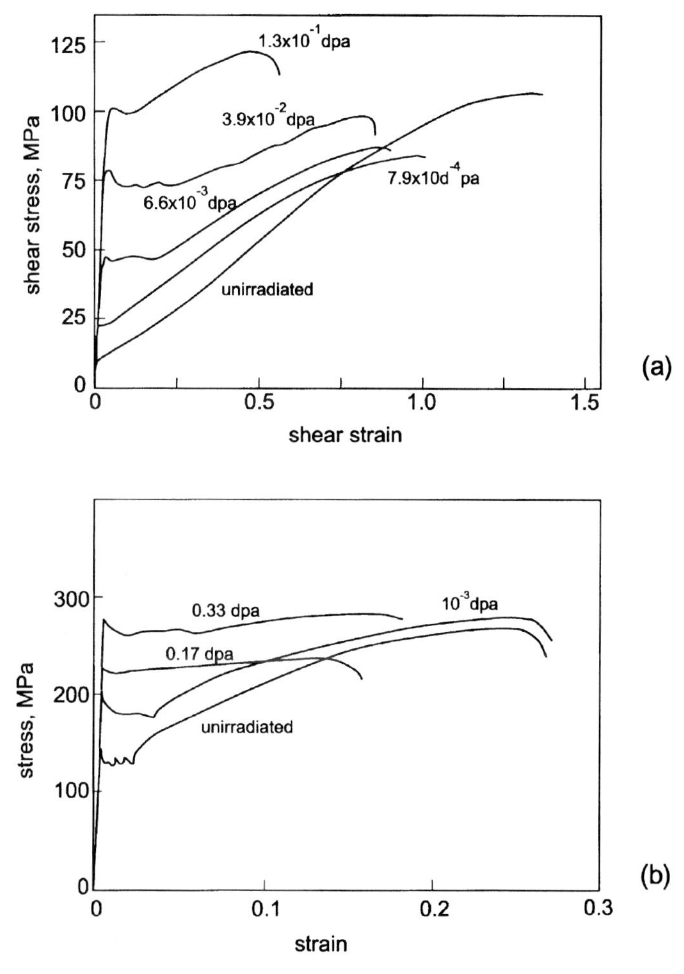

Figure 10.17 Stress–strain curves for (a) copper single crystals (⟨110⟩ orientation) irradiated by 590 Me V protons and (b) polycrystalline iron irradiated by fast neutrons to different doses at room temperature. The damage dose is given as the average number of times each atom is displaced by the radiation (displacements per atom (dpa)). (Reprinted from Victoria, Baluc, Bailat, Dai, Luppo, Schaublin and Singh, *J. Nucl. Mater.* **276**, 114, 2000, with permission from Elsevier Science.)

α-iron, for example, where it is large ($m \simeq 40$), equation (10.37) shows that ρ_m has to increase by 10^4 times at yield for the stress to drop by 20 per cent. That is, if $\rho_{ml} = 10^7\ \text{cm}^{-2}$, $\rho_{mu} = 10^3\ \text{cm}^{-2}$. Since the *total* dislocation density initially might be typically 10^6 to $10^7\ \text{cm}^{-2}$, locking plays an important role in ensuring that the initial mobile density ρ_{mu} is small.

The physical interpretation of this analysis is as follows. When a crystal containing a low density of dislocations which are free to glide is strained using a constant crosshead speed s, the dislocations cannot move fast enough at low stresses to produce sufficient strain in the specimen. The stress in the specimen (and spring K) therefore rises, and as it does so the dislocations multiply rapidly and move faster, as implied by relation (10.35). The stress stops rising (point E on Fig. 10.1(c)) when $\text{d}\tau/\text{d}l = 0$, which from equation (10.34) occurs when $\dot{\varepsilon}_p = b\rho_m\bar{v} = s/l_0$, i.e. the strain rate of the specimen equals the applied strain rate. However, multiplication continues with increasing strain, producing more than enough dislocations and $b\rho_m\bar{v} > s/l_0$. The stress therefore drops until $\bar{v}$ is slow enough for the two strain rates to match. In general, three factors favour this phenomenon. First, the initial mobile dislocation ρ_{mu} must be small; second, the dislocation velocity must not increase too rapidly with increasing stress, i.e. small m; and third, the dislocations must multiply rapidly. The first is the most important.

Irradiation Hardening and Dislocation Channelling

A phenomenon that illustrates several features discussed in this section is found to occur in metals irradiated with energetic atomic particles, e.g. fast neutrons in a fission reactor. The incident particles cause *radiation damage* by displacing many atoms from their lattice sites by collisions, thereby creating supersaturations of vacancies and self-interstitial atoms. Many collect together, forming interstitial and vacancy dislocation loops or vacancy stacking-fault tetrahedra (sections 3.7, 5.5, 5.7, 6.2 and 6.3), and others diffuse to dislocations, causing climb, or to other defect sinks such as grain boundaries. The loops and small clusters that remain in the lattice act as obstacles to dislocation motion in a similar way to that described for alloys in the next section, and raise the stress required for yield and subsequent plastic flow, as seen in Fig. 10.17(a) and (b) for proton-irradiated single crystals of copper and neutron-irradiated polycrystalline iron, respectively. This is *irradiation hardening*.

As noted in section 6.3, small loops consisting of clusters of self-interstitials may be quite mobile and so they will be able to migrate to attractive regions where their energy is lowered just below the core of edge dislocations. They can form atmospheres, locking the dislocations and causing a yield drop in metals where it is not usually observed, e.g. copper at high irradiation dose as in Fig. 10.17(a). In some circumstances, dislocation motion at yielding creates regions almost free of obstacles, as seen in the example of Fig. 10.18. The mechanisms

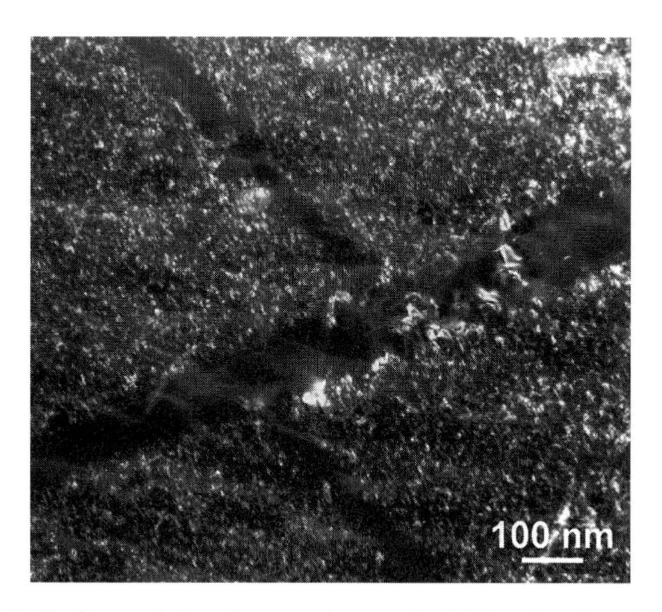

Figure 10.18 Transmission electron micrograph of dislocation channelling in a palladium single crystal irradiated by 590 MeV protons to a damage dose of 0.12 dpa. The sample contains a high density of point defect clusters at this dose and when yielding occurs channels parallel to {111} slip planes are created that are almost free of defects. (Reprinted from Victoria, Baluc, Bailat, Dai, Luppo, Schaublin and Singh, *J. Nucl. Mater.* **276**, 114, 2000, with permission from Elsevier Science.)

involved in this *dislocation channelling* are not understood, but may be associated with the ability of gliding dislocations to absorb by climb the self-interstitial atoms and vacancies in the clusters they encounter during glide, so that channelling can occur by repeated glide and climb.

10.6 The Flow Stress for Random Arrays of Obstacles

The ability to confer increased strength on a crystal of one element by the deliberate addition of atoms of other elements is one of the important achievements of materials technology. It is possible to raise the yield strength of metals such as aluminium, copper and nickel, for example, up to 100 or more times the value for the pure elements by suitable choice of alloying additions and heat treatment. Clearly, alloying introduces barriers to the motion of dislocations, but the mechanism by which dislocations are hindered depends on the form the solute atoms adopt within the solvent lattice. These microstructural features are assessed in section 10.7. In the present section, general expressions for the flow stress at 0 K are derived, and it is sufficient to know that the obstacles exert forces on a dislocation as it moves on its slip plane. The theory of obstacle strengthening can be complicated but the important features can be distinguished with the simple approach adopted here.

An obstacle may be classed as *strong* or *weak* depending on whether or not a dislocation bends through a large angle in its vicinity. Also, an obstacle may be *localised* or *diffuse* depending on whether or not the force it exerts on the line is confined to a small part of the line or is distributed approximately uniformly over a long length. The latter situation is considered first.

Diffuse Forces

Each obstacle is assumed to create an internal shear stress τ_i in the matrix proportional to the misfit parameter δ (sections 8.4 and 10.4). This produces a force on a dislocation of maximum value $\tau_i b$ per unit length. It is further assumed that τ_i is constant over a region of size Λ, the spacing of the obstacles. An obstacle can therefore bend a dislocation against the opposition of its line tension to a radius of curvature R given by (equation (4.30))

$$R = \alpha Gb/\tau_i \simeq Gb/2\tau_i \tag{10.38}$$

Strong forces result in $R \lesssim \Lambda$ and weak ones in $R \gg \Lambda$. *Strong* forces therefore bend the line sufficiently for it to follow the contours of minimum interaction energy with each obstacle, as shown in Fig. 10.19(a), and the flow stress is that required to overcome the mean internal stress:

$$\tau_{\text{diff}}(\text{strong}) \simeq \frac{2}{\pi}\tau_i \tag{10.39}$$

Line tension prevents the dislocation bending round individual *weak* obstacles, and it samples regions of both high and low interaction energy, trying to avoid the former (Fig. 10.19(b)). During plastic flow, the line advances by segments of length L ($\gg\Lambda$) moving independently of each other from one position of low energy to another. Over the length L, there are $n \simeq L/\Lambda$ obstacles, and in a random array, there will be an excess $\sqrt{n}$ acting in one direction on the line. The flow stress is

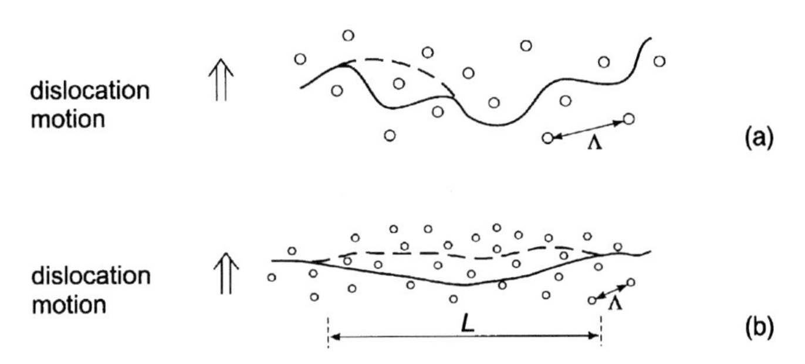

Figure 10.19 Dislocation on a slip plane in which obstacles of mean spacing Λ exert diffuse forces. It advances from the full to the dashed line shape as indicated. The obstacle forces are strong in (a) and weak in (b). (After Nabarro (1975).)

therefore that required for the segment L to overcome the mean internal stress of $\sqrt{n}$ obstacles, each of which acts over length Λ, i.e.

$$\tau bL \simeq \sqrt{n}\frac{2}{\pi}\tau_i b\Lambda$$

therefore

$$\tau \simeq \frac{2\tau_i}{\pi}\left(\frac{\Lambda}{L}\right)^{1/2} \tag{10.40}$$

It has been recognised from the time of the earliest theories that it is difficult to estimate L. Here, it is assumed that the segment L has radius of curvature R ($\gg L$) equal to $Gb/2\tau$, and that its midpoint moves a distance Λ at each advance forward. Then, from the geometry of a circle:

$$\frac{\Lambda}{2} = R - [R^2 - (L/2)^2]^{1/2} \simeq \frac{L^2}{8R} \tag{10.41}$$

Replacing R by $Gb/2\tau$ to find L and substituting in equation (10.40), the flow stress to move a dislocation through an array of weak, diffuse forces is found to be

$$\tau_{\text{diff}}(\text{weak}) \simeq \frac{2}{\pi^{4/3}}\tau_i\left[\frac{\tau_i\Lambda}{Gb}\right]^{1/3} \tag{10.42}$$

Other assumptions change both the numerical constant and the exponent in this expression.

Localised Forces

In this situation, the dislocation experiences no internal stress in regions between the obstacles, and the resistance offered by each obstacle is represented by a force K acting at a point on the line. Under an applied shear stress τ, the line bows between neighbouring obstacles to a radius of curvature $R \simeq Gb/2\tau$, as shown in Fig. 10.20(a). From equilibrium between the line tension forces T at an obstacle and K:

$$K = 2T\cos\phi \simeq Gb^2\cos\phi \tag{10.43}$$

If the obstacle spacing is l (Fig. 10.20(a)), each obstacle resists the forward force τbl on a segment of line of initial length l. The dislocation breaks away from an obstacle when τbl equals the maximum resisting force K_{max}. The flow stress for a regular square away of localised obstacles is therefore

$$\tau_{\text{loc}} = \frac{K_{\text{max}}}{bl} = \frac{2T}{bl}\cos\phi_c \simeq \frac{Gb}{l}\cos\phi_c \tag{10.44}$$

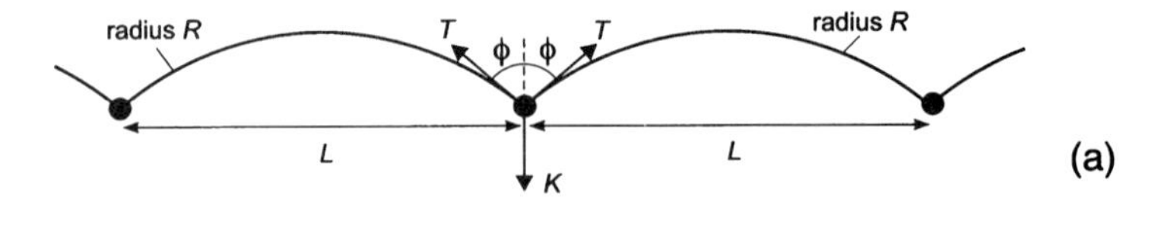

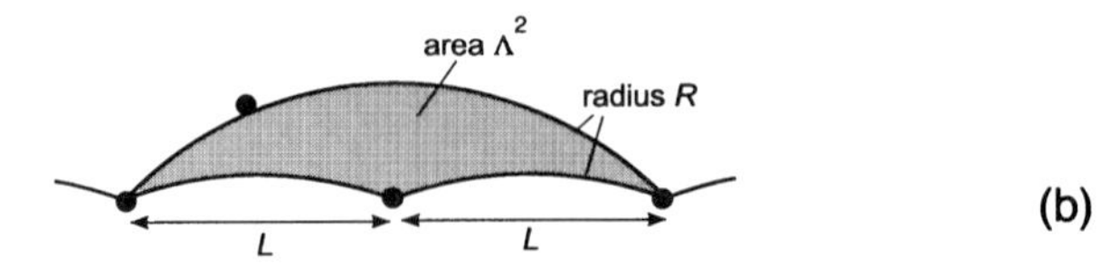

Figure 10.20 (a) Dislocation on a slip plane in which each obstacle exerts a localised glide resistance force K, balanced in equilibrium by line tension forces T. (b) The area used to calculate the effective obstacle spacing l for weak forces.

where the obstacle strength is characterised by the critical angle $\phi_c = \cos^{-1}(K_{\max}/2T)$ at which breakaway occurs.

In random arrays of *weak* obstacles, for which $\phi_c \simeq \pi/2$, a dislocation advances by breaking from a few widely-spaced obstacles before unpinning along the entire length. It retains a roughly straight shape. The effective obstacle spacing l along an almost straight line is given by the *Friedel relation*, as follows. In steady state at the flow stress, when a dislocation unpins from one obstacle it moves forward to encounter one other. The area swept in so doing is therefore the area of glide plane per obstacle Λ^2, where Λ is the average spacing of obstacles in the plane. From Fig. 10.20(b), this area is simply the area of one large segment of a circle of radius R minus that of two others, and is

$$\Lambda^2 \simeq \frac{2l^3}{3R} - \frac{l^3}{6R} \quad (\text{if } l \ll R)$$
$$\simeq \frac{l^3\tau}{Gb} \tag{10.45}$$

From equation (10.44) the effective obstacle spacing l is therefore $\Lambda/(\cos\phi_c)^{1/2}$, so that using the same equation, the flow stress for weak localised obstacles is

$$\tau_{\text{loc}}\,(\text{weak}) \simeq \frac{Gb}{\Lambda}(\cos\phi_c)^{3/2} \tag{10.46}$$

Dislocations in random arrays of *strong* obstacles ($\phi_c \simeq 0$) move in regions where the obstacles are relatively widely spaced (and the glide resistance correspondingly small). This increases the effective obstacle spacing and the flow stress is

$$\tau_{\text{loc}}(\text{strong}) \simeq 0.84\frac{Gb}{\Lambda}(\cos\phi_c)^{3/2} \tag{10.47}$$

10.7 The Strength of Alloys

To apply the analyses of the previous section to real alloys, it is necessary to consider the relative values of the flow stress given by equations (10.39), (10.42), (10.46) and (10.47). First, however, a simplified account of the effects of alloying in obstacle structure is presented.

Solutions, Precipitates and Ageing

There is usually a maximum concentration to which the atoms of element B can be dissolved in solution in a crystal of element A. For concentrations below this *solubility limit*, any strengthening effect is due to the resistance to dislocation motion created by the atoms of B dispersed randomly in the A lattice. Important examples include alpha-brass (Cu alloyed with Zn), the titanium alloy Ti–6Al–4V and the 5000 series aluminium alloys (Al–Mg).

For concentrations of solute above the solubility limit, the excess B atoms tend to precipitate within particles of either element B or a compound of A and B. The solubility limit varies with temperature, however. It is therefore possible to have an alloy composition which is less than the limit at one temperature but exceeds it at another. An example afforded by the aluminium–copper system is shown by the equilibrium phase diagram in Fig. 10.21. The α-phase is a *solid solution* of copper atoms dissolved substitutionally in aluminium. At 550 °C, the solubility limit is almost 6 wt per cent, but at room temperature it is less than 0.1 wt per cent. Thus, the alloy 'Duralumin', which contains 4 wt per cent copper, is a homogeneous solid solution at the *solution treatment* temperature of 550 °C. But on slowly cooling the alloy, the second phase starts to precipitate out at the temperature indicated by point x, and at 20 °C the alloy consists of large $CuAl_2$ precipitates (θ-phase) in equilibrium with the aluminium-rich matrix. However, if the solution-treated alloy is quenched to room temperature, insufficient time is allowed for precipitation and a *super-saturated* solid solution is obtained. On subsequent annealing at 150 °C, say, there is just sufficient thermal energy available for precipitate *nucleation* to occur. Initially, a fine dispersion of numerous small precipitates is formed, but with increasing ageing time at the elevated temperature, the dispersion coarsens. The number of precipitates decreases and their size and spacing increases, and eventually, after very long times, the equilibrium structure is established.

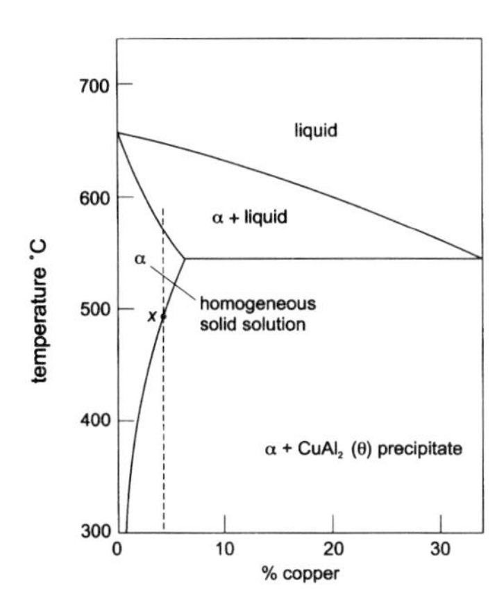

Figure 10.21 Section of the aluminium–copper equilibrium diagram.

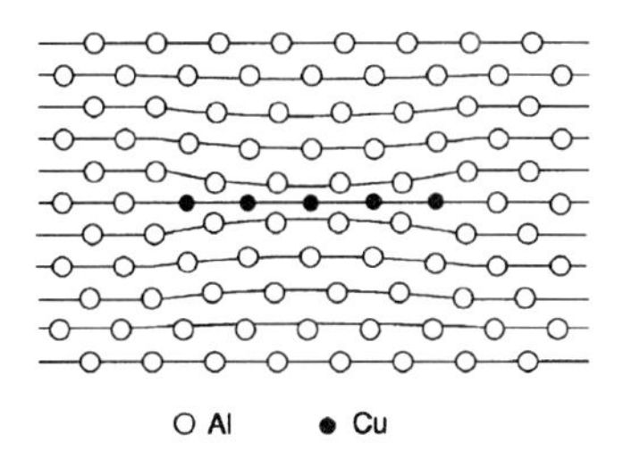

Figure 10.22 Schematic representation of a {100} section through a GP zone in aluminium–copper illustrating the coherency strains. The zones in real alloys are larger than that shown here.

In many alloys, such as Al–Cu, metastable phases occur during the transition of the precipitates to equilibrium. The structure of the interface between these zones and the matrix governs the kinetics of their formation and growth. The first zones to form are small and *coherent*, in the sense defined in section 9.6. In Al–Cu, for example, the copper atoms nucleate platelets (known as GP zones) of about 10 nm diameter on the {100} aluminium planes, as shown schematically in Fig. 10.22. Since the copper atoms are smaller than the aluminium atoms, the surrounding

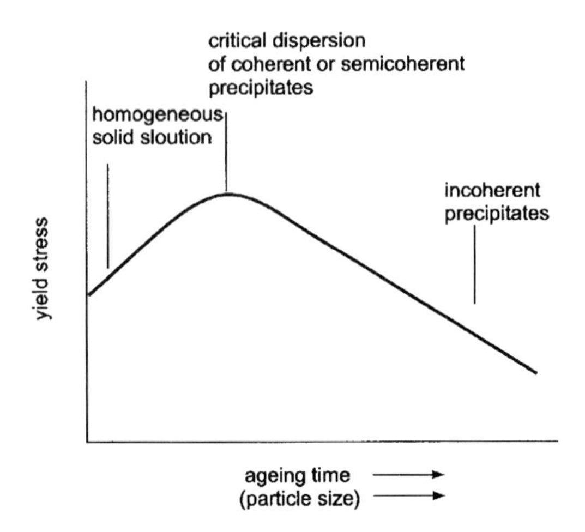

Figure 10.23 Variation of yield stress with ageing time typical of an age-hardening aluminium alloy.

matrix is strained, as indicated. In some other alloys, the zones are oversized. As the zones grow by bulk diffusion, they thicken (θ''-phase) and the large elastic energy associated with the coherency strains is reduced by the transition to semicoherent zones (θ'-phase) with misfit dislocations (section 9.6) in the interface. These precipitates have crystal structures which may be coherent with the matrix on some faces but not all. Eventually, after further growth, coherency is lost completely and precipitates of the equilibrium phase – $CuAl_2$ (θ-phase) in an Al–Cu alloy – are produced. Ageing therefore results in a transition from (a) a solid solution of misfitting atoms to (b) a fine dispersion of coherent precipitates surrounded by elastic strain to (c) a coarse dispersion of particles with incoherent interfaces and negligible lattice strain. These structures offer differing resistance to dislocation motion, and so the yield strength of the alloy changes with time, as illustrated in Fig. 10.23. The solid solution is stronger than the pure metal, but by ageing it is possible to further increase the strength. A peak strength is encountered corresponding to a critical dispersion of coherent or semicoherent precipitates, and beyond this the strength falls as *overageing* occurs. Selection of the optimum heat-treatment conditions is therefore crucial to maximising the strength of alloys. Important examples include the aluminium alloy series 2000 (Al–Cu), 6000 (Al–Mg–Si) and 7000 (Al–Zn).

The behaviour of a dislocation under applied stress is different in the three situations (a)–(c) described above, and the flow stress for the three cases is treated separately below using the analysis of section 10.6.

Solution Strengthening

In the *diffuse*-force model of a solid solution, there are c/b^3 obstacles per unit volume, where b is the interatomic spacing and c the atomic-fraction concentration of solute, so that the mean spacing of neighbouring solute atoms is

$$\Lambda = b/c^{1/3} \tag{10.48}$$

The maximum internal stress averaged over the space of radius $\Lambda/2$ around each solute is

$$\tau_i \simeq G|\delta|c\ln(1/c) \tag{10.49}$$

In the *local*-force model, on the other hand, only those solute atoms in the two planes immediately adjacent to the slip plane contribute. There are $2c/b^2$ of them per unit area of plane, so that in this case

$$\Lambda = b/(2c)^{1/2} \tag{10.50}$$

The maximum resisting force exerted by each of these solute atoms is found from equation (10.27) (and the subsequent discussion) to be

$$K_{\max} \simeq \frac{1}{5}Gb^2|\delta| \tag{10.51}$$

The quantities τ_i and $K_{\max}$ are seldom large enough for strong-force conditions to apply. Equations (10.42) and (10.46) based on weak forces are therefore appropriate, and the latter gives much the largest flow stress when the parameters of equations (10.48) to (10.51) are substituted. The flow stress for a solid solution is thus

$$\tau_{\text{loc}}(\text{weak}) \simeq \sqrt{2}Gc^{1/2}(\cos\phi_c)^{3/2} \tag{10.52}$$

where $\cos\phi_c \simeq K_{\max}/Gb^2$ is given by equation (10.51): the stress increases as the square root of the concentration. For the largest misfit expected for substitutional solutes, i.e. $|\delta| \simeq 0.15$ giving $\cos\phi_c \simeq 0.03$, the flow stress will be raised to about $G/400$ when $c = 10$ per cent, in reasonable agreement with experiment. Since the solute atom represents an energy barrier of magnitude $\simeq bK_{\max}$ to dislocation motion, it may be anticipated from the analysis of section 10.2 that the flow stress will decrease with increasing temperature above 0 K. This is found to be so experimentally, the thermal component of the flow stress being given by equation (10.10) with $\Delta F \simeq bK_{\max}$, $p \simeq 2/3$ and $q \simeq 3/2$.

The analysis leading to equation (10.52) neglects several features known to exist in real alloys. First, real obstacles have a finite range of interaction with a dislocation and characteristic force–distance profiles. These details have been incorporated in several theories and some suggest a $c^{2/3}$, rather than $c^{1/2}$, dependence of the flow stress, and for many alloys a more complicated dependence probably exists. Second, more than one type of obstacle is frequently present. If τ_1 is the flow stress for one species and τ_2 that for a second, the flow stress when both are present is well fitted by

$$\tau = (\tau_1^2 + \tau_2^2)^{1/2} \tag{10.53}$$

Precipitate Strengthening

When precipitates nucleate and grow, they intersect slip planes in a random fashion. A gliding dislocation must either cut through the precipitates or penetrate the array by bowing between the obstacles. It will adopt the mechanism offering the *lowest* resistance. The stress corresponding to this short-range mechanism must be compared with that required to overcome the internal stress τ_i which coherent precipitates produce in the surrounding lattice. The flow stress for the alloy is then the *larger* of the two.

Consider first the flow stress for *diffuse* forces governed by τ_i. With the assumption that the precipitates are spherical, τ_i is given by equation (10.49), where c now represents the volume fraction of precipitate. If each precipitate contains N atoms, equation (10.48) for the mean obstacle spacing is modified to

$$\Lambda = b(N/c)^{1/3} \tag{10.54}$$

In the early stages of ageing, when Λ and N are small, the weak-force conditions of section (10.6) apply and from equation (10.42) the flow stress is

$$\tau_{\text{diff}}(\text{weak}) \simeq 0.4\, N^{1/9} G|\delta|^{4/3} c^{11/9} [\ln(1/c)]^{4/3} \tag{10.55}$$

In the intermediate stages of ageing, Λ and N increase until the dislocation can bend around individual precipitates, i.e. $R \simeq \Lambda$ (Fig. 10.19(a)). Then, from equations (10.39) and (10.49), the flow stress is

$$\tau_{\text{diff}}(\text{strong}) \simeq \frac{2}{\pi} G|\delta| c \ln(1/c) \tag{10.56}$$

When overageing occurs, precipitate coherency is lost, and τ_i falls to zero and makes no contribution to the flow stress.

The *localised* forces offer two distinct resistances, as discussed above. A dislocation cutting through a precipitate experiences a resistance force K_{max} (Fig. 10.20(a)), and in the early stages of ageing equation (10.46) is appropriate. The Friedel spacing Λ for precipitates is found by noting that the area fraction of precipitate on a plane equals the volume fraction c. Since the average area of obstacle on a plane intersecting precipitates of volume Nb^3 randomly is $(\pi/6)^{1/3}(Nb^3)^{2/3}$, then

$$c = (\pi/6)^{1/3}(Nb^3)^{2/3}/\Lambda^2$$

i.e.

$$\Lambda \simeq bN^{1/3}/c^{1/2} \tag{10.57}$$

Various factors can determine K_{max}. For example, (a) the precipitate may be disordered and the dislocation has to create an antiphase boundary, (b) the Peierls stress may be high in the precipitate, and (c) the step of height b the cutting dislocation creates around the obstacle periphery may have a high energy. The first two give K_{max} proportional to the obstacle diameter $bN^{1/3}$ and typical values are of the order of $Gb^2N^{1/3}/100$. Thus, the critical cusp angle ϕ_c is approximately $\cos^{-1}(N^{1/3}/100)$, and weak-force conditions apply only when $N < 10^5$. The flow stress from equations (10.46) and (10.57) is then

$$\tau_{loc}(\text{weak}) \simeq \frac{1}{1000}N^{1/6}Gc^{1/2} \tag{10.58}$$

As ageing proceeds, N and K_{max} increase further and dislocations find it easier to bow between the obstacles than to pass through them, as shown schematically in Fig. 10.24. Under these strong-force conditions when $K_{max} \simeq Gb^2$ and $\cos\phi_c \simeq 1$, the flow stress is obtained from equations (10.47) and (10.57). It is known as the *Orowan stress* and is

$$\tau(\text{Orowan}) \simeq 0.84\,Gb/\Lambda = 0.84\,N^{-1/3}Gc^{1/2} \tag{10.59}$$

In the overaged state, dislocations cannot penetrate the incoherent interface and equation (10.59) still applies.

It is now possible to explain the variation of the flow stress throughout the ageing process (Fig. 10.23). In the underaged condition, dislocations cut through the precipitates, the flow stress increasing with N when either long-range internal stress (equation (10.55)) or short-range local force (equation (10.58)) determines the strength. The two effects are comparable in size and which is the dominant one depends on parameters such as $|\delta|$ and antiphase boundary energy for the alloy in question. When the precipitates are large enough for strong-force conditions to apply ($N \gtrsim 10^5$), the flow stress given by equation (10.56) is generally larger than the Orowan stress (equation (10.59)). Thus, unless coherency is lost at an early stage, further ageing results in a flow stress controlled by large coherency strains (equation (10.56)). It is independent of N and corresponds to the optimum heat-treatment. In this state, a dislocation must first overcome τ_i before cutting a precipitate. When coherency is lost by overageing, a dislocation moves through the obstacles by bowing between them. The flow stress (equation (10.59)) then decreases with increasing N. Except in the very earliest stages of ageing, the energy barriers offered to dislocation motion by precipitates are generally too large in comparison with kT for thermal activation to be significant, i.e. $\Delta F \gtrsim Gb^3$ in equation (10.10). The flow stress is almost athermal.

It can be seen from Fig. 10.24 that when the spacing of incoherent precipitates is not large in comparison with their size, a correction must be effected in equation (10.59) by replacing Λ by $(\Lambda - D)$. It is also seen that a dislocation moving through obstacles by the Orowan mechanism leaves a ring of dislocation around each one. These *Orowan loops* may affect the stress for subsequent dislocations by reducing the effective spacing $(\Lambda - D)$. The nature of Orowan loops and the ways in which they are affected by cross slip are discussed in section 7.5 (Fig. 7.12).

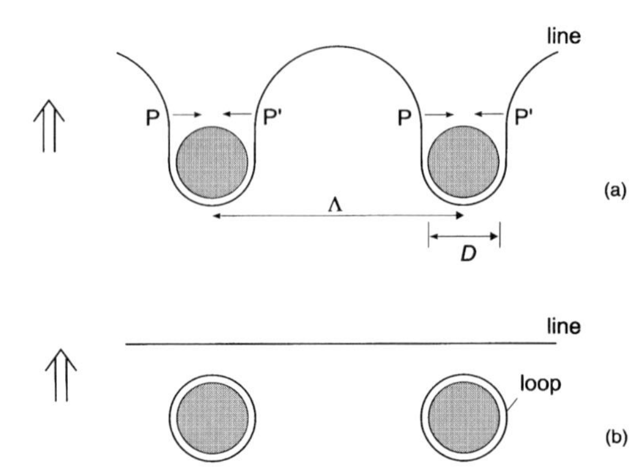

Figure 10.24 The Orowan mechanism. (a) The line bows between the obstacles until the segments at P and P' are parallel ($\phi_c = 0$). They then attract and meet. (b) The line moves forward leaving Orowan loops around the obstacles.

The Orowan stress is also the flow stress for *dispersion strengthening*, a process in which incoherent obstacles such as oxide particles are deliberately added to a softer metal matrix. Examples include thoria-dispersed (TD) nickel ($Ni-ThO_2$) and sintered aluminium powder (SAP) ($Al-Al_2O_3$).

10.8 Work Hardening

The mechanisms discussed in the preceding sections of this chapter describe the glide of dislocations in single crystals where the resistance encountered does not change with increasing plastic strain. They neglect the increase in glide resistance that occurs when dislocations move, interact and change their distribution and density. This results in *work hardening*, which can often be exploited to advantage by raising the strength of solids shaped by plastic deformation. Taylor (1934) recognised in the earliest days of dislocation theory that work-hardening results from the interaction of dislocations. It was postulated that the flow stress τ is just the stress required to force two edge dislocations on parallel slip planes of spacing l past each other against their elastic interaction. Hence, from Fig. 4.12,

$$\tau = \alpha Gb/l \tag{10.60}$$

where α is a constant or order 0.1. The line density ρ of a network of dislocations of spacing l is $\rho \simeq l^{-2}$. With the assumption that each dislocation moves a distance x before being stopped by a network of other dislocations, it is seen from equations (3.9) and (10.60) that

$$\tau = \alpha G\left(\frac{b}{x}\right)^{1/2} \varepsilon^n \tag{10.61}$$

where n is 0.5 in the Taylor theory. Although such a parabolic relation between stress and strain is a feature of the plastic deformation of many polycrystalline materials, this is not the case for single crystals, for which equation (10.61) is derived. Also, it is known that dislocations from sources do not move as isolated defects but as groups creating slip bands. Furthermore, from the form of the stress–strain curve, it is known that barriers to dislocation motion are actually created by deformation. The incorporation of these aspects into a theory which can explain how the dislocation arrangement changes with increasing plastic strain and predict the flow stress for a given dislocation state has proved difficult, for many effects are involved.

It is now established that single crystals of a variety of metallic and non-metallic crystal structures exhibit *three-stage* behaviour: the stress–strain curve for a face-centred cubic metal, for example, is shown schematically in Fig. 10.1(d) and a real example is reproduced in Fig. 10.17(a). The extent of each stage is a function of the crystallographic orientation of the tensile axis, the purity of the sample and the temperature of the test. In the body-centred cubic and prism-slip hexagonal metals, similar behaviour (but at higher stress) is observed at intermediate temperatures, and in the basal-slip hexagonal metals stage I may extend almost to fracture. In all cases, the rate of work-hardening (i.e. the slope θ of the τ versus ε curve in Fig. 10.1(d)) is low in *stage I*, typically $\theta_I \simeq 10^{-4} G$, where G is the shear modulus. This linear region (often called *easy glide*) is followed by another, *stage II*, in which the slope θ is much higher, typically $\theta_{II} \simeq (3-10)\theta_I$. In *stage III, dynamic recovery* occurs, the τ versus ε curve is parabolic and θ decreases with increasing strain. The onset of stage III is promoted by an increase in temperature. Numerous investigations using transmission electron microscopy and measurement of surface slip steps have shown how dislocation behaviour contributes to these stages.

The plastic strain in *stage I* results from dislocations moving on the slip system with the highest resolved shear stress. On this *primary system*, dislocation sources operate at the critical resolved shear stress and dislocations move over large distances (typically between $\sim 100\,\mu m$ and the crystal diameter). Many tens of dislocations contribute to slip lines observed on the surface. Electron microscopy of specimens deformed only in stage I reveals edge dislocation dipoles (see section 4.6). They are formed by the elastic interaction of dislocations of opposite sign moving in opposite directions from sources on different (but parallel) slip planes, as shown schematically in Fig. 10.25. Screw dislocations are seldom seen, because screws of opposite sign attract each other (section 4.6) and can annihilate by cross slip. The exception to this occurs in crystals of low stacking-fault energy, for then the constriction required for cross slip is difficult. Also, in overaged or dispersion-strengthened alloys (section 10.7), prismatic loops generated by cross slip (section 7.5) interact with screw dislocations to form helices.

Very few dislocations are formed on *secondary* (i.e. non-primary) slip systems in stage I, but in *stage II* there is a considerable increase in multiplication and movement of secondary dislocations. The onset of

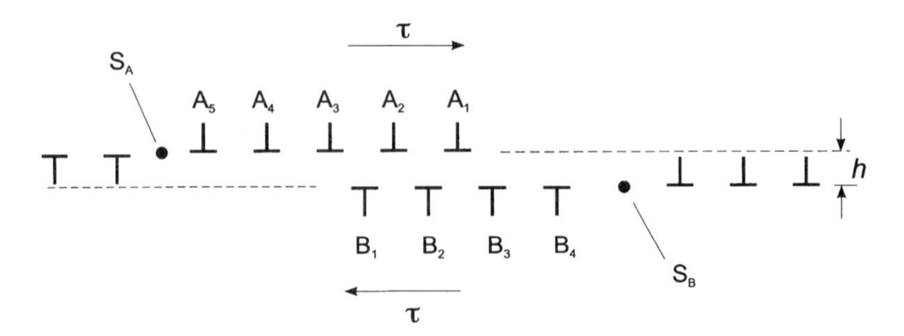

Figure 10.25 Dislocation sources S_A and S_B operate under resolved shear stress τ, and dislocations A_1, A_2, A_3... interact with B_1, B_2, B_3... of opposite sign to form an array of dipoles known as a *multipole*.

this secondary activity brings stage I to an end. It occurs when the applied load (aided by internal stress from primary dislocations) and crystal orientation are such that the critical resolved shear stress on the secondary system is reached. Thus, the extent of stage I depends on crystal structure and orientation. The densities of primary and secondary dislocations are approximately equal in stage II, although the primaries contribute most to the plastic strain. Interactions between primary and secondary dislocations lead to the formation of barriers to slip, so that although new primary slip lines appear as strain increases, the slip line length decreases with increasing strain. (The formation mechanism of Lomer–Cottrell barriers in face-centred cubic metals is described in section 5.6.) Transmission electron microscopy reveals that the density of secondary dislocations is highest in regions where the primary density is high, and the two sets, together with their reaction products, form sheets of dislocations parallel to the primary planes.

A variety of models has been proposed to explain the value of the flow stress τ for a particular dislocation distribution. To this must be added the lattice and alloying contributions discussed in section 10.3 and 10.7. In conformity with equation (10.9), τ contains a part τ_G, which is temperature-independent, and a part τ^*, which is assisted by thermal activation and is temperature-dependent. τ_G arises from the resistance to dislocation motion caused by long-range elastic interactions and has the form

$$\tau_G \propto Gb/l \qquad (10.62)$$

where length l depends on the controlling mechanism. For example, if τ_G is the stress required to make parallel dislocations of opposite sign pass on parallel slip planes of spacing h (Fig. 10.25), then $l = h$, as seen by putting $y = h$ in Fig. 4.12. If τ_G is required to overcome the internal stress caused by dislocations piled up at barriers, then l is the distance from the gliding dislocation to the centre of the pile-up (section 9.8). When forest dislocations of spacing l_f intersecting the slip plane are considered, three contributions to τ arise. The first is a long-range term (section 7.7) and gives relation (10.62) with $l = l_f$. The other two are independent of internal stress. One is the applied stress required to create a jog by forest intersection (section 7.7) and the other the stress necessary to move a jogged dislocation by vacancy generation (section 7.3). The latter two mechanisms contribute to τ^* and, from Chapter 7, have the form at 0 K of

$$\tau^*(0) \propto Gb/l_f \tag{10.63}$$

If the dislocation arrangement specified by a spacing l is random, then the density of those dislocations, ρ, is approximately l^{-2}. Thus, both τ_G and $\tau^*(0)$ have the form

$$\tau = \alpha Gb\sqrt{\rho} \tag{10.64}$$

Experiments show that α in τ_G falls in the range 0.05 to 1, but it is not clear whether τ_G in stages I and II correlates better with equation (10.64) when ρ is taken as the density of primary dislocations, ρ_p, or forest dislocations, ρ_f. There is cause for uncertainty because the dislocation arrangements observed are not random. However, from the constancy with strain of the ratio of τ at two different temperatures in pure face-centred cubic metals, it has been suggested that either τ_G and τ^* are both controlled by ρ_f or the contributions of ρ_p and ρ_f to τ_G remain in constant proportion during deformation. In alloys and body-centred cubic crystals, solid-solution and Peierls-stress effects may dominate τ^*, which is larger than τ_G at low to moderate temperatures.

The equations above do not give a work-hardening rate θ, nor do they predict the form of the dislocation arrangement. There is, as yet, no completely statisfactory theory of stage I. Most models have considered the applied stress τ_G for one dislocation to pass other, parallel dislocations in dipole/multipole configurations, which, as explained above, are characteristic of this stage. Unfortunately, multipoles are probably very mobile unless they contain exactly equal numbers of dislocations of opposite sign, so that it is necessary to invoke some (unexplained) trapping mechanism for multipoles to predict the observed θ. It is possible that forest dislocations play a role here. In stage I, the long-range internal stress field from dislocations in dipole pairs is weak and θ correspondingly small. In stage II, secondary slip leads to strong obstacles and pile-ups of primary dislocations. The large internal stress that results leads to further secondary slip and the formation of stable dislocation boundaries with long-range stress fields. New dislocations are generated in relatively soft regions of the crystal, before being blocked themselves. The slip line length therefore decreases with increasing strain and τ_G increases rapidly. Again, however, there is no completely satisfactory theory which accounts quantitatively for the net contributions from the forest and primary dislocations.

Stage II ceases and *stage III* begins when dislocations leave their original slip planes. Except for high temperatures at which edge dislocations can climb, this occurs by the cross slip of screw dislocations, as

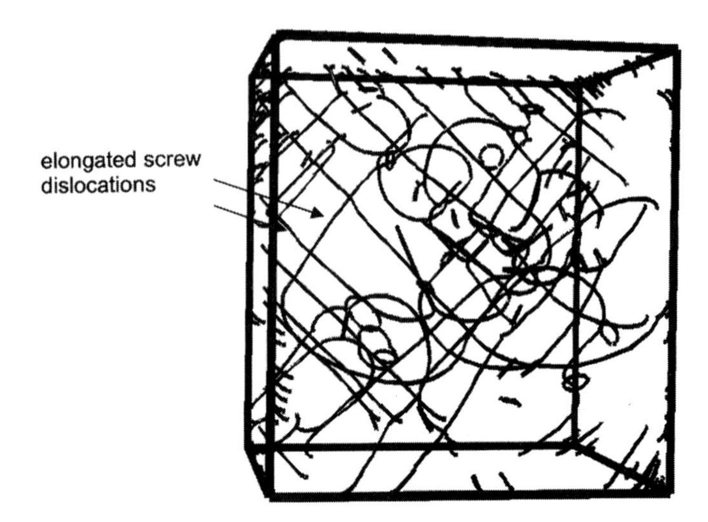

Figure 10.26 Simulation cell (size 30 μm) of a dislocation dynamics simulation of a molybdenum crystal pulled in a ⟨100⟩ direction at a strain rate of $10\,s^{-1}$. The dislocation structure has a large number of ⟨111⟩ screws. (Reprinted from Zhib, Diaz de la Rubia, Rhee and Hirth, *J. Nucl. Mater.* **276**, 154, 2000, with permission from Elsevier Science.)

confirmed by cross-slip traces seen by surface examination. The attraction and annihilation of screws of opposite sign leads to a reduction in dislocation density (and internal stress) in stage III, and the edge dislocations rearrange to form low-angle boundaries. This change of structure is promoted by the applied stress and is thus known as dynamic recovery. Since cross slip is hindered by low stacking-fault energy (section 5.3) and assisted by thermal vibrations, stage III dominates the stress–strain curve for high stacking-fault energy crystals at moderate to high temperatures, as illustrated by the dashed curve in Fig. 10.1(d). The stress at which stage III starts decreases exponentially with increasing temperature and stacking-fault energy.

As noted above, analytical theory has not explained entirely all the features of work hardening. Answers to some of the questions that remain are starting to emerge from large-scale computer simulation of the behaviour of dislocation arrangements that mimic those in real crystals. A dislocation is treated in the elasticity approximation (Chapter 4) as a series of flexible segments that are free to move to obtain a balance of the force on each due to their neighbouring segments, other dislocations and the applied stress. A stress-dependent dislocation velocity can be employed, with differences between edges and screws (section 3.5), so that the dynamics of dislocation behaviour are allowed for. Furthermore, the resistance to dislocation motion due to nodes and jogs formed by intersection of dislocations can be incorporated with reasonable precision. An example of a simulation cell for a crystal of molybdenum, a body-centred cubic metal, loaded along a ⟨100⟩ axis at

300 K to a plastic strain of 0.1% is shown in Fig. 10.26. The mobility of edge dislocations was taken to be much higher than that of screws, in conformity with experiment (section 3.5), so that the substructure is seen to be dominated by long screw dislocations. The calculated stress–strain curve is plotted in Fig. 10.27, together with the dislocation density and number of jogs. The yield stress is in good agreement with experimental values and the hardening is seen to be associated with the increasing formation of jogs as the dislocation density increases.

10.9 Deformation of Polycrystals

There are important differences between plastic deformation in single-crystal and polycrystalline materials. In the latter, the individual crystals have different orientations, and the applied resolved shear stress for slip

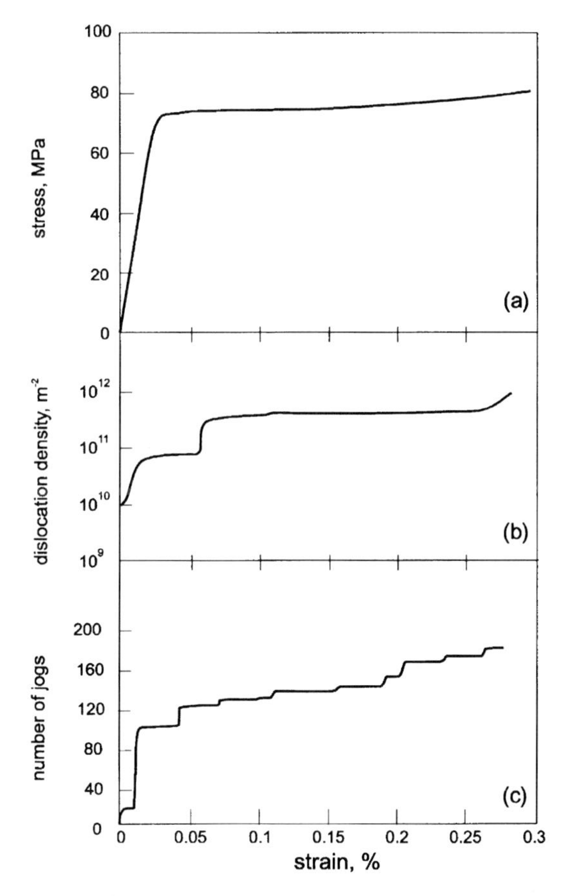

Figure 10.27 Dependence of (a) stress, (b) dislocation density and (c) number of jogs on the strain for the simulation of Fig. 10.26. (Reprinted from Zhib, Diaz de la Rubia, Rhee and Hirth, *J. Nucl. Mater.* **276**, 154, 2000, with permission from Elsevier Science.)

(equation (3.1)) varies from grain to grain. A few grains yield first, followed progressively by the others. The grain boundaries, being regions of considerable atomic misfit, act as strong barriers to dislocation motion, so that unless the average grain size is large, the stage I easy-glide exhibited by single crystals does not occur in polycrystals. The stress–strain curve is therefore not simply a single-crystal curve averaged over random orientations. Furthermore, the internal stresses around piled-up groups of dislocations at the boundaries of grains that have yielded may cause sources in neighbouring grains to operate. Thus, the macroscopic yield stress at which all grains yield depends on grain size. Finally, a grain in a polycrystal is not free to deform plastically as though it were a single crystal, for it must remain in contact with, and accommodate the shape changes of, its neighbours. An inability to meet this condition leads to a small strain to failure, as typified by Fig. 10.1(b).

Considering the last point first, it is seen from section 4.2 that when volume remains constant, as it does during plastic deformation of crystals, a general elastic shape change is fully specified by five independent components of strain. Extending this to the ability to undergo general *plastic* shape changes, *five independent slip systems* must be able to operate, a result known as the *von Mises condition*. An independent system produces a shape change which cannot be obtained by combinations of other systems. In the face-centred cubic metals, the twelve $\langle 1\bar{1}0\rangle$ $\{111\}$ slip systems provide five independent systems and satisfy the condition. So, too, do the slip systems in body-centred cubic metals. In other structures (see Chapter 6) there are often insufficient systems, except at high temperature. For example, hexagonal metals in which basal slip is strongly preferred only have three slip systems of which two are independent – neglecting any contribution from deformation twinning – and show little ductility at low temperatures.

Equation (3.1) shows that the tensile yield stress σ_y and the critical resolved shear stress τ_c for a single crystal which slips on one plane in one direction are related by the Schmid factor $(\cos\phi\cos\lambda)$. It may be rewritten

$$\sigma_y = m\tau_c \tag{10.65}$$

For polycrystals in which the grains have random orientations, this may be generalised to

$$\sigma_y = M\tau_c \tag{10.66}$$

where M is known as the *Taylor factor*. By averaging the stress over the grains and considering the most-favoured slip systems, it has been shown that $M \simeq 3$ for the face-centred and body-centred cubic metals.

From experimental measurement of the yield stress of polycrystalline aggregates in which grain size d is the only material variable, it has been found that the *Hall–Petch relationship* is satisfied:

$$\sigma_y = \sigma_0 + k_y d^{-n} \tag{10.67}$$

where exponent n is approximately 0.5, k_y is a material constant and σ_0 is a constant stress of uncertain origin. The flow stress beyond yield follows a similar form. Hence, at cold-working temperatures, where grain boundaries do not contribute to creep, high yield and flow stresses are favoured by small grain size. One rationalisation of equation (10.67) is that a pile-up at a grain boundary in one grain can generate sufficiently large stresses to operate sources in an adjacent grain at the yield stress. For example, the stress τ_1 on the leading dislocation generated by source S_1 in grain 1 under resolved shear stress τ (Fig. 10.28) is seen from equations (9.27) and (9.28) to be

$$\tau_1 = \frac{d}{A}\tau^2 \tag{10.68}$$

where $A \simeq Gb/2\pi$. If it is assumed that source S_2 in grain 2 operates when τ_2 reaches a critical value τ_1^*, then

$$\sigma_y = k_y d^{-1/2} \tag{10.69}$$

where k_y is $m(A\tau_1^*)^{1/2}$, and m converts resolved shear stress to tensile stress. The Hall–Petch relationship follows by assuming the dislocations from S_2 encounter a friction stress or internal back-stress σ_0. Although such a pile-up model was first used to explain the relationship (Petch, 1953), other interpretations, such as that based on grain-boundary dislocations (Li and Chou, 1970), are possible.

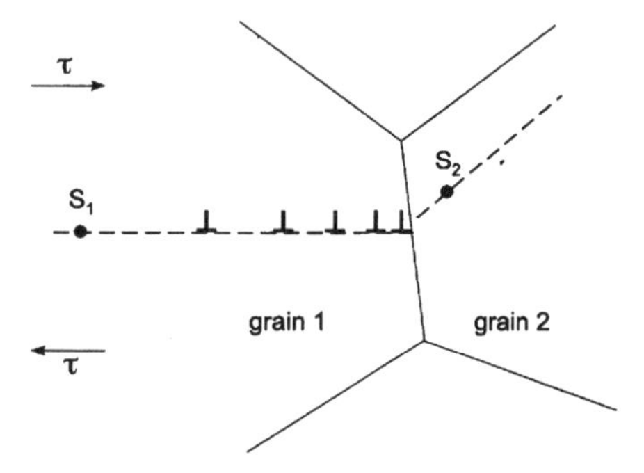

Figure 10.28 Schematic illustration of a pile-up formed in grain 1 under an applied resolved shear stress τ. S_2 is a source in grain 2. The trace of the preferred slip plane in each grain is marked by a dashed line.

10.10 Dislocations and Fracture

Surface irregularities and cracks are particularly potent sources of dislocations in solids under stress. Indeed, the generation of dislocations in the vicinity of crack tips plays a vital role in determining whether material exhibits brittle or ductile behaviour when fracture occurs, i.e. whether crack propagation is preceded by significant dislocation emission at the crack tip or not. In ductile metals, yielding at the tip of a crack can blunt it and crack propagation is accompanied by extensive plastic flow. However, many materials, e.g. ceramics, semiconductors and refractory metals, exhibit a *brittle-to-ductile transition* as their temperature is raised. Some, such as silicon, show a very sharp transition over a narrow temperature range, but most do not and their transition is soft, occurring over a wide temperature range of typically 50–100°. Dislocations influence the behaviour by affecting the stress field near the crack tip before the cleavage occurs. Experiments show that dislocations glide and multiply near the crack tip over the temperature range of the transition. A soft transition is assisted by a high density of available dislocation sources along the crack front.

To a first approximation, if material under an applied tensile stress σ contains a crack of length c perpendicular to the stress axis, the stress field near the tip of the crack is proportional to $\sigma\sqrt{c/r_c}$, where r_c is the radius of curvature of the crack tip. Hence, the applied stress is strongly magnified in the tip region of sharp cracks ($r_c \ll c$). It is convenient to use the *stress intensity factor* K to describe the stress around the crack:

$$K = \alpha\sigma\sqrt{\pi c} \tag{10.70}$$

where α is a constant approximately equal to 1. The stress field near the tip varies as $K/\sqrt{r}$, where r is the distance from the tip. Local fracture occurs at a critical value K_f of the stress intensity factor corresponding to there being sufficient strain energy in the crack tip stress field to create new fracture surfaces. For a given crack length, the fracture stress and K_f are given by equation (10.70). In ideally brittle crystals, K_f equals the critical value K_c as given by the Griffith theory for pure cleavage. Dislocation activity near the tip of a crack can *shield* the tip from the full effects of the applied stress and result in a local stress intensity factor that is lower than that given by equation (10.70). Models based on this effect have been developed to explain the role of dislocations in the transition from brittle to ductile behaviour.

A simple explanation of the effect of dislocations emitted from a source near a crack tip under an applied stress is as illustrated schematically in Fig. 10.29. The first dislocation is emitted at K well below K_c and will be able to move away from the source if the stress it experiences due to the crack tip stress field exceeds that to operate dislocation sources near the crack tip: this might include, for example, an image stress (section 4.8) due to the crack surface. Subsequent dislocations experience an additional back stress from the earlier ones, and that also has to be overcome. The sign of these dislocations is such that they shield the material at the crack tip from the full stress intensity. Thus, the applied stress intensity K at fracture is greater than K_c because of the shielding effect of the emitted dislocations. The dynamics of the situation depend on the time the shielding dislocations remain in the vicinity of the crack before gliding away and the rate of increase of applied load. This can be modelled by treating dislocation movement using a realistic dependence of the velocity of a dislocation on the local stress (section 3.5) and is found to provide a good description of the rise in K during the soft transitions of simple materials. A comparison of model and experimental results for a single crystal of molybdenum is plotted in Fig. 10.30 for two different values of the rate of loading $\mathrm{d}K/\mathrm{d}t$.

The sharp transition phenomenon is modelled in this framework by assuming that the spacing of the dislocation sources along the tip of the crack is large and that they only begin to operate in a typical fracture experiment when K is close to K_c, i.e. shielding does occur but only after

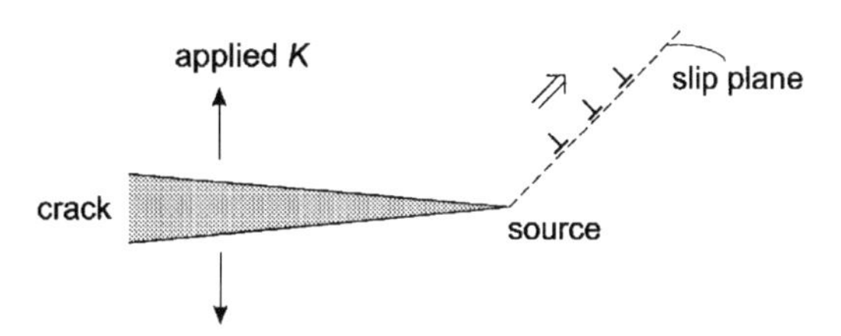

Figure 10.29 Schematic illustration of the generation of shielding dislocations from a source just ahead of a crack.

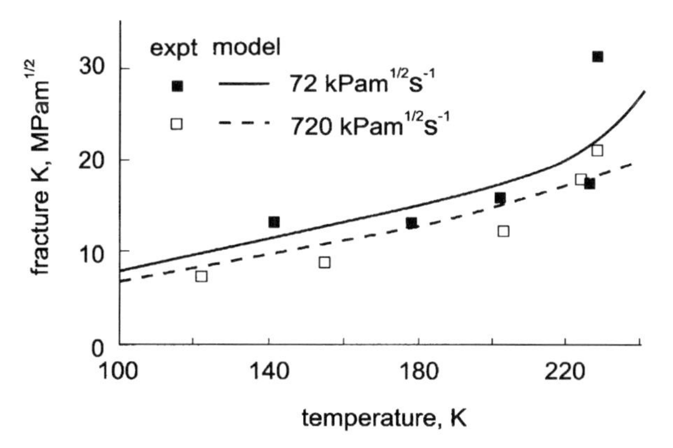

Figure 10.30 Experimental and modelled brittle-to-ductile transition data for molybdenum at two different loading rates $\mathrm{d}K/\mathrm{d}t$. For the model, $K_c = 6\,\mathrm{MPa\,m}^{-1/2}$, the exponent m (equation 3.3b) in the dislocation velocity rises from 5.6 to 4.1 as temperature increases from 150 to 250 K, and dislocations sources operate at $0.2K_c$. (After Fig. 9 of Roberts (1996), *Computer Simulation in Materials Science*, p. 409, Kluwer, with permission from Kluwer Academic Publishers.)

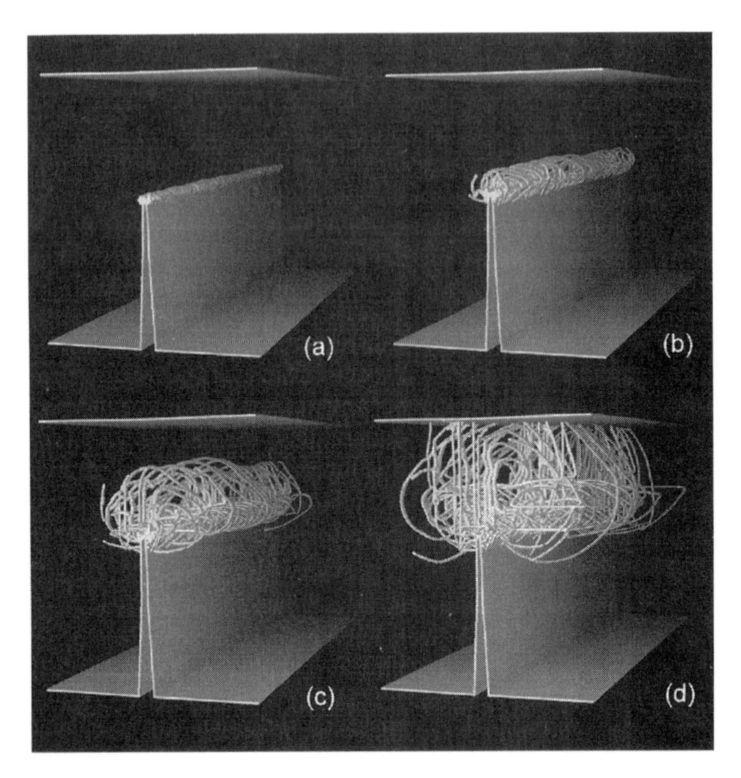

Figure 10.31 Four stages in the behaviour of a crack in xenon as observed in a computer simulation of a crystal containing 100 million atoms. At very high crack velocity (approach one third of the speed of surface sound waves), the crack tip roughens and dislocations are emitted, thus blunting the crack and arresting its motion. (Reprinted from Abraham, Schneider, Land, Lifka, Skovira, Gerner and Rosenkrantz, *J. Mech. Phys. Solids* **45**, 1461, 1997, with permission from Elsevier Science.)

dislocations are emitted suddenly. This results in a rapid increase in crack tip shielding and the abrupt increase in K_f with increasing temperature.

The models above assume that the dislocation sources near the crack tip pre-exist. Most evidence points to the difficulty of actually creating sufficient dislocations by nucleation at sites along the crack front, e.g. at ledges. However, there may be conditions under which homogeneous nucleation can occur at the crack tip. Computer simulation (section 2.7) of crystals containing tens of millions of atoms confirms that crack propagation in the absence of dislocations occurs by planar cleavage at low crack velocity. At very high crack velocity approaching one third of the speed of surface sound waves, however, the crack tip roughens on the atomic scale. This instability results in the prolific emission of dislocations and the cessation of crack motion by crack tip blunting. Four images from a computer simulation of a model of xenon containing 100 million atoms are shown in Fig. 10.31. Only atoms at the surface or along the core of dislocation lines are visualised. In (a) the crack has run from the lower surface of the model by pure cleavage, but has reached a velocity (300 ms^{-1}) where dislocations have started to be generated along the crack tip. They have multiplied and moved away from the crack in (b) and (c), and have reached the upper surface of the specimen in (d). The elapsed time between (a) and (d) is 400 ps.

Further Reading

Ashby, M. F. and Jones, D. R. H. (1996) *Engineering Materials*, vols 1 and 2, Butterworth-Heinemann.

Basinski, S. J. and Basinski, Z. S. (1979) 'Plastic deformation and work hardening', *Dislocations in Solids*, vol. 4, p. 261 (ed. F. R. N. Nabarro), North-Holland.

Bilby, B. A. (1950) 'On the interaction of solute atoms and dislocations', *Proc. Phys. Soc.* **A63**, 191.

Brown, L. M. and Ham, R. K. (1971) 'Dislocation-particle interactions', *Strengthening Methods in Crystals*, p. 12 (eds. A. Kelly and R. B. Nicholson), Elsevier.

Christian, J. W. (1975) *The Theory of Transformations in Metals and Alloys*, Pergamon.

Cottrell, A. H. (1954) 'Interaction of dislocations and solute atoms', *Relation of Properties to Microstructure*, p. 131, Amer. Soc. Metals, Cleveland, Ohio.

Cottrell, A. H. and Bilby, B. A. (1949) 'Dislocation theory of yielding and strain ageing in iron', *Proc. Phys. Soc.* **A62**, 49.

de Batist, R. (1972) *Internal Friction of Structural Defects in Crystalline Solids*, North-Holland.

Edington, J. W., Melton, K. N. and Cutler, C. P. (1976) 'Superplasticity', *Prog. Mater. Sci.* **21**, 61.

Eshelby, J. D. (1956) 'The continuum theory of lattice defects', *Solid State Phys.* **3**, 79.

Eshelby, J. D. (1957) 'Determination of the elastic field of an ellipsoidal inclusion', *Proc. Roy. Soc.* **A241**, 376.

Eshelby, J. D. (1975) 'Point defects', *The Physics of Metals: 2 Defects*, p. 1, (ed. P. B. Hirsch), Cambridge University Press.

Friedel, J. (1964) *Dislocations*, Pergamon.

Frost, H. J. and Ashby, M. F. (1982) *Deformation-Mechanism Maps*, Pergamon.

Gerold, V. (1979) 'Precipitation hardening', *Dislocations in Solids*, vol. 4, p. 219, (ed. F. R. N. Nabarro), North-Holland.

Gifkins, R. C. (Ed.) (1983) *Strength of Metals and Alloys* (3 vols.), Pergamon.

Haasen, P. (1979) 'Solution hardening in f.c.c. metals', *Dislocations in Solids*, vol. 4, p. 155, (ed. F. R. N. Nabarro), North-Holland.

Hertzberg, R. W. (1996) *Deformation and Fracture Mechanics of Engineering Materials*, John Wiley.

Hirsch, P. B. (1975) 'Work hardening', *The Physics of Metals: 2 Defects*, p. 189, (ed. P. B. Hirsch), Cambridge University Press.

Hirth, J. P. and Lothe, J. (1982) *Theory of Dislocations*, Wiley.

Honeycombe, R. W. K. (1968) *The Plastic Deformation of Metals*, Edward Arnold.

Johnston, W. G. (1962) 'Yield points and delay times in single crystals', *J. Appl. Phys.* **33**, 2716.

Kelly, A., Groves, G. W. and Kidd, P. (2000) *Crystallography and Crystal Defects*, John Wiley.
Kocks, U. F., Argon, A. S. and Ashby, M. F. (1975) 'Thermodynamics and kinetics of slip', *Prog. in Mater. Sci.* **19**.
Li, J. C. M. and Chou, Y. T. (1970) 'The role of dislocations in the flow stress grain size relationship', *Met. Trans.* **1**, 1145.
Martin, J. W. (1998) *Precipitation Hardening*, Butterworth-Heinemann.
Mott, N. F. (1952) 'Mechanical strength and creep in metals', *Imperfections in Nearly Perfect Crystals*, p. 173, Wiley.
Mott, N. F. and Nabarro, F. R. N. (1948) *Strength of Solids*, p. 1, Phys. Soc. London.
Nabarro, F. R. N. (1947) 'Dislocations in a simple cubic lattice', *Proc. Phys. Soc.* **59**, 256.
Nabarro, F. R. N. (1967) *The Theory of Crystal Dislocations*, Oxford University Press.
Nabarro, F. R. N. (1975) 'Solution and precipitation hardening', *The Physics of Metals: 2 Defects*, p. 152, (ed. P. B. Hirsh), Cambridge University Press.
Nowick, A. S. and Berry, B. S. (1972) *Anelastic Relaxation in Crystalline Solids*, Academic Press.
Peierls, R. (1940) 'The size of a dislocation', *Proc. Phys. Soc.* **52**, 34.
Petch, N. J. (1953) 'The cleavage strength of polycrystals', *J. Iron Steel Inst.* **174**, 25.
Ritchie, I. G. and Fantozzi, G. (1992) 'Internal friction due to the intrinsic properties of dislocations', *Dislocations in Solids*, vol. 9, chap. 45 (ed. F. R. N. Nabarro), North-Holland.
Smith, E. (1979) 'Dislocations and cracks', *Dislocations in Solids*, vol. 4, p. 363, (ed. F. R. N. Nabarro), North-Holland.
Taylor, G. I. (1934) 'Mechanism of plastic deformation of crystals', *Proc. Roy. Soc.* **A145**, 362.

The SI System of Units

Base-units

metre (m)	— length
kilogram (kg)	— mass
second (s)	— time
ampere (A)	— electric current
kelvin (K)	— thermodynamic temperature
candela (cd)	— luminous intensity

Some derived units

	name of SI unit	SI base-units
frequency	hertz (Hz)	$1\,Hz = 1\,s^{-1}$
force	newton (N)	$1\,N = 1\,kg\,m\,s^{-2}$
work, energy	joule (J)	$1\,J = 1\,N\,m$

Multiplication factors

	Prefix	Symbol
10^{12}	tera	T
10^{9}	giga	G
10^{6}	mega	M
10^{3}	kilo	k
10^{-3}	milli	m
10^{-6}	micro	μ
10^{-9}	nano	n
10^{-12}	pico	p
10^{-15}	femto	f
10^{-18}	atto	a

Some useful conversions

1 angstrom	$= 10^{-10}\,m = 100\,pm$ or $0.1\,nm$
1 kgf	$= 9.8067\,N$
1 atm	$= 101.33\,kN\,m^{-2}$
1 torr	$= 133.32\,N\,m^{-2}$
1 bar	$= 0.1\,MN\,m^{-2}$
1 dyne	$= 10^{-5}\,N$
$1\,dyne\,cm^{-1}$	$= 1\,mN\,m^{-1}$
$1\,dyne\,cm^{-2}$	$= 0.1\,N\,m^{-2}$
1 cal	$= 4.1868\,J$
1 erg	$= 10^{-7}\,J = 0.1\,\mu J$

$1\ \mathrm{erg\,cm^{-2}}$	$= 1\ \mathrm{mJ\,m^{-2}} = 1\ \mathrm{mN\,m^{-1}}$
1 inch	$= 25.4\ \mathrm{mm}$
1 lb	$= 0.4536\ \mathrm{kg}$
1 lbf	$= 4.4482\ \mathrm{N}$
$1\ \mathrm{lbf\,in^{-2}}$	$= 6.895\ \mathrm{kN\,m^{-2}}$
$1\ \mathrm{lb\,in^{-1}}$	$= 175\ \mathrm{N\,m^{-1}}$
1 eV	$= 1.6 \times 10^{-19}\ \mathrm{J} = 0.16\ \mathrm{aJ}$

Index